# Lecture Notes in Physics

Founding Editors: W. Beiglböck, J. Ehlers, K. Hepp, H. Weidenmüller

## The Lecture Notes in Physics

The series Lecture Notes in Physics (LNP), founded in 1969, reports new developments in physics research and teaching – quickly and informally, but with a high quality and the explicit aim to summarize and communicate current knowledge in an accessible way. Books published in this series are conceived as bridging material between advanced graduate textbooks and the forefront of research and to serve three purposes:

- to be a compact and modern up-to-date source of reference on a well-defined topic

- to serve as an accessible introduction to the field to postgraduate students and nonspecialist researchers from related areas

- to be a source of advanced teaching material for specialized seminars, courses and schools

Both monographs and multi-author volumes will be considered for publication. Edited volumes should, however, consist of a very limited number of contributions only. Proceedings will not be considered for LNP.

Volumes published in LNP are disseminated both in print and in electronic formats, the electronic archive being available at springerlink.com. The series content is indexed, abstracted and referenced by many abstracting and information services, bibliographic networks, subscription agencies, library networks, and consortia.

Proposals should be sent to a member of the Editorial Board, or directly to the managing editor at Springer:

Christian Caron
Springer Heidelberg
Physics Editorial Department I
Tiergartenstrasse 17
69121 Heidelberg / Germany
christian.caron@springer.com

J. Gemmer
M. Michel
G. Mahler

# Quantum Thermodynamics

Emergence of Thermodynamic Behavior Within Composite
Quantum Systems

Second Edition

 Springer

Jochen Gemmer
Universität Osnabrück
FB Physik
Barbarastr. 7
49069 Osnabrück
Germany

M. Michel
Univ. of Surrey
Dept. of Physics
School of Electronics/Physical
Sciences
Guildford, Surrey
United Kingdom GU2 7XH

Günter Mahler
Universität Stuttgart
Inst. Theoretische Physik und Synergetik
Pfaffenwaldring 57
70569 Stuttgart
Germany
mahler@theo1.physik.uni-stuttgart.de

Gemmer, J. et al., *Quantum Thermodynamics: Emergence of Thermodynamic Behavior Within Composite Quantum Systems*, Lect. Notes Phys. 784 (Springer, Berlin Heidelberg 2009), DOI 10.1007/978-3-540-70510-9

Lecture Notes in Physics ISSN 0075-8450     e-ISSN 1616-6361
ISBN 978-3-642-26099-5                        e-ISBN 978-3-540-70510-9
DOI 10.1007/978-3-540-70510-9
Springer Heidelberg Dordrecht London New York

*Cover design*: Integra Software Services Pvt. Ltd., Pondicherry

Printed on acid-free paper

Springer is part of Springer Science+Business Media (www.springer.com)

# Preface 2nd Edition

More than 4 years have passed since the first edition of this book appeared. This time period has seen a number of exciting developments in the emerging field of quantum thermodynamics. Meanwhile it is becoming a popular keyword, and quite a number of researchers and students alike are being attracted by the fact that it is here, where one finds plenty of room for new thoughts, for the design of new theoretical (as well as experimental) tools, for new concepts, for the revision of seemingly well-established interpretations.

The reduction of thermodynamics to an underlying theory has challenged physicists for more than a century. So far such attempts have met with partial success at most. In this context an interesting and consistent route is offered by quantum thermodynamics.

Historically there has been a strong link between quantum mechanics and thermodynamics, right from the start. It was Planck's profound insight in thermodynamics, which originally gave birth to quantum theory.

Quantum thermodynamics intends to show that decoherence is far from being just a technical nuisance but a generic phenomenon of partite quantum systems giving rise to some of the most dominating, if apparently classical, features of closed (finite) quantum systems: thermal equilibrium.

Meanwhile, our focus has shifted from these foundational issues towards applications, in particular, non-equilibrium phenomena and thermodynamic processes in the nano-domain. Looking back, we felt a strong need to revise also the presentation of the equilibrium section. We tried hard to improve readability, to reduce technical details as far as we could. As a result a large fraction of this book has thoroughly been redone.

Even today, the central parts of this book are not yet "mainstream". We hope the prospective reader will find this updated version useful. We apologize for printing errors, which may have escaped our attention. We take the responsibility for misleading statements or even misunderstandings on our side: Work is in progress.

Stuttgart                                                                                      G. Mahler
October 2008

# Preface 1st Edition

This monograph views thermodynamics as an *incomplete description* of many free-dom quantum systems. Left unaccounted for may be an environment with which the system of interest interacts; closed systems can be described incompletely by focussing on any subsystem with fewer particles and declearing the remainder as the environment. Any interaction with the environment brings the open system to a mixed quantum state, even if the closed compound state is pure. Moreover, observables (and sometimes even the density operator) of an open system may relax to equilibrium values while the closed compound state keeps evolving unitarily á la Schrödinger forever.

The view thus taken can hardly be controversial for our generation of physicists. And yet, the authors offer surprises. Approach to equilibrium, with equilibrium characterized by maximum ignorance about the open system of interest, does not require excessively many particles: some dozens suffice! Moreover, the precise way of partitioning which might reflect subjective choices is immaterial for the salient features of equilibrium and equilibration. And what is nicest, quantum effects are at work in bringing about universal thermodynamic behavior of modest size open systems. von Neumann's concept of entropy thus appears as much more widely useful than sometimes feared, way beyond truly macroscopic systems in equilibrium.

The authors have written numerous papers on their quantum view of thermodynamics, and this monograph is a most welcome coherent review.

Essen,                                                      *Fritz Haake*
June 2004

# Acknowledgments

The authors thank Dr. Michael Hartmann (now Technische Universität Munich) and Dipl. Phys. Thomas Jahnke (Universität Stuttgart) for contributing some sections. Some chapters are based on work to which Dr. Robin Steinigeweg (Universität Osnabrück), Dipl. Phys. Hannu Wichterich (University College London), and last but not at all least P.D. Dr. Heinz-Peter Breuer (Albert-Ludwigs-Universität Freiburg) contributed substantially. We have profited a lot from fruitful discussions with Dipl. Phys. Heiko Schröder, Dipl. Phys. Jens Teifel, Dipl. Phys. Hendrik Weimer (Universität Stuttgart), Prof. Jan Birjukov (Ural State Technical University, Jekaterinburg), Prof. Armen Allahverdyan (Yerevan Physics Institute, Yerevan), Prof. Dr. Heinz-Jürgen Schmidt, Dipl. Phys. Christian Bartsch, Dipl. Phys. Mehmet Kadiroglu (Universität Osnabrück), Prof. Jürgen Schnack, Prof. Peter Reimann (Universität Bielefeld), and Dr. Fabian Heidrich-Meisner (RWTH Aachen). We are indebted to Prof. Joel L. Lebowitz for a comment on a draft version of this book. It is a pleasure to thank Springer Verlag, especially Dr. Christian Caron, for continuous encouragement and excellent cooperation. Financial support by the "Deutsche Forschungsgesellschaft," the "Landesstiftung Baden-Württemberg," and the "Deutsche Akademische Austausch Dienst (DAAD)" is gratefully acknowledged.

# Contents

# Part I
# Background

# Chapter 1
# Introduction

*Over the years enormous effort was invested in proving
ergodicity, but for a number of reasons, confidence in the
fruitfulness of this approach has waned.*
— Y. Ben-Menahem and I. Pitowsky [1]

**Abstract** The basic motivation behind the present text is threefold: To give a
new explanation for the emergence of thermodynamics, to investigate the interplay
between quantum mechanics and thermodynamics, and to explore possible extensions of the common validity range of thermodynamics.

Originally, thermodynamics has been a purely phenomenological science. Early scientists (Galileo, Santorio, Celsius, Fahrenheit) tried to give definitions for quantities
which were intuitively obvious to the observer, like pressure or temperature, and
studied their interconnections. The idea that these phenomena might be linked to
other fields of physics, like classical mechanics, e.g., was not common in those days.
Such a connection was basically introduced when Joule calculated the heat equivalent in 1840 showing that heat was a form of energy, just like kinetic or potential
energy in the theory of mechanics.

At the end of the 19th century, when the atomic theory became popular,
researchers began to think of a gas as a huge amount of bouncing balls inside a box.
With this picture in mind it was tempting to try to reduce thermodynamics entirely
to classical mechanics. This was exactly what Boltzmann tried to do in 1866 [2],
when he connected entropy, a quantity which so far had only been defined phenomenologically, to the volume of a certain region in phase space, an object defined
within classical mechanics. This was an enormous step forward, especially from a
practical point of view. Taking this connection for granted one could now calculate
all sorts of thermodynamic properties of a system from its Hamilton function. This
gave rise to modern thermodynamics, a theory the validity of which is beyond any
doubt today. Its results and predictions are a basic ingredient for the development of
all kinds of technical apparatuses ranging from refrigerators to superconductors.

Boltzmann himself, however, tried to prove the conjectured connection between
the phenomenological and the theoretical entropies, but did not succeed without
invoking other assumptions like the famous ergodicity postulate or the hypothesis
of equal "a priori probabilities." Later on, other physicists (Gibbs [3], Birkhoff [4],
Ehrenfest [5], von Neumann [6], etc.) tried to prove those assumptions, but none

Gemmer, J. et al.: *Introduction*. Lect. Notes Phys. **784**, 3–6 (2009)
DOI 10.1007/978-3-540-70510-9_1

of them seems to have solved the problem satisfactorily. It has been pointed out, though, that there are more properties of the entropy to be explained than its mere equivalence with the region in phase space, before thermodynamics can be reduced to classical mechanics, thus the discussion is still ongoing [1]. The vast majority of the work done in this field is based on classical mechanics.

Meanwhile, quantum theory, also initially triggered by the atomic hypothesis, has made huge progress during the last century and is today believed to be more fundamental than classical mechanics. At the beginning of the 21st century it seems highly unlikely that a box with balls inside could be anything more than a rough caricature of what a gas really is. Furthermore, thermodynamic principles seem to be applicable to systems that cannot even be described in classical phase space. Those developments make it necessary to rethink the work done so far, whether it has led to the desired result (e.g., demonstration of ergodicity) or not. The fact that a basically classical approach apparently did so well may even be considered rather surprising.

Of course, there have been suggestions of how to approach the problem on the basis of quantum theory [7–14], but again, none of them seems to have established the emergence of thermodynamics from quantum mechanics as an underlying theory in a conclusive way.

Roughly 4 years after the first edition of this book had appeared, we have now the opportunity to present an updated and considerably revised version. The presentation has been changed, and there are now four parts instead of three, which also means increased page numbers. Nevertheless, the original motivation for this text has largely remained unchanged; it consists primarily of three interrelated objectives:

(i) To give somewhat new explanation for the emergence of thermodynamic behavior. This point of view definitely leaves one question open: whether or not all macroscopic thermodynamic systems belong to the class of systems that will be examined in the following. The answer to this question is beyond the scope of this text.

(ii) To stimulate the delicate interplay between quantum mechanics and thermodynamics. This could possibly shed new light on some interpretational problems within quantum mechanics: with the "exorcism" of subjective ignorance as a guiding principle underlying thermodynamic states, the general idea of quantum states representing subjective knowledge might lose much of its credibility.

(iii) To ask whether the principles of thermodynamics might be applicable to systems other than the pertinent large, many-particle systems. It is a main conclusion of this work that the answer has to be positive. For it turns out that a large class of small quantum systems without any restriction concerning size or particle number show thermodynamic behavior with respect to an adequately defined set of thermodynamic variables. This behavior ("nano-thermodynamics") requires some embedding but is nevertheless established entirely on the basis of the Schrödinger equation.

This book is not intended to be a review. Related work includes, in particular, the so-called decoherence theory. We cannot do justice to the numerous investigations; we merely give a few references [15–18]. It might be worth mentioning here that decoherence has, during the last years, mainly been discussed as one of the main obstacles in the implementation of large-scale quantum computers [19], possibly neglecting other aspects of the phenomenon.

We close these introductory remarks with a short "manual" for the reader:

The first part of this book is simply a collection of well-established concepts in the field of thermodynamics and quantum mechanics. Chapters 2 and 3 are not meant to serve as a full-fledged introduction, more as a reminder of the central topics in quantum mechanics and thermodynamics. They may very well be skipped by a reader familiar with these subjects. Chapter 4 is a collection of historical approaches to thermodynamics with a focus on their subtleties rather than their undoubtable brilliance. Again this chapter is not imperative for the understanding of Part II.

In Part II the emergence of equilibrium thermodynamical behavior from Schrödinger-type quantum dynamics is investigated. To those ends Chap. 5 lists the properties of thermodynamic variables that do not follow in an obvious way and thus require careful explanation on the basis of an underlying theory (quantum mechanics). These explanations are then given in the remainder of Part II. In Chap. 6 the central ideas of the present quantum approach to thermodynamics which are "typicality," "objective lack of knowledge," etc., are explained in a rather general way. Their concrete implementation is then given in the remainder of Part II. Section 7.3 gives a flavor of how the frequently appearing Hilbert space averages are mathematically done. However, full explanation of the respective mathematics is given in Appendices A, B, and C. Wherever needed in Part II results from the Appendix concerning Hilbert space averages are simply quoted. Apart from being relevant in the context of a "derivation of thermodynamics," Chaps. 11, 12, and 15 comprise their messages in a self-contained way and may thus be read without prior reading of previous sections.

Part III is dedicated to linear non-equilibrium thermodynamics. The focus is not on traditional linear response theory but on the relaxation dynamics of non-equilibrium states. Chapter 17 is again a very basic review of well-established relaxation theories. Chapter 18 is somewhat technical and comments on the interrelations between Hilbert space averaging and projection techniques and the implications thereof. The remainder of Part III is then based on projection techniques and may be read without any notion of Hilbert space averages or typicality. In Chaps. 19 and 20 the relaxation dynamics of systems coupled to finite baths and of non-equilibrium density distributions in spatially structured systems are discussed. Chapter 21 investigates transport induced by reservoirs featuring different temperatures. The latter is self-contained and may be read without reference to other parts.

Part IV, eventually, is concerned with some specific applications. Chapter 22 gives a brief, but nonetheless illustrative example for "typical" state features to be expected from a product Hilbert space – independent of any physical constraints. Chapters 23 and 24 try to explore the measurement problem in thermodynamics: How to get information about thermodynamical variables, despite the fact that they are not observables in the usual sense. Thermodynamic processes (Chap. 25), while not in the main focus of this book, nevertheless constitute a central connection between theory and applications.

# References

1. Y. Ben-Menahem, I. Pitowsky, Stud. Hist. Phil. Mod. Phys. **32**, 503 (2001)
2. L. Boltzmann, *Lectures on Gas Theory* (University of California Press, Los Angeles, 1964)
3. J.W. Gibbs, *Elementary Principles in Statistical Mechanics* (Dover Publications, New York, 1960)
4. G. Birkhoff, Proc. Natl. Acad. Sci. USA **17**, 656 (1931)
5. P. Ehrenfest, T. Ehrenfest, *The Conceptual Foundations of the Statistical Approach in Mechanics* (Cornell University Press, Ithaca, 1959)
6. J.V. Neumann, Proc. Natl. Acad. Sci. USA **18**, 70 (1932)
7. L. Landau, I. Lifshitz, *Quantum Mechanics, Course of Theoretical Physics*, vol. 3, 3rd edn. (Pergamon Press, Oxford, 1977)
8. J.V. Neumann, Z. Phys. **57**, 30 (1929)
9. W. Pauli, Z. Fierz, Z. Phys. **106**, 572 (1937)
10. L. Landau, E. Lifshitz, *Statistical Physics, Part 1, Course of Theoretical Physics*, vol. 5, 3rd edn. (Pergamon Press, Oxford, 1980)
11. E. Schrödinger, *Statistical Thermodynamics*. A course of seminar lectures, del. in Jan.–March 1944 (Cambridge University Press, Cambridge, 1948)
12. G. Lindblad, *Non-equilibrium Entropy and Irreversibility, Mathematical Physics Studies*, vol. 5 (Reidel, Dordrecht, 1983)
13. W. Zurek, Phys. Today **41**(10), 36 (1991)
14. W. Zurek, J. Paz, Phys. Rev. Lett. **72**, 2508 (1994)
15. D. Giulini, E. Joos, C. Kiefer, J. Kupsch, I.O. Stamatescu, H. Zeh, *Decoherence and the Appearance of a Classical World in Quantum Theory*, 2nd edn. (Springer, Berlin, Heidelberg, 2003)
16. D. Braun, F. Haake, W.T. Strunz, Phys. Rev. Lett. **86**, 2913 (2001)
17. W.H. Zurek, Rev. Mod. Phys. **75**, 715 (2003)
18. L. Hackermüller, K. Hornberger, B. Brezger, A. Zeilinger, M. Arndt, Nature **427**, 711 (2004)
19. M. Nielsen, I. Chuang, *Quantum Computation and Quantum Information* (Cambridge University Press, Cambridge, 2000)

# Chapter 2
# Basics of Quantum Mechanics

*Those who are not shocked when they first come across*
*quantum theory cannot possibly have understood it.*
— N. Bohr [1]

**Abstract** Before we can start with the quantum mechanical approach to thermodynamics we have to introduce some fundamental terms and definitions of standard quantum mechanics for later reference. This chapter should introduce the reader only to some indispensable concepts of quantum mechanics necessary for the text at hand, but is far from being a complete overview of this subject. For a complete introduction we refer to standard textbooks [2–7].

## 2.1 Introductory Remarks

The shortcomings of classical theories had become apparent by the end of the 19th century. Interestingly enough, one of the first applications of quantum ideas has been within thermodynamics: Planck's famous formula for black body radiation was based on the hypothesis that the exchange of energy between the container walls and the radiation field should occur in terms of fixed energy quanta only. Later on, this idea has been put on firmer ground by Einstein postulating his now well-known rate equations [8].

Meanwhile quantum mechanics has become a theory of unprecedented success. So far, its predictions have always been confirmed by experiment.

Quantum mechanics is usually defined in terms of some loosely connected axioms and rules. Such a foundation is far from the beauty of, e.g., the "principles" underlying classical mechanics. Motivated, in addition, by notorious interpretation problems, there have been numerous attempts to modify or "complete" quantum mechanics.

A first attempt was based on so-called "hidden variables" [9]. Its proponents essentially tried to expel the non-classical nature of quantum mechanics. More recent proposals intend to "complete" quantum mechanics not within mechanics, but on a higher level: by means of a combination with gravitation theory [10], with psychology [11], or with (quantum-) information theory [12, 13].

While the emergence of classicality from an underlying quantum substrate has enjoyed much attention recently, it has so far not been appreciated that the under-

Gemmer, J. et al.: *Basics of Quantum Mechanics*. Lect. Notes Phys. **784**, 7–22 (2009)
DOI 10.1007/978-3-540-70510-9_2

standing of quantum mechanics may benefit also from subjects like quantum thermodynamics.

## 2.2 Operator Representations

In quantum mechanics we deal with systems (Hamilton models), observables, and states. They all are represented by Hermitian operators. Their respective specification requires data (parameters), which have to be defined with respect to an appropriate reference frame. These frames are operator representations. Let us in the following consider some aspects of these operator representations in detail. First we will concentrate on simple systems and their representation.

### 2.2.1 Transition Operators

If we restrict ourselves to systems living in a finite and discrete Hilbert space $\mathscr{H}$ (a complex vector space of dimension $n_{\text{tot}}$), we may introduce a set of orthonormal state vectors $|i\rangle \in \mathscr{H}$. From this orthonormal and complete set of state vectors with

$$\langle i | j \rangle = \delta_{ij} , \quad i, j = 1, 2, \ldots, n_{\text{tot}} , \tag{2.1}$$

we can define $n_{\text{tot}}^2$ transition operators (in general non-Hermitian)

$$\hat{P}_{ij} = |i\rangle\langle j| , \quad \hat{P}_{ij}^{\dagger} = \hat{P}_{ji} . \tag{2.2}$$

These operators are, again, orthonormal in the sense that

$$\text{Tr}\left\{ \hat{P}_{ij} \hat{P}_{i'j'}^{\dagger} \right\} = \delta_{ii'}\delta_{jj'} , \tag{2.3}$$

where $\text{Tr}\{\ldots\}$ denotes the trace operation. Furthermore, they form a complete set in so-called Liouville space, into which any other operator $\hat{A}$ can be expanded,

$$\hat{A} = \sum_{i,j} A_{ij} \hat{P}_{ij} , \tag{2.4}$$

$$A_{ij} = \text{Tr}\left\{ \hat{A} \hat{P}_{ij}^{\dagger} \right\} = \langle i|\hat{A}|j\rangle . \tag{2.5}$$

The $n_{\text{tot}}^2$ parameters are, in general, complex ($2n_{\text{tot}}^2$ real numbers). For Hermitian operators we have, with

$$\hat{A}^\dagger = \sum_{i,j} A_{ij}^* \hat{P}_{ij}^\dagger = \sum_{i,j} A_{ji} \hat{P}_{ji} = \hat{A} \;, \tag{2.6}$$

$$A_{ij}^* = A_{ji} \;, \tag{2.7}$$

i.e., we are left with $n_{\text{tot}}^2$-independent real numbers. All these numbers must be given to uniquely specify any Hermitian operator $\hat{A}$.

## 2.2.2 Pauli Operators

There are many other possibilities to define basis operators, besides the transition operators. For $n_{\text{tot}} = 2$ a convenient set is given by the so-called Pauli operators $\hat{\sigma}_i$ ($i = 0, \ldots, 3$). The new basis operators can be expressed in terms of transition operators

$$\hat{\sigma}_1 = \hat{P}_{12} + \hat{P}_{21} \;, \tag{2.8}$$

$$\hat{\sigma}_2 = \mathrm{i}(\hat{P}_{12} - \hat{P}_{21}) \;, \tag{2.9}$$

$$\hat{\sigma}_3 = -\hat{P}_{11} + \hat{P}_{22} \;, \tag{2.10}$$

$$\hat{\sigma}_0 = \hat{\mathbb{1}} \;. \tag{2.11}$$

These operators are Hermitian and – except for $\hat{\sigma}_0$ – traceless. The Pauli operators satisfy several important relations: $(\hat{\sigma}_i)^2 = \hat{\mathbb{1}}$ and $[\hat{\sigma}_1, \hat{\sigma}_2] = 2\mathrm{i}\hat{\sigma}_3$ and their cyclic extensions. Since the Pauli operators form a complete orthonormal operator basis, it is possible to expand any operator in terms of these basis operators. Furthermore we introduce raising and lowering operators, in accordance with

$$\hat{\sigma}_+ = 1/2(\hat{\sigma}_1 + \mathrm{i}\hat{\sigma}_2) \;, \quad \hat{\sigma}_- = 1/2(\hat{\sigma}_1 - \mathrm{i}\hat{\sigma}_2) \;. \tag{2.12}$$

Also for higher dimensional cases, $n_{\text{tot}} \geq 2$, one could use as a basis the Hermitian generators of the $\text{SU}(n_{\text{tot}})$ group.

## 2.2.3 State Representation

The most general way to define the state of a quantum mechanical system is by its density matrix, $\rho_{ij}$, which specifies the representation of the density operator,

$$\hat{\rho} = \sum_{i,j} \rho_{ij} \hat{P}_{ij} \tag{2.13}$$

subject to the condition

$$\text{Tr} \{\hat{\rho}\} = \sum_i \rho_{ii} = 1 \ . \tag{2.14}$$

The expectation value for some observable $\hat{A}$ in state $\hat{\rho}$ is now given by

$$\langle A \rangle = \text{Tr} \{\hat{A}\hat{\rho}\} = \sum_{i,j} A_{ij}\rho_{ij} \ . \tag{2.15}$$

The density matrix $\rho_{ij} = \langle i|\hat{\rho}|j\rangle$ is a positive definite and Hermitian matrix. The number of independent real numbers needed to specify $\hat{\rho}$ is thus $d = n_{\text{tot}}^2 - 1$. For the density operator of an arbitrary pure state $|\psi\rangle$ we have $\hat{\rho} = |\psi\rangle\langle\psi|$. In the eigenrepresentation one finds, with $W_i = \rho_{ii}$,

$$\hat{\rho} = \sum_i W_i \hat{P}_{ii} \ , \tag{2.16}$$

which may be interpreted as a "mixture" of pure states $\hat{P}_{ii} = |i\rangle\langle i|$ with the statistical weight $W_i$. From this object the probability $W(|\psi\rangle)$ to find the system in an arbitrary pure state, expanded in the basis $|i\rangle$

$$|\psi\rangle = \sum_i \psi_i|i\rangle \ , \tag{2.17}$$

can be calculated as

$$W(|\psi\rangle) = \langle\psi|\hat{\rho}|\psi\rangle = \sum_i |\psi_i|^2 W_i \ . \tag{2.18}$$

To measure the distance of two arbitrary, not necessarily, pure states given by $\hat{\rho}$ and $\hat{\rho}'$ we define a "distance measure"

$$D^2_{\hat{\rho}\hat{\rho}'} = \text{Tr} \left\{(\hat{\rho} - \hat{\rho}')^2\right\} \ . \tag{2.19}$$

This commutative measure (sometimes called Bures metric) has a number of convenient properties: $D^2_{\hat{\rho}\hat{\rho}'} \geq 0$ with the equal sign holding if and only if $\hat{\rho} = \hat{\rho}'$; the triangle inequality holds as expected for a conventional distance measure; for pure states

$$D^2_{|\psi\rangle|\psi'\rangle} = 2\left(1 - |\langle\psi|\psi'\rangle|^2\right) \leq 2 \tag{2.20}$$

and $D^2$ is invariant under unitary transformations. A second measure of distance is the fidelity defined by [14]

$$F_{\hat{\rho}\hat{\rho}'} = \text{Tr} \left\{\left(\sqrt{\hat{\rho}}\, \hat{\rho}' \sqrt{\hat{\rho}}\right)^{1/2}\right\} \ . \tag{2.21}$$

For pure states $F$ is just the modulus of the overlap: $F_{|\psi\rangle|\psi'\rangle} = |\langle\psi|\psi'\rangle|$.

### 2.2.4 Purity and von Neumann Entropy

For a pure state in eigenrepresentation all matrix elements in (2.13) of the density matrix are zero except $\rho_{ii} = 1$, say, i.e., the density operator $\hat{\rho} = \hat{P}_{ii}$ is a projection operator. Obviously in this case $\hat{\rho}^2 = \hat{\rho}$, due to the properties of the projection operator, so that the so-called purity becomes

$$P = \text{Tr}\{\hat{\rho}^2\} = 1 . \tag{2.22}$$

In general, we have

$$P(\hat{\rho}) = \sum_{i,j}\sum_{i',j'} \rho_{ij}\,\rho_{i'j'}\,\text{Tr}\left\{\hat{P}_{ij}\,\hat{P}_{i'j'}^\dagger\right\} = \sum_{i,j}|\rho_{ij}|^2 > \frac{1}{n_{\text{tot}}} . \tag{2.23}$$

Because of the Cauchy–Schwarz relation

$$|\rho_{ij}|^2 \le \rho_{ii}\,\rho_{jj} , \tag{2.24}$$

we conclude that $P \le 1$. The equality sign holds for pure states only. $P$ can be calculated for any density matrix without prior diagonalization. In the diagonal representation (cf. (2.16)) the purity is simply the sum of the squares of the probabilities $W_i$ to find the system in a respective eigenstate,

$$P(\{W_i\}) = \text{Tr}\{\hat{\rho}^2\} = \sum_i W_i^2 . \tag{2.25}$$

Note that the purity itself is invariant with respect to unitary transformations. Its value does not depend on the representation chosen.

Furthermore, a very important quantity is another measure called the von Neumann entropy [14]. Also this measure is defined for any state $\hat{\rho}$ as

$$S(\hat{\rho}) = -k_B \text{Tr}\{\hat{\rho}\,\ln\hat{\rho}\} \ge 0 , \tag{2.26}$$

where $k_B$ denotes a proportional constant, the Boltzmann constant. (At this point the inclusion of $k_B$ is arbitrary and not yet meant to anticipate any connection to thermodynamics.) For a pure state the minimum entropy $S = 0$ is reached. The maximum entropy obtains for

$$\rho_{ij} = \frac{1}{n_{\text{tot}}}\,\delta_{ij} , \quad i, j = 1, 2, \ldots, n_{\text{tot}} , \tag{2.27}$$

i.e., for a density matrix proportional to the normalized unit matrix, with the entropy

$$S_{\text{max}} = k_B \ln n_{\text{tot}} \, . \tag{2.28}$$

In the same limit the purity $P$ is minimal,

$$P_{\text{min}} = \frac{1}{n_{\text{tot}}} \, . \tag{2.29}$$

The maximum entropy (or minimum purity) is thus found for the broadest possible probability distribution, the equipartition over all pure states (remember (2.27)). Therefore $S$ and $P$ are both measures for the "broadness" of the distribution.

The purity can be expressed as a rather simple function of the full state, the evaluation of which does not require the diagonalization of a matrix, as opposed to the calculation of the von Neumann entropy. We will thus mainly consider $P$ rather than $S$.

In general, though, these two measures do not uniquely map onto each other. Nevertheless in the limits of maximum $S$ (minimum $P$) and maximum $P$ (minimum $S$) they do. The formal approximation $\ln \hat{\rho} \approx \hat{\rho} - \hat{1}$ leads to the "linearized" entropy

$$S_{\text{lin}} = k_B(1 - P) \geq 0 \, . \tag{2.30}$$

Since, as will be shown in Sect. 2.4, $S$ is a constant of motion, the question for the possible origin of $S > 0$ arises. One interpretation is essentially classical and traces a finite $S$ back to subjective ignorance. In the eigenrepresentation of the density operator (see (2.16)) the density operator can be seen as a "mixture" of pure states $\hat{P}_{ii} = |i\rangle\langle i|$ and the entropy then reads

$$S = -k_B \sum_i W_i \ln W_i \, . \tag{2.31}$$

Alternatively, a nonpure state may result from the system under consideration being entangled with another system, while the total state is pure. In this case $S$ indicates a principal uncertainty. It is always possible to find such an embedding, as will be discussed in the next section.

### 2.2.5 Bipartite Systems

Systems typically consist of subsystems. In the case of a bipartite system, the total Hilbert space can be decomposed into a product space

$$\mathcal{H} = \mathcal{H}^{(1)} \otimes \mathcal{H}^{(2)} \, , \tag{2.32}$$

with dimension $n_{\text{tot}} = n^{(1)} \cdot n^{(2)}$. A complete set of orthonormal vectors is then given by the product states ($\otimes$ means tensor product of the vectors involved)

$$|ij\rangle = |i\rangle \otimes |j\rangle , \tag{2.33}$$

with $i = 1, 2, \ldots, n^{(1)}$ numbering the states in $\mathscr{H}^{(1)}$ and $j = 1, 2, \ldots, n^{(2)}$ in $\mathscr{H}^{(2)}$. The states fulfill the orthonormality relation

$$\langle ij|i'j'\rangle = \delta_{ii'}\, \delta_{jj'} . \tag{2.34}$$

Based on this we can define the transition operators

$$\hat{P}_{ij|i'j'} = |ij\rangle\langle i'j'| = \hat{P}_{ii'}^{(1)} \otimes \hat{P}_{jj'}^{(2)} , \tag{2.35}$$

where $\hat{P}_{ii'}^{(\mu)}$ is a transition operator in the subspace of the subsystem $\mu = 1, 2$. These, again, form a complete orthogonal set such that any operator $\hat{A}$ can be expanded in the form

$$\hat{A} = \sum_{i,j} \sum_{i',j'} A_{ij|i'j'}\, \hat{P}_{ij|i'j'} . \tag{2.36}$$

For a pure state

$$|\psi\rangle = \sum_{i,j} \psi_{ij}\, |i\rangle \otimes |j\rangle \tag{2.37}$$

the density operator $\hat{\rho} = |\psi\rangle\langle\psi|$ has the matrix representation

$$\rho_{ij|i'j'} = \psi_{ij}\, \psi_{i'j'}^* . \tag{2.38}$$

If we are interested in the state of one of the subsystems alone we have to trace over the other subsystem. The reduced density operator of the system of interest is now given by

$$\hat{\rho}^{(1)} = \mathrm{Tr}_2\{\hat{\rho}\} = \sum_{i,i'} \sum_{j} \langle ij|\hat{\rho}|i'j\rangle\, |i\rangle\langle i'|$$
$$= \sum_{i,i'} \rho_{ii'}\, \hat{P}_{ii'}^{(1)} , \tag{2.39}$$

with $\rho_{ii'} = \sum_{j} \rho_{ij|i'j}$. Here $\mathrm{Tr}_2\{\ldots\}$ means trace operation within Hilbert space $\mathscr{H}^{(2)}$. The result for subsystem 2 is obtained by exchanging the indices of the two subsystems.

The expectation value for any local operator $\hat{A}^{(1)} \otimes \hat{1}^{(2)}$ can be calculated from

$$\langle \hat{A}^{(1)} \rangle = \mathrm{Tr}_1\{\hat{A}^{(1)}\hat{\rho}^{(1)}\} . \tag{2.40}$$

The corresponding purity, say, for the reduced state of the first subsystem, is

$$P(\hat{\rho}^{(1)}) = \sum_{i,i'} |\rho_{ii'}|^2 = \sum_{i,i'} \sum_{j,j'} \rho_{ij|i'j} \; \rho_{i'j'|ij'} \; . \tag{2.41}$$

Furthermore, the reduced von Neumann entropies are given by

$$S(\hat{\rho}^{(\mu)}) = -k_B \mathrm{Tr}_\mu \left\{ \hat{\rho}^{(\mu)} \ln \hat{\rho}^{(\mu)} \right\} \; , \quad \mu = 1, 2 \; . \tag{2.42}$$

One easily convinces oneself that for

$$\hat{\rho} = \hat{\rho}^{(1)} \otimes \hat{\rho}^{(2)} \tag{2.43}$$

the total entropy is additive,

$$S = S(\hat{\rho}^{(1)}) + S(\hat{\rho}^{(2)}) \; . \tag{2.44}$$

In general, the theorem by Araki and Lieb [15] tells us that

$$|S(\hat{\rho}^{(1)}) - S(\hat{\rho}^{(2)})| \le S \le S(\hat{\rho}^{(1)}) + S(\hat{\rho}^{(2)}) \; . \tag{2.45}$$

This theorem implies that if the total system is in a pure state ($S = 0$) then $S(\hat{\rho}^{(1)}) = S(\hat{\rho}^{(2)})$, no matter how the system is partitioned. Under the same condition $P(\hat{\rho}^{(1)}) = P(\hat{\rho}^{(2)})$. Then if $S(\hat{\rho}^{(1)}) = S(\hat{\rho}^{(2)}) > 0$, it follows that (2.44) does not apply and the total (pure) state cannot be written in a product form. This is interpreted to result from "entanglement," for which the local entropies $S(\hat{\rho}^{(1)}) = S(\hat{\rho}^{(2)})$ thus constitute an appropriate measure.

Such pure entangled states have been of central interest now for almost 70 years. They can have properties that seem to contradict intuition. If a local measurement on one subsystem is made, i.e., a projection of only one subsystem state is performed, the local state of the other subsystem can be severely affected, which has raised the question of whether quantum mechanics could be valid at all [16]. Nevertheless, these states can theoretically be shown to result from product states, if the subsystems are allowed to interact for a while. On a small scale such a buildup has been demonstrated experimentally; it is a widespread belief that entanglement as a fundamental quantum mechanical property should show up mainly between very small objects.

### 2.2.6 Multipartite Systems

Alternatively, one may consider a network of $N$ subsystems of dimension $n$ each. Then $n_{\mathrm{tot}} = n^N$. As a consequence of the direct product structure, the number of parameters required to specify a density operator then grows exponentially with $N$

$$d = n^{2N} - 1 \; . \tag{2.46}$$

For the classical system of $N$ point particles we would need $6N$ real parameters, i.e., we would just have to specify position and momentum of each individual particle. This so-called phase space is the direct sum of the individual particle spaces. The analog in the quantum case would be to specify the local states of the $N$ subsystems, for which we would need $(n^2 - 1)N$ parameters. (This was the dimension of the direct sum of subsystem Liouville spaces.) Defining

$$\gamma = \frac{d}{(n^2 - 1)N} ,  \qquad (2.47)$$

we see that for $n = 2$, $N = 3$, $\gamma = 7$, but for $N = 10$, $\gamma \approx 30\,000$. The tremendous information needed over the local parameters is due to the fact that correlations (entanglement) dominate, in general. For product states $\gamma = 1$.

The blowup of $\gamma$ is a typical quantum property, closer to the heart of quantum mechanics than the famous Heisenberg uncertainty relation. Both are due to the non-commutativity of the underlying operators, though.

The number of parameters needed to specify a Hamilton model typically grows only polynomially with $N$. This is because direct interactions are usually restricted to finite clusters, e.g., up to pairs.

## 2.3 Dynamics

So far, we have considered some properties of Hilbert spaces, the basis operators and appropriate states. We turn now to some dynamical aspects of quantum systems.

The unitary dynamics of a closed system generated by a Hamilton operator $\hat{H}$ is given by the Schrödinger equation

$$i\hbar \frac{\partial}{\partial t} |\psi(t)\rangle = \hat{H}(t)|\psi(t)\rangle  \qquad (2.48)$$

for the time-dependent pure state $|\psi(t)\rangle$. This is the fundamental equation specifying the so-called Schrödinger picture: here the state vectors $|\psi(t)\rangle$ carry all dynamics, while the basic operators are time independent. But note that the Hamiltonian could include explicitly time-dependent potentials. These would render the system non-autonomous, though.

From the Schrödinger equation one can easily derive the evolution equation directly for the density operator. This is the Liouville–von Neumann equation

$$i\hbar \frac{\partial \hat{\rho}}{\partial t} = [\hat{H}, \hat{\rho}] ,  \qquad (2.49)$$

with $[\hat{A}, \hat{B}] = \hat{A}\hat{B} - \hat{B}\hat{A}$ defining the commutator. This equation can be written in the form

$$\frac{\partial \hat{\rho}}{\partial t} = \hat{\mathcal{L}} \hat{\rho} \, , \qquad (2.50)$$

where $\hat{\mathcal{L}}$ is a so-called superoperator acting (here) on the operator $\hat{\rho}$ to produce the new operator

$$\hat{\mathcal{L}} \hat{\rho} = -\frac{i}{\hbar} [\hat{H}, \hat{\rho}] \, . \qquad (2.51)$$

Modified superoperators control the dynamics of open quantum systems, which we will consider in detail in Sect. 4.8.

The Liouville–von Neumann equation can formally be solved by

$$\hat{\rho}(t) = \hat{U}(t) \, \hat{\rho}(0) \, \hat{U}^{\dagger}(t) \, , \qquad (2.52)$$

where the unitary time evolution operator, $\hat{U}^{\dagger}\hat{U} = \hat{U}\hat{U}^{\dagger} = \hat{1}$, also obeys the Schrödinger equation,

$$i\hbar \frac{\partial}{\partial t} \hat{U}(t) = \hat{H}(t) \, \hat{U}(t) \, . \qquad (2.53)$$

For $\partial \hat{H}/\partial t = 0$, i.e., no explicit time-dependent Hamiltonian, it has the formal solution

$$\hat{U}(t) = e^{-i\hat{H}t/\hbar} \, . \qquad (2.54)$$

When represented with respect to a specific set of basis operators, the Liouville–von Neumann equation is equivalent to

$$i\hbar \frac{\partial}{\partial t} \rho_{ij}(t) = \text{Tr} \left\{ [\hat{H}, \hat{\rho}(t)] \, \hat{P}_{ij}^{\dagger} \right\} \, . \qquad (2.55)$$

This equation determines the evolution of the matrix elements of the density operator. The solution $\rho_{ij}(t)$, subject to the condition $\sum_i \rho_{ii} = 1$, can thus be visualized as a deterministic quantum trajectory in Liouville space, controlled by the Hamiltonian and by the initial state $\hat{\rho}(0)$.

In the Heisenberg picture, the dynamics is carried by time-dependent observables

$$\hat{A}_{\text{H}}(t) = \hat{U}^{\dagger}(t) \, \hat{A} \, \hat{U}(t) \, , \qquad (2.56)$$

while the states are constant, $\hat{\rho}_{\text{H}}(t) = \hat{\rho}(0)$. If $\partial \hat{A}/\partial t = 0$ in the Schrödinger picture, the corresponding evolution equation for the now time-dependent operators reads

$$i\hbar \frac{d}{dt} \hat{A}_{\text{H}} = -[\hat{H}, \hat{A}_{\text{H}}] \, . \qquad (2.57)$$

In either picture the time dependence of the expectation value of an operator $\langle \hat{A} \rangle = \mathrm{Tr}\left\{\hat{A}\,\hat{\rho}\right\} = \mathrm{Tr}\left\{\hat{A}_{\mathrm{H}}\,\hat{\rho}(0)\right\}$ is given by

$$i\hbar \frac{\partial}{\partial t}\langle \hat{A} \rangle = \mathrm{Tr}\left\{[\hat{H}, \hat{\rho}]\,\hat{A}\right\} = -\mathrm{Tr}\left\{[\hat{H}, \hat{A}_{\mathrm{H}}]\,\hat{\rho}(0)\right\}$$

$$= -\langle[\hat{H}, \hat{A}_{\mathrm{H}}]\rangle , \tag{2.58}$$

which is known as the "Ehrenfest theorem." Since this evolution equation is similar to the classical equation of motion based on the Poisson bracket, this theorem can be interpreted to state that "the classical equations of motion are valid for expectation values in quantum mechanics."

## 2.4 Invariants

According to the Heisenberg equation of motion (2.57), conserved quantities are those which commute with the system Hamiltonian $\hat{H}$. In eigenrepresentation $\hat{H}$ can be written as

$$\hat{H} = \sum_{i=1}^{n_{\mathrm{tot}}} E_i\,\hat{P}_{ii} . \tag{2.59}$$

As a consequence, the projectors commute with the Hamiltonian itself,

$$[\hat{P}_{jj}, \hat{H}] = 0 , \quad j = 1, \ldots, n_{\mathrm{tot}} . \tag{2.60}$$

Since commutators are invariant under unitary transformations, the above relation thus holds in the Schrödinger as well as in the Heisenberg pictures. For the change of the energy distribution we find

$$i\hbar \frac{\partial}{\partial t} W_j = i\frac{\partial}{\partial t}\mathrm{Tr}\left\{\hat{P}_{jj}\,\hat{\rho}\right\} = -\mathrm{Tr}\left\{[\hat{H}, \hat{P}_{jj}^{(\mathrm{H})}]\,\hat{\rho}(0)\right\} = 0 , \tag{2.61}$$

i.e., the energy distribution, the probability of finding the system in state $j$, is a constant of motion.

Furthermore, defining the expectation value of an arbitrary function of the density operator $\hat{\rho}$

$$\langle f(\hat{\rho}) \rangle = \mathrm{Tr}\left\{\hat{\rho}\,f(\hat{\rho})\right\} , \tag{2.62}$$

one infers that

$$i\hbar\frac{\partial}{\partial t}\langle f(\hat{\rho})\rangle = i\hbar\operatorname{Tr}\left\{\frac{\partial}{\partial t}\hat{\rho}f(\hat{\rho})\right\} + i\hbar\operatorname{Tr}\left\{\hat{\rho}\frac{\partial}{\partial t}f(\hat{\rho})\right\}$$
$$= \operatorname{Tr}\left\{[\hat{H},\hat{\rho}]f(\hat{\rho})\right\} + \operatorname{Tr}\left\{\hat{\rho}[\hat{H},f(\hat{\rho})]\right\}. \tag{2.63}$$

Here we have made use of the Liouville equation (2.49) and its variant

$$i\hbar\frac{\partial}{\partial t}f(\hat{\rho}) = [\hat{H},f(\hat{\rho})]. \tag{2.64}$$

Observing the invariance of the first trace term in (2.63) under cyclic permutations, we see that the right-hand side cancels,

$$\frac{\mathrm{d}}{\mathrm{d}t}\langle f(\hat{\rho})\rangle = 0. \tag{2.65}$$

Taking now $f(\hat{\rho}) = \hat{\rho}$, the term $\langle f(\hat{\rho})\rangle = \operatorname{Tr}\left\{\hat{\rho}^2\right\}$ is just the purity, so that

$$\frac{\mathrm{d}}{\mathrm{d}t}P = 0. \tag{2.66}$$

For $f(\hat{\rho}) = \ln\hat{\rho}$ one concludes that the von Neumann entropy is invariant, too. In fact, any moment $\operatorname{Tr}\left\{(\hat{\rho})^k\right\}$ is a constant of motion in closed quantum systems. But note that the local reduced von Neumann entropy of a part of the system defined in (2.42) is not necessarily conserved under a unitary time evolution of the full system (see Sect. 6.1).

For later reference we finally investigate a bipartite system with the total Hamiltonian $\hat{H}$. Here we may encounter a situation for which

$$[\hat{A}^{(1)},\hat{H}] = 0,\quad [\hat{B}^{(2)},\hat{H}] = 0, \tag{2.67}$$

where the operator

$$\hat{A}^{(1)} = \sum_i A_i\,\hat{P}_{ii}^{(1)} \tag{2.68}$$

acts only on subsystem 1, and

$$\hat{B}^{(2)} = \sum_j B_j\,\hat{P}_{jj}^{(2)} \tag{2.69}$$

acts only on subsystem 2. As a consequence,

$$[(\hat{A}^{(1)})^k,\hat{H}] = \sum_i A_i^k[\hat{P}_{ii}^{(1)},\hat{H}] = 0. \tag{2.70}$$

As this has to hold for any $k$, we conclude that

$$[\hat{P}_{ii}^{(1)}, \hat{H}] = 0 , \tag{2.71}$$

and correspondingly,

$$[\hat{P}_{jj}^{(2)}, \hat{H}] = 0 . \tag{2.72}$$

According to these considerations the expectation value

$$W_{ij}^{(12)} = \mathrm{Tr}\left\{\hat{P}_{ii}(1)\,\hat{P}_{jj}(2)\,\hat{\rho}\right\} \tag{2.73}$$

is thus a conserved quantity, too. This expectation value is the joint probability for finding subsystem 1 in state $i$ and subsystem 2 in state $j$.

## 2.5 Time-Dependent Perturbation Theory

Just as in classical mechanics or any other theoretical framework, there are very few examples that allow us to achieve an exact analytical solution. In quantum mechanics even numerical solutions are seriously constrained by exponential explosion of state parameters (cf. Sect. 2.2.6). Many, quite powerful, approximation schemes have been developed. For later reference we summarize here the basics of standard perturbation theory.

To consider time-dependent phenomena it is often very helpful – if not unavoidable – to use a perturbation theory instead of a full solution of the time-dependent problem. To outline such a theory, we use in addition to the Schrödinger and the Heisenberg pictures the interaction or Dirac picture.

### 2.5.1 Interaction Picture

In the interaction picture, both observables and states are time dependent. We consider the Hamilton operator

$$\hat{H} = \hat{H}_0 + \hat{V}(t) , \tag{2.74}$$

where $\hat{H}_0$ represents the unperturbed Hamiltonian and $\hat{V}(t)$ the time-dependent perturbation. According to the unitary transformation

$$\hat{U}_0(t, t_0) = \exp\left(-\frac{i}{\hbar}\hat{H}_0(t - t_0)\right) , \tag{2.75}$$

where $t_0$ is the time at which the perturbation is switched on, one can transform the states as well as the operators of the Schrödinger picture into the interaction picture (index I)

$$|\psi(t)\rangle = \hat{U}_0(t, t_0) |\psi_\mathrm{I}(t)\rangle \,, \tag{2.76}$$

$$\hat{A}_\mathrm{I} = \hat{U}_0^\dagger(t, t_0) \, \hat{A} \, \hat{U}_0(t, t_0) \,. \tag{2.77}$$

Based on these transformations, the Schrödinger equation reads

$$i\hbar\left(\frac{\partial}{\partial t}\hat{U}_0\right)|\psi_\mathrm{I}(t)\rangle + i\hbar\hat{U}_0\frac{\partial}{\partial t}|\psi_\mathrm{I}(t)\rangle = (\hat{H}_0 + \hat{V})\hat{U}_0|\psi_\mathrm{I}(t)\rangle \,. \tag{2.78}$$

Observing that

$$i\hbar \frac{\partial}{\partial t}\hat{U}_0 = \hat{H}_0 \, \hat{U}_0 \tag{2.79}$$

and

$$\hat{U}_0 \hat{U}_0^\dagger = \hat{1} \,, \tag{2.80}$$

the above equation reduces to an effective Schrödinger equation for $|\psi_\mathrm{I}(t)\rangle$

$$i\hbar \frac{\partial}{\partial t}|\psi_\mathrm{I}(t)\rangle = \hat{V}_\mathrm{I}(t) |\psi_\mathrm{I}(t)\rangle \,, \tag{2.81}$$

identifying $\hat{V}_\mathrm{I}(t) = \hat{U}_0^\dagger \hat{V}(t)\hat{U}_0$. This equation has the formal solution

$$|\psi_\mathrm{I}(t)\rangle = \hat{U}_\mathrm{I}(t, t_0) |\psi_\mathrm{I}(t_0)\rangle, \tag{2.82}$$

with the evolution equation

$$i\hbar \frac{\partial}{\partial t}\hat{U}_\mathrm{I}(t, t_0) = \hat{V}_\mathrm{I}(t) \, \hat{U}_\mathrm{I}(t, t_0) \,. \tag{2.83}$$

The corresponding dynamics for observables in the interaction picture (remember (2.77)) is then controlled by

$$\frac{\mathrm{d}\hat{A}_\mathrm{I}}{\mathrm{d}t} = \frac{1}{i\hbar}[\hat{A}_\mathrm{I}(t), \hat{H}_0] + \hat{U}_0^\dagger \frac{\partial \hat{A}}{\partial t} \, \hat{U}_0 \,. \tag{2.84}$$

## 2.5.2 Series Expansion

The formal solution (2.82) of the effective Schrödinger equation (2.81) may be written as

$$\hat{U}_I(t, t_0) = \hat{1} - \frac{i}{\hbar} \int_{t_0}^t \mathrm{d}t_1 \, \hat{V}_\mathrm{I}(t_1) \, \hat{U}_\mathrm{I}(t_1, t_0) \,. \tag{2.85}$$

This integral equation can be solved for $\hat{U}_I(t, t_0)$ by iteration,

$$\hat{U}_I(t, t_0) = \hat{1} + \frac{i}{\hbar} \int_{t_0}^{t} dt_1 \, \hat{V}_I(t_1) + \left(\frac{i}{\hbar}\right)^2 \int_{t_0}^{t} dt_1 \int_{t_0}^{t_1} dt_2 \, \hat{V}_I(t_1) \, \hat{V}_I(t_2) + \cdots$$

$$= \sum_{n=0}^{\infty} \left(\frac{i}{\hbar}\right)^n \int_{t_0}^{t} dt_1 \int_{t_0}^{t_1} dt_2 \cdots \int_{t_0}^{t_{n-1}} dt_n \, \hat{V}_I(t_1) \cdots \hat{V}_I(t_n) , \qquad (2.86)$$

which is called the Dyson series expansion. In first order the transition probability due to $\hat{V}_I(t)$ is given by

$$W_{ij}(t) = \left| \delta_{ij} + \frac{1}{i\hbar} \int_{t_0}^{t} dt_1 \, \langle j | \hat{V}_I(t_1) | i \rangle \right|^2 . \qquad (2.87)$$

For $i \neq j$ and going back to the Schrödinger picture, we find

$$W_{ij}(t) = \frac{1}{\hbar^2} \left| \int_{t_0}^{t} dt_1 \, \exp\left(\frac{i(E_j - E_i)t_1}{\hbar}\right) \langle j | \hat{V}(t_1) | i \rangle \right|^2 . \qquad (2.88)$$

Let the time-dependent perturbation be

$$\hat{V}(t) = \begin{cases} 0 \\ \hat{V} \end{cases} \text{ for } \begin{array}{l} t \leq 0 \\ t > 0 \end{array} \qquad (2.89)$$

and

$$\frac{E_j - E_i}{\hbar} = \omega_{ji} . \qquad (2.90)$$

Then we find for the transition probability

$$W_{ij}(t) = \frac{1}{\hbar^2} \left| \frac{e^{i\omega_{ji}t} - 1}{\omega_{ji}} \langle j | \hat{V} | i \rangle \right|^2 \qquad (2.91)$$

$$= \frac{1}{\hbar^2} \left| \frac{\sin(\omega_{ji} t/2)}{\omega_{ji}/2} \right|^2 |\langle j | \hat{V} | i \rangle|^2 , \qquad (2.92)$$

which gives Fermi's golden rule for large times

$$W_{ij}(t) = t \, \frac{2\pi}{\hbar} \, \delta(E_j - E_i)|\langle j | \hat{V} | i \rangle|^2 , \qquad (2.93)$$

i.e., a constant transition rate.

# References

1. N. Bohr, quoted in W. Heisenberg, *Physics and Beyond* (Harper and Row, New York, 1971), p. 206
2. L.I. Schiff, *Quantum Mechanics*, 3rd edn. (McGraw-Hill, Duesseldorf, 1968)
3. J.V. Neumann, *Mathematischen Grundlagen der Quantenmechanik*. Die Grundlehren der Mathematischen Wissenschaften (Springer, Berlin, Heidelberg, New York, 1968)
4. L.B. Ballentine, *Quantum Mechanics*, 2nd edn. (World Scientific, Singapore, 1998)
5. L. Landau, I. Lifshitz, *Quantum Mechanics, Course of Theoretical Physics*, vol. 3, 3rd edn. (Pergamon Press, Oxford, 1977)
6. W.R. Theis, *Grundzüge der Quantentheorie* (Teubner, Stuttgart, 1985)
7. G. Mahler, V.A. Weberruß, *Quantum Networks*, 2nd edn. (Springer, Berlin, Heidelberg, 1998)
8. R. Loudon, *The Quantum Theory of Light* (Oxford University Press, Oxford, 1985)
9. F.J. Belinfante, *A Survey of Hidden-Variables Theories* (Pergamon Press, Oxford, 1973)
10. R. Penrose, in *Mathematical Physics*, ed. by A. Fokas, A. Grigoryan, T. Kibble, B. Zegarlinski (Imperial College, London, 2000)
11. H.P. Stapp, *Mind, Matter, and Quantum Mechanics*, 2nd edn. (Springer, Berlin, Heidelberg, 2004)
12. C.A. Fuchs, quant-ph/0106166 (2001)
13. M.C. Caves, C.A. Fuchs, R. Schack, Phys. Rev. A **65**, 022305 (2002)
14. M. Nielsen, I. Chuang, *Quantum Computation and Quantum Information* (Cambridge University Press, Cambridge, 2000)
15. H. Araki, E.H. Lieb, Commun. Math. Phys. **18**, 160 (1970)
16. A. Einstein, B. Podolsky, N. Rosen, Phys. Rev. **47**, 777 (1935)

# Chapter 3
# Basics of Thermodynamics and Statistics

*Not knowing the 2nd law of thermodynamics is like never having read a work of Shakespeare.*

— C. P. Snow [1]

**Abstract** After having introduced some central concepts, results, and equations from quantum mechanics, we will now present the main definitions and laws of phenomenological thermodynamics and of thermostatistics. The aim is not at all to give a complete overview of the concepts of classical thermodynamics, but a brief introduction and summary of this old and useful theory. For a complete exposition of the subject we refer to some standard textbooks [2–5]. (A timeline of notable events ranging from 1575 to 1980 can be found in [6].)

## 3.1 Phenomenological Thermodynamics

In spite of the fact that all physical systems are finally constructed out of basic subunits, the time evolution of which follows complicated coupled microscopical equations, the macrostate of the whole system is typically defined only by very few macroscopic observables like temperature, pressure, volume. From a phenomenological point of view one finds simple relations between these macroscopic observables, essentially condensed into four main statements – the fundamental laws of thermodynamics. These central statements are based on experience and cannot be founded within the phenomenological theory.

### 3.1.1 Basic Definitions

A physical system is understood to be an operationally separable part of the physical world. Microscopically such a system can be defined by a Hamilton function (classically) or a Hamilton operator (quantum mechanically), whereas in the macroscopic domain of thermodynamics systems are specified by state functions. Such a state function, like the internal energy $U$, is defined on the space of so-called macrostates.

Such a macrostate of the system is defined by a complete and independent set of state variables (macrovariables) $Z_i$, where $i = 1, 2, \ldots, n_{var}$. The dimension $n_{var}$ is small compared to the number of microscopic variables for the system under consideration. The macroscopic state variables $Z_i$ come in two variants: *extensive*

Gemmer, J. et al.: *Basics of Thermodynamics and Statistics*. Lect. Notes Phys. **784**, 23–39 (2009)
DOI 10.1007/978-3-540-70510-9_3      © Springer-Verlag Berlin Heidelberg 2009

*variables*, $X_i$ (e.g., volume $V$, entropy $S$), which double if the system is doubled, and *intensive variables*, $\xi_i$ (e.g., pressure $P$, temperature $T$), which remain constant under change of system size. For each extensive variable $X_i$ there is a conjugate intensive variable $\xi_i$, with $i = 1, 2, \ldots, n_{\mathrm{var}}$. Starting from an all-extensive macrostate, $Z_i = X_i$, one can get different representations by replacing $X_j$ by $\xi_j$ (for some given $j$). For the all-extensive macrostate one usually chooses the internal energy $U(X_i)$ as the appropriate state function. (Another choice would be the entropy, see below.) The coordinate transformation to other state variables, or more precisely, Legendre transformation, leads to new state functions.

There are no isolated macrosystems: system and environment constitute the most fundamental partition of the physical world underlying any physical description. In the thermodynamic regime the environment can be used to fix certain state variables like volume $V$, temperature $T$, pressure $P$. The system proper is usually classified according to the allowed exchange processes with the environment. "Completely closed" means no matter exchange, no energy exchange; "closed" means no exchange of matter; otherwise the system is termed "open."

The existence of equilibrium states is taken as a fundamental fact of experience. After a certain relaxation time any completely closed macrosystem approaches an equilibrium state (stationary state), which the system will then not leave anymore spontaneously. The number of independent state variables becomes a minimum in equilibrium given by $n_{\mathrm{var}}$. There are $n_{\mathrm{var}}$ state equations, relations between extensive and intensive macrovariables in equilibrium, which help to specify the experimentally verifiable properties of the system.

As a thermodynamic process we consider a sequence of state changes defined in the state space of macrovariables of the system and its environment. A reversible process must consist of equilibrium states only: relaxation from a non-equilibrium state to an equilibrium state is always irreversible by definition.

Moderate deviations from global equilibrium are based on the concept of local equilibrium. In this case the macrosystem can further be partitioned into macroscopic subsystems, which, by themselves, are still approximately in equilibrium. The local state would thus be time independent, if isolated from the other neighboring parts.

Conventional thermodynamics is sometimes also called thermostatics, as the state changes are studied here without explicit reference to time. As a phenomenological theory thermodynamics cannot define its own range of validity. In particular, it does not give any criteria, according to which a given system should be expected to behave thermodynamically or not.

### 3.1.2 Fundamental Laws

To consider thermodynamic phenomena in detail, we often need, besides the macrovariables, some additional quantities $A$, which are functions of the independent macrovariables $Z_i$, $i = 1, 2, \ldots, n_{\mathrm{var}}$. In thermodynamic processes the total change of $A$ over a closed cycle may not be independent of the path, i.e.,

$$\oint \delta A \neq 0 . \tag{3.1}$$

Such a quantity is non-integrable and is said to have no complete differential. Nevertheless, it is possible to define an infinitesimal change,

$$\delta A = \sum_{i=1}^{n_{var}} \frac{\partial A}{\partial Z_i} dZ_i , \tag{3.2}$$

for which, however,

$$\frac{\partial}{\partial Z_j} \frac{\partial A}{\partial Z_i} \neq \frac{\partial}{\partial Z_i} \frac{\partial A}{\partial Z_j} . \tag{3.3}$$

Sometimes one can introduce an integrating factor for the quantity $A$ such that the last relation is fulfilled and $A$ becomes integrable. Furthermore two non-integrable quantities may add up to form an integrable one. State functions are always integrable. In the following $dA$ will denote a complete differential, $\delta A$ an infinitesimal change (not necessarily a complete differential), and $\Delta A$ a finite change of $A$.

Zeroth Law:

For a thermodynamic system there exists an empirical temperature $T$ such that two systems are in thermal equilibrium, if $T^{(1)} = T^{(2)}$. Any monotonic function $f(T)$ of $T$ can also be used as an empirical temperature.

First Law:

For any thermodynamic system the total internal energy $U$ is an extensive state function. In a completely closed system $U$ is constant in time,

$$\delta U = 0 . \tag{3.4}$$

$U$ may change only due to external energy transfer: $\delta U = \delta U^{ext}$. Examples are as follows:

- Change of volume $V$: $\delta U^{ext} = -P \, dV$ ($P$: pressure).
- Change of magnetization $\mathbf{M}$: $\delta U^{ext} = \mathbf{B} \, d\mathbf{M}$ ($\mathbf{B}$: magnetic field).
- Change of particle number $N$: $\delta U^{ext} = \mu \, dN$ ($\mu$: chemical potential).

The total contribution has the general form

$$\delta A = \sum_{i=1}^{n_{var}-1} \xi_i \, dX_i , \tag{3.5}$$

where $\delta A$ is called the total applied work, $X_i$ an extensive work variable (excluding entropy), and $\xi_i$ the conjugate intensive variable to $X_i$ (excluding temperature). Why just these variables $X_i$, no others? The answer is that there exist environments (i.e., some appropriate apparatus) such that these energy changes can actually be carried out in a controlled fashion.

For thermodynamic systems we may have, in addition, a "heat contribution" $\delta Q$. The *first law of thermodynamics* thus reads explicitly

$$dU = \delta Q + \delta A = \delta Q + \sum_{i=1}^{n_{\text{var}}-1} \xi_i \, dX_i \, , \tag{3.6}$$

$\delta Q$ and $\delta A$ do not constitute complete differentials by themselves, but their sum does. For any closed path in macrostate space we thus have

$$\oint dU = 0 \, , \tag{3.7}$$

which constitutes a form of energy conservation; there is no perpetual motion machine (perpetuum mobile) of the first kind, i.e., there is no periodic process in which work is extracted without supplying energy or heat. Periodic means that the machine is exactly in the same state after each cycle (ready for the next one), which is not necessarily true for the environment.

Second Law:

The first law guarantees that each process conserves the energy of the whole system (system and environment together). However, there are processes that we typically do not observe even though they would not violate the first law. According to Clausius:

> Heat never flows spontaneously from a cold body to a hotter one.

An important alternative formulation of the second law makes use of the concept of the perpetuum mobile of second kind (Thomson's formulation):

> It is impossible to construct a periodically operating machine, which does nothing else but transforms heat of a single bath into work.

Experience tells us that the above two formulations of the second law of thermodynamics are fulfilled, in general. However, it is not possible to prove this law within the phenomenological theory of thermodynamics. In statistical mechanics there have been numerous attempts to do just this. We will introduce some of them later in Chap. 4.

For reversible processes one finds that

$$dS = \frac{\delta Q}{T} \, . \tag{3.8}$$

In this case $1/T$ is an integrating factor for $\delta Q$. $S$ is called entropy. If irreversible processes participate, the quantity is not integrable anymore. In general, for arbitrary (reversible and irreversible) processes we have

$$\delta S \geq \frac{\delta Q}{T} \, . \tag{3.9}$$

As long as irreversible processes take place, entropy is increased until the system reaches equilibrium. In equilibrium $S$ takes on a maximum value, usually constrained by some conservation laws.

The entropy of a system can thus change due to internal entropy production and external entropy transfer:

$$\delta S = \delta S^{\text{int}} + \delta S^{\text{ext}} \, . \tag{3.10}$$

The second law states that

$$\delta S^{\text{int}} \geq 0 \, . \tag{3.11}$$

A system is called adiabatically closed, if

$$\delta S^{\text{ext}} = \frac{\delta Q}{T} = 0 \quad \Longrightarrow \quad \delta S \geq 0 \, . \tag{3.12}$$

In the case of a reversible process we need to have $dS^{\text{tot}} = dS^{\text{g}} + dS^{\text{c}} = 0$, where $dS^{\text{g}}$ is the entropy change of the system, $dS^{\text{c}}$ the entropy change of the environment. Only under this condition can a process run backward without violating the second law, i.e., after the reverse process everything, the system as well as the environment, is in exactly the same state as before.

It is important to note that entropy changes are measurable. For this purpose we couple the system to an external system (bath at temperature $T^{\text{c}}$) and perform a reversible process with $\Delta S^{\text{tot}} = \Delta S^{\text{g}} + \Delta S^{\text{c}} = 0$. For fixed $T = T^{\text{c}}$ we can identify

$$\Delta S^{\text{g}} = \int \frac{\delta Q}{T^{\text{c}}} \, . \tag{3.13}$$

$\Delta S^{\text{g}}$ can thus be measured via the reversible heat exchange.

Remark: The measurability of entropy changes for any individual thermodynamic system has far-reaching consequences. It excludes the possibility to consistently interpret $S$ in terms of subjective ignorance (though one may still use this metaphor in a pragmatic way). Furthermore, in so far as quantum physical uncertainties can be shown to give rise to thermodynamic entropy, this may shed new light on the question whether quantum mechanical states could be interpreted as representing our subjective knowledge or ignorance, indicating that this is not the case. Quantum thermodynamics, as treated in this book, should thus help to clarify ongoing controversial disputes.

Third Law (Nernst):

For $T \to 0$ we have for systems without "frozen-in disorder,"

$$S \to 0 \tag{3.14}$$

independent of $X_i$ or $\xi_i$, respectively. As a consequence, specific heats, e.g., go to zero for $T \to 0$. This is interpreted to imply that the zero point of the absolute temperature $T$ cannot be reached. In support of the above remark, $S = 0$ cannot mean that we have "complete knowledge" of the respective ground state; this is hardly ever the case.

### 3.1.3 Gibbsian Fundamental Form

The so-called Gibbsian fundamental form now follows as a combination of the first and the second law (for reversible processes i.e. remaining in equilibrium state space)

$$dU = T \, dS - \sum_{i=1}^{n_{\text{var}}-1} \xi_i \, dX_i \, . \tag{3.15}$$

The "natural" independent macrovariables for $U$ are thus $S$ and $X_i$, which are all extensive. Euler's homogeneity relation (definition of a complete differential)

$$dU = \frac{\partial U}{\partial S} \, dS + \sum_{i=1}^{n_{\text{var}}-1} \frac{\partial U}{\partial X_i} \, dX_i \tag{3.16}$$

allows us to identify

$$T(S, X_i) = \left( \frac{\partial U}{\partial S} \right)_{X_i} , \tag{3.17}$$

$$\xi_j(S, X_i) = \left( \frac{\partial U}{\partial X_j} \right)_{S, X_i \neq X_j} . \tag{3.18}$$

The absolute temperature $T$ thus has the property of an empirical temperature as defined in the zeroth law, it is the conjugate variable to $S$.

### *3.1.4 Thermodynamic Potentials*

So far we have restricted ourselves to the internal energy $U(S, X_i)$ of the system as a thermodynamic state function or potential (see (3.15)). Instead of $U$ we may alternatively consider the entropy function, $S(U, X_i)$. Both of these basic state functions are functions of extensive variables only. Rewriting the Gibbsian fundamental form we get

$$dS = \frac{1}{T} dU - \sum_{i=1}^{n_{var}-1} \frac{\xi_i}{T} dX_i \qquad (3.19)$$

from which, comparing with the Euler equation (cf. (3.16)), we read

$$\left( \frac{\partial S}{\partial U} \right)_{X_i} = \frac{1}{T} , \qquad (3.20)$$

$$\left( \frac{\partial S}{\partial X_j} \right)_{U, X_i \neq X_j} = -\frac{\xi_j}{T} . \qquad (3.21)$$

However, for concrete physical situations, e.g., special contact conditions of system and environment, it is more appropriate to use a different set of independent variables. This set should be better adapted to the considered situation. The method to perform this coordinate transformation is called the *Legendre transformation*.

We start from the energy function $U(S, X_i)$, and restrict ourselves to simple fluid systems ($n_{var} = 2$) with the single work variable $X = V$. For this volume $V$ the conjugate intensive variable is the pressure $P = -\partial U / \partial V$. The free energy $F$ (or Helmholtz free energy) results from a Legendre transformation of the function $U(S, V)$ replacing $S$ by its conjugate $T$:

$$F(T, V) = U(S(T, V), V) - \frac{\partial U}{\partial S} S = U - TS , \qquad (3.22)$$

$$dF = dU - T\,dS - S\,dT = -S\,dT - P\,dV . \qquad (3.23)$$

For the enthalpy $H$ we replace $V$ by the conjugate variable $P$,

$$H(S, P) = U - \frac{\partial U}{\partial V} V = U + PV , \qquad (3.24)$$

$$dH = dU + V\,dP + P\,dV = T\,dS + V\,dP . \qquad (3.25)$$

Finally, the Gibbs free energy (or free enthalpy) results, if we replace $S$ and $V$ by their respective conjugate variables,

$$G(T, P) = U - \frac{\partial U}{\partial S} S - \frac{\partial U}{\partial V} V = U - TS + PV \,, \tag{3.26}$$

$$dG = -S\,dT + V\,dP \,. \tag{3.27}$$

All these thermodynamic potentials are equivalent. Of course there are additional potentials, if there are more work variables, e.g., magnetic variables, exchange of particle numbers.

What is the use of these thermodynamic potentials? According to the second law the entropy reaches its maximum value in equilibrium. As a consequence these thermodynamic potentials will reach a minimum value for the equilibrium state of the system under specific conditions. This allows us to use these potentials to compute the properties of the system if a calculation based on the entropy is impossible. Additionally we need these potentials in the statistical theory, as will be seen below.

## 3.2 Linear Non-equilibrium Thermodynamics

Up to this point all considerations and discussions referred to equilibrium situations, i.e., situations that are reached after sufficient time if systems are left alone. These equilibrium states are time independent, thus the theory developed so far excludes, technically speaking, all phenomena that feature a time evolution. Of course, relations like the Gibbs fundamental form (see Sect. 3.1.3) or the formulation of thermodynamic potentials are essentially meant to describe processes like adiabatic expansion, isothermal compression that surely do have a time dependence, but all these processes are driven by the change of external parameters like volume, temperature. These processes are called "quasi-static" since it is assumed that the system will be immediately at rest the very moment in which the external parameter stops changing. Thus, this theory does not include any process in which a system develops without an external influence.

Such processes happen on the way to equilibrium. They are thus irreversible and much harder to deal with because the internal energy is no longer a state function of the extensive quantities. Therefore Gibbs' fundamental form of (3.15) is not necessarily valid and one needs more and more parameters to specify the state of a system.

There is, however, a class of irreversible processes that happen close to equilibrium which are, with some additional assumptions, accessible from a slightly enlarged theory called "linear irreversible thermodynamics."

The first assumption is that in such processes equilibrium thermodynamics remains locally valid, i.e., it is assumed that it is possible to divide the system into spatial cells to each of which equilibrium thermodynamics applies, only the thermodynamic quantities may now vary from cell to cell. Regarding the level of description one does not need entirely new quantities, one only needs the standard quantities for each cell and these are assumed to be small enough so that the quantities may be given in the form of smooth space (and time)-dependent functions.

So, from extensive quantities one goes to densities (e.g., $U \rightarrow u(\mathbf{q}, t)$, where $\mathbf{q}$ is the vector of position coordinates) and from intensive quantities to fields (e.g., $T \rightarrow T(\mathbf{q}, t)$). For the entropy density one gets (see Sect. 3.1.4)

$$s(\mathbf{q}, t) = s\big(u(\mathbf{q}, t), X_i(\mathbf{q}, t)\big) , \tag{3.28}$$

or, specializing in situations in which no extensive quantities other than energy and entropy vary,

$$s(\mathbf{q}, t) = s\big(u(\mathbf{q}, t)\big) . \tag{3.29}$$

To describe the evolution of the system one has to introduce the "motion" of the energy – the energy current $\mathbf{j}_u$. Since the overall energy is conserved, the current is connected with the energy density by a continuity equation,

$$\frac{\partial u}{\partial t} + \nabla \mathbf{j}_u = 0 . \tag{3.30}$$

If one had an equation connecting $\mathbf{j}_u$ to the functions describing the thermal state of the system like $u(\mathbf{q}, t)$ or $T(\mathbf{q}, t)$, one could insert this equation into (3.30) getting an autonomous equation describing the behavior of the system. The problem is that such an equation depends on the material and could, in principle, take on the most complicated forms. The aim of the considerations at hand is to show that, under some assumptions, this equation can just assume a form into which the properties of the specific material enter via a few constants.

Since equilibrium thermodynamics is supposed to be locally valid, one finds for the differential of the entropy density, with (3.20),

$$ds = \frac{\partial s}{\partial u} du = \frac{1}{T} du \tag{3.31}$$

and thus for the entropy current $\mathbf{j}_s$ connected with the energy current,

$$\mathbf{j}_s = \frac{1}{T} \mathbf{j}_u . \tag{3.32}$$

Entropy is no longer a conserved quantity, so a local entropy production rate $\dot{s}$ enters,

$$\dot{s} = \frac{\partial s}{\partial t} + \nabla \mathbf{j}_s . \tag{3.33}$$

Now, substituting (3.31) and (3.32) into (3.33), one finds, exploiting (3.30)

$$\dot{s} = -\frac{1}{T^2} \nabla T \mathbf{j}_u . \tag{3.34}$$

Demanding that $\dot{s}$ has to be positive at any point and any time, (3.34) sets restrictions on the above-mentioned equation for $\mathbf{j}_u$.

A basic restriction deriving from another concept is the Markov assumption, i.e., the assumption that the $\mathbf{j}_u(\mathbf{q}, t)$ should only depend on the state of the system at the actual time $t$ and not at the configurations at former times $t' < t$. This dependence thus has to be local in time. It could nevertheless, in principle, be non-local in space. $\mathbf{j}_u(\mathbf{q}, t)$ could depend on the values of, say $T(\mathbf{q}', t)$ at all $\mathbf{q}' \neq \mathbf{q}$ or spatial derivatives of $T$ of arbitrarily high order. This, however, is forbidden by (3.33). $\mathbf{j}_u$ can only depend on first-order derivatives (no higher, no lower order) of $T$, otherwise the positivity of $\dot{s}$ could not be guaranteed. Specializing now in cases with very small temperature gradients, one can neglect all terms in which the first-order derivatives enter other than linearly, eventually finding

$$\mathbf{j}_u(\mathbf{q}, t) = -\kappa \, \nabla T(\mathbf{q}, t) \; . \tag{3.35}$$

The explicit form of $\kappa$ depends on the material. This is the above-mentioned equation that allows, together with (3.30), for a closed description of linear irreversible processes (equation of heat conduction). Equation (3.35) is also known as Fourier's lawand has turned out to be appropriate for describing a huge class of experiments that proceed close to equilibrium.

This concept can be generalized: the external forces $F_i$, like gradients of electric potentials, chemical potentials, or temperature, are taken to be responsible for the respective currents $\mathbf{j}_i$: heat currents, diffusion currents, as well as energy currents, which are not independent of each other. In general, we expect

$$\mathbf{j}_i = \sum_j \mathsf{L}_{ij} \, F_j \; , \tag{3.36}$$

where $\mathsf{L}_{ij}$ is the matrix of transport coefficients. Due to the Onsager theorem, the matrix $\mathsf{L}_{ij}$ should be symmetric ($\mathsf{L}_{ij} = \mathsf{L}_{ji}$). Since the current $\mathbf{j}_i$ also induces an entropy flow through the system, we have to be very careful in choosing currents and forces. However, if we choose these quantities ensuring that the entropy density increases, $\dot{s} \geq 0$, while the currents $\mathbf{j}_i$ flow, it follows from the continuity equation for the entropy that

$$\dot{s} = \sum_i \mathbf{j}_i \, F_i \tag{3.37}$$

and, as a further consequence, we find Onsager's theorem fulfilled (see [7]).

## 3.3 Statistics

All considerations so far were aimed at a macroscopic description of large systems in contact with different types of environments. To investigate the behavior of such systems, we have introduced the central laws of thermodynamics – phenomeno-logical restrictions for the development of a system, without any fundamental jus-tification. These laws, and especially the second one, are exclusively founded on experience obtained in a large variety of different thermodynamic experiments. This phenomenological theory of thermodynamics does not allow for the calculation of a thermodynamic potential from a microscopic picture; instead, the potential must be found from experiment.

The connection of the macroscopic behavior of a given system to the micro-scopic time evolution, according to a classical Hamilton function or to a quantum mechanical Hamiltonian, did not enter our considerations yet, and will be the main subject of this book. Nevertheless, even if it might not provide any explanation, the theory of statistical physics provides a "recipe" for how to calculate an equilibrium entropy from a microscopic picture. This recipe together with the above-mentioned phenomenological theory enables us to calculate all sorts of equilibrium behavior. This way of dealing with thermodynamic phenomena has sometimes been com-pared to "driving a car without knowing how an engine works"; nevertheless, it is an important technique and therefore briefly described in the following.

### 3.3.1 Boltzmann's Principle, A Priori Postulate

Boltzmann postulated the following connection between the thermodynamic entropy and the microstate of an isolated system (all extensive state variables, internal energy $U$, volume $V$, and particle number $N$, are fixed from the outside) as

$$S = k_B \ln m(U, V, N) , \qquad (3.38)$$

where $k_B$ is the so-called Boltzmann constant and $m$ is the number of microstates accessible for the system under the given restrictions. The number of accessible microstates $m$ is often also called statistical weight or sometimes thermodynamic weight, and we will evaluate this quantity below.

However, let us first consider a macrostate of a given system. From phenomenol-ogy we have learned that the equilibrium state must have maximum entropy (the second law, see Sect. 3.1.2) and thus, according to Boltzmann, this state should also belong to a maximum number of microstates. For illustration, think of a gas in a container: the states of maximum entropy are states where the gas particles are equally distributed over the whole volume and, of course, the number of such states is very large in comparison to the number of states, where all gas particles are in one corner of the container, say.

The entropy defined above is an extensive quantity in the sense that two systems with statistical weights $m^{(1)}$ and $m^{(2)}$ have the joint weight $m^{(1)} \cdot m^{(2)}$ and the total entropy of both systems $S = k_B(\ln m^{(1)} + \ln m^{(2)})$.

Within statistical mechanics of isolated systems we have another very important postulate – the assumption of equal a priori probabilities of finding the system in any one of the $m$ possible microstates belonging to the respective macrostate. As a postulate, this statement is not provable either, but as, e.g., the energy of a gas in a volume $V$ does not depend on the position of the gas particles within the container, each of these "microstates" might, indeed, be expected to be equally likely.

This idea of assuming certain probabilities for microstates rather than calculating them led to yet another way of describing a macrostate of a system, which is the so-called *statistical ensemble*. This ensemble consists of $m$ identical virtual systems for each accessible microstate and each is represented by a point in phase space. This concept has been supported by the claim that the thermodynamic system should be *quasi-ergodic*, i.e., its trajectory would come infinitesimally close to every possible microstate within its time evolution, thus one would be allowed to replace the time average by the ensemble average. Later we will discuss in more detail the ideas behind this quasi-ergodic theorem and the problems we have to face after its introduction (see Sect. 4.2).

We can now describe the state of a system by the density of points in phase space belonging to the statistical ensemble. This density $W(\mathbf{q}, \mathbf{p}, t)$ contains the probability of finding a point in phase space at position $(\mathbf{q}, \mathbf{p})$ at time $t$. According to the a priori postulate this probability should be constant within the respective energy shell (see below, (3.40)) and elsewhere zero

$$
W(\mathbf{q}, \mathbf{p}, t) = \begin{cases} \frac{1}{m} = \text{const.} & E < H(\mathbf{q}, \mathbf{p}) < E + \Delta E \\ 0 & \text{else} \end{cases} . \tag{3.39}
$$

The statistical ensemble defined by this special density is called the *microcanonical ensemble* (see Fig. 3.1b).

### 3.3.2 Microcanonical Ensemble

The microscopic behavior of any $N$ particle system is described by the respective Hamilton function $H(\mathbf{q}, \mathbf{p})$, dependent on all generalized coordinates of the system. A microstate of the system is then represented by a point in the systems phase space, the $6N$-dimensional space spanned by all position and momentum coordinates of the $N$ particles. For an isolated system, a system which does not exchange any extensive variable like energy, volume with the environment, the Hamilton function defines an energy surface $H(\mathbf{q}, \mathbf{p}) = U$ in the phase space. The state evolution is therefore constrained to this hypersurface in phase space. Since the total isolation of a system is a very idealized restriction, let us consider in the following not completely isolated systems, for which the internal energy is fixed only within a small interval

$$
E < H(\mathbf{q}, \mathbf{p}) = U < E + \Delta E . \tag{3.40}
$$

The representing trajectory of the system is then restricted to an energy shell of the thickness $\Delta E$ in phase space, contrary to the restriction to an energy surface in the case of total isolation.

To exploit Boltzmann's postulate we need to know the number of microstates $m$ in such an energy shell of the respective phase space. Usually we divide the phase space into cells of the size $h^{3N}$, arguing that in each cell there is exactly one microstate of the system. This assertion is reminiscent of a quantum state in phase space, due to the uncertainty relation. However, we could also have introduced an abstract division into cells. In any case, the number of states should be the phase space volume of the respective energy shell in phase space divided by the cell size.

The total volume of the phase space below the energy surface $H(\mathbf{q}, \mathbf{p}) = E$ is given by the volume integral

$$\Omega(E) = \iint_{H(\mathbf{q},\mathbf{p}) \leq E} \prod_{\mu=1}^{N} d\mathbf{q} d\mathbf{p} \, , \tag{3.41}$$

and the volume of the energy shell by $\Omega(E + \Delta E) - \Omega(E)$. The latter can directly be evaluated by the volume integral

$$\Omega(E + \Delta E) - \Omega(E) = \iint_{E < H(\mathbf{q},\mathbf{p}) < E + \Delta E} \prod_{\mu=1}^{N} d\mathbf{q} d\mathbf{p} \, . \tag{3.42}$$

For further reference we also define here an infinitesimal quantity, the state density (cf. Fig. 3.1a)

**Fig. 3.1a** State density $G(E)$

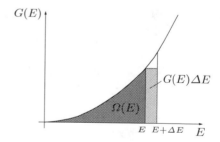

**Fig. 3.1b** Probability $W(p, q)$ for a microcanonical ensemble

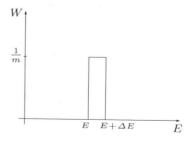

$$G(E) := \lim_{\Delta E \to 0} \frac{\Omega(E + \Delta E) - \Omega(E)}{\Delta E} \tag{3.43}$$

$$= \frac{d\Omega(E)}{dE} . \tag{3.44}$$

Finally the number of microstates in the respective energy shell, according to the above argumentation, is

$$m = \frac{\Omega(E + \Delta E) - \Omega(E)}{h^{3N}} . \tag{3.45}$$

We thus find in linear approximation for small $\Delta E$

$$m \approx \frac{\Delta E\, G(E)}{h^{3N}} , \tag{3.46}$$

where $G(E)$ is the state density (3.43) at the energy surface $H(\mathbf{q}, \mathbf{p}) = E$, which we have assumed does not change much in the small interval $\Delta E$. The Boltzmann entropy now reads

$$S = k_B \ln m = k_B \ln \frac{\Delta E\, G(E)}{h^{3N}} . \tag{3.47}$$

In most cases $\Delta E$ can be considered a constant independent of $E$. As explained later this is not true in all cases (see Sect. 4.5).

From (3.20) and the entropy definition we are then able to define a temperature of an isolated system in equilibrium by the state density at the energy $E$,

$$\frac{1}{T} = \frac{\partial S}{\partial E} = k_B \frac{\partial}{\partial E} \ln \frac{\Delta E}{h^{3N}} G(E) = \frac{k_B}{G(E)} \frac{\partial G(E)}{\partial E} . \tag{3.48}$$

Due to this result the statistical temperature corresponds to the relative change of the state density with the energy.

So far we have restricted ourselves to a microcanonical ensemble where all possible microstates are equally likely. Because of the assumed isolation of the system this ensemble is not very well adapted for a variety of experimental situations. Therefore we extend our considerations to a more general exchange concept, with more detailed information about the microstate of the system.

### 3.3.3 Statistical Entropy, Maximum Principle

First, we define a new quantity

$$S'(W_i) = -k_B \sum_i W_i \ln W_i , \tag{3.49}$$

where distinguishable states of the system, whatever they might be, are labeled by $i$ and $W_i$ is the probability of finding the system in state $i$. Originally, this definition was proposed by the information theoretician Shannon, who intended to measure lack of knowledge by this function. In thermostatistics the probabilities $W_i$ are the probabilities for finding the system in a microstate with energy $E$.

To find the best guess about the probability distribution, provided one knows some property of the system for sure, one has to compute the maximum of $S'$ with respect to the $W_i$s under the restriction that the resulting description of the system has to feature the known property. This maximum of $S'$ then is the entropy $S$. This scheme is often referred to as Jaynes' principle introduced in [8, 9].

Again, if all extensive quantities, like volume and internal energy of a system, are fixed (microcanonical situation), one has to maximize $S'$ over all states featuring this energy and volume, which are all states from the accessible region. The only macrocondition to meet is the normalization of the distribution

$$\sum_i W_i - 1 = 0 . \tag{3.50}$$

As expected from our former considerations, in this case (isolated situation) a uniform distribution (see (3.39)) over all those states results, as claimed in the a priori postulate. Therefore definition (3.49) meets the definition of Boltzmann in the case of a microcanonical situation, if we introduce (3.39) as the respective probability distribution of the microcanonical ensemble.

Much more interesting are other contact conditions, e.g., canonical ones. In a canonical situation energy can be exchanged between system and environment – the system is in contact with a heat bath. As an additional macrocondition we require that the mean value of the energy is equivalent to the internal energy of the system $U$, given by

$$\sum_i W_i E_i = U . \tag{3.51}$$

Of course the normalization condition (3.50) should also be obeyed. Now, we maximize $S'$ with respect to the $W_i$s under observance of both these conditions required for the variation

$$\delta\left( \sum_i W_i \ln W_i + \alpha \left( \sum_i W_i - 1 \right) + \beta \left( \sum_i W_i E_i - U \right) \right) = 0 , \tag{3.52}$$

with the Lagrange multipliers $\alpha$ and $\beta$. From this variation we find the probability distribution

$$W_i = \frac{1}{Z} e^{-\beta E_i} \quad \text{with} \quad Z = e^{1+\alpha} , \tag{3.53}$$

called the Boltzmann distribution, where we have introduced the partition function $Z$ instead of the Lagrange multiplier $\alpha$.

It remains to evaluate the two Lagrange multipliers. By introducing the result of the variation in condition (3.50), we find

$$Z = \sum_i e^{-\beta E_i} . \tag{3.54}$$

For the second Lagrange multiplier we start from the entropy definition introducing the distribution

$$S = -k_B \sum_i W_i \ln W_i = k_B \ln Z \underbrace{\sum_i W_i}_{=1} + k_B \beta \underbrace{\sum_i W_i E_i}_{=U} . \tag{3.55}$$

We thus get

$$U - \frac{1}{k_B \beta} S = -\frac{1}{\beta} \ln Z . \tag{3.56}$$

The left-hand side is the free energy $F$ (see (3.22)), if we identify $\beta$ and $Z$, respectively, as

$$\beta = \frac{1}{k_B T} \quad \text{and} \quad F = -k_B T \ln Z . \tag{3.57}$$

Note that there is no way to avoid this ad hoc identification if one wants to get a connection to phenomenological thermodynamics. In the same way other extensive quantities allowed for exchange can be handled, again yielding results which are in agreement with experiments.

We have thus found a recipe to evaluate the thermodynamic potentials and therefore the entropy of the system only by microscopic properties. These properties are specified by the Hamilton function of the system entering the partition function. If one is able to evaluate the logarithm of the partition function, all other thermodynamic properties, state equations, intensive parameters, etc., follow from phenomenological considerations.

However, a complete derivation of thermodynamics from microscopical theories is still missing. As already mentioned the above statistical considerations are only recipes for concrete evaluation of thermodynamic behavior. A detailed introduction to some approaches to thermodynamics from a microscopical theory will be given in the following chapter.

# References

1. C.P. Snow, *The Two Cultures and the scientific revolution* (Cambridge University Press, New York, 1959)
2. H.B. Callen, *Thermodynamics and an Introduction to Thermostatistics*, 2nd edn. (Wiley, New York, 1985)
3. H. Römer, T. Filk, *Statistische Mechanik* (VCH, Weinheim, New York, Basel, Cambridge, Tokyo, 1994)
4. W. Weidlich, *Thermodynamik und statistische Mechanik* (Akademische Verlagsgesellschaft, Wiesbaden, 1976)
5. H.J. Kreuzer, *Nonequilibrium Thermodynamics and its statistical foundation* (Clarandon Press, Oxford, 1981)
6. J. Biggus, http://history.hyperjeff.net/statmech.html (2002)
7. G.D. Mahan, *Many-Particle Physics*, 3rd edn. (Plenum Press, New York, London, 2000)
8. E. Jaynes, Phys. Rev. **106**, 620 (1957)
9. E. Jaynes, Phys. Rev. **108**, 171 (1957)

# Chapter 4
# Brief Review of Pertinent Concepts

*Given the success of Ludwig Boltzmann's statistical approach in explaining the observed irreversible behavior of macroscopic systems . . . , it is quite surprising that there is still so much confusion about the problem of irreversibility.*
— J. L. Lebowitz [1]

*Boltzmann's ideas are as controversial today, as they were more than hundred years ago, yet they are still defended (Lebowitz 1993). Boltzmann's H-Theorem is based on the unjustifiable assumption that the motions of particles are uncorrelated before collision.*

— H. Primas [2]

**Abstract** This chapter is meant to provide a neither complete nor irreducible overview of the ideas and arguments which have been brought forward during the search for a derivation of thermodynamics. A central issue in these efforts has been the irreversibility which is most likely absent in any underlying theory. Among others, we give a brief summary of central classical concepts such as the Boltzmann equation, $H$-Theorem, ergodicity, the ensemble as introduced by Gibbs and Ehrenfest's view on the subject. Furthermore we discuss the Shannon entropy, the von Neumann entropy, and the theory of open quantum systems within the quantum mechanical regime. Any "decision" in favor of any of those approaches is left to the reader. Here, only the most prominent ideas that are or have been around are briefly introduced.

Even though phenomenological thermodynamics works very well, as outlined in the previous chapter, many scientists felt and feel an urge to "derive" the laws of thermodynamics from an underlying theory [3].

Almost all approaches of this type focus on the irreversibility that seems to be present in thermodynamic phenomena, but is not intrinsically part of any underlying theory. So to a large extent these approaches intend to prove the second law of thermodynamics in terms of this irreversibility. They try to formulate entropy as a function of quantities, the dynamics of which can be calculated within a microscopic picture in such a way that the entropy would eventually increase during any evolution, until a maximum is reached. This maximum value should be proportional to the logarithm of the volume of the accessible phase space (energy shell); see (3.47).

Gemmer, J. et al.: *Brief Review of Pertinent Concepts*. Lect. Notes Phys. **784**, 41–62 (2009)
DOI 10.1007/978-3-540-70510-9_4　　　　　© Springer-Verlag Berlin Heidelberg 2009

Only if this limit is reached the identification of the "microscopical entropy" with the phenomenological entropy will eventually yield state equations that are in agreement with experiment. It has not been appreciated very much that there are further properties of the entropy that remain to be shown, even after the above behavior has been established (see Sect. 4.5 and Sect. 5.1).

One problem of all approaches based on Hamiltonian mechanics is the applicability of classical mechanics itself. To illustrate this, let us consider a gas consisting of atoms or molecules. In principle, such a system should, of course, be described by quantum mechanics. Nevertheless, for simplicity, one could possibly treat the system classically, if it were to remain in the Ehrenfest limit (see Sect. 2.3), i.e., if the spread of the wave packages were small compared to the structure of the potentials which the particles encounter. Those potentials are generated by the particles themselves, which basically repel each other. If we take the size of those particles to be roughly some $10^{-10}$ m, we have to demand that the wave packages should have a width smaller than $10^{-10}$ m in the beginning. Assuming particle masses between some single and some hundred proton masses and plugging those numbers into the corresponding formulas [4], we find that the spread of such wave packages will be on the order of some meters to 100 m after 1 s, which means the system leaves the Ehrenfest limit on a timescale much shorter than the one typical for thermodynamic phenomena. If we demand the packages to be smaller in the beginning, their spreading gets even worse. Considering this, it is questionable whether any explanation based on Hamiltonian dynamics in phase space or $\mu$-space (cf. Sect. 4.1) can ever be a valid foundation of thermodynamics at all. This insufficiency of the classical picture becomes manifest at very low temperatures (freezing out inner degrees of freedom) and it is not obvious why it should become valid at higher temperatures even if it produces good results.

Nevertheless a short, and necessarily incomplete overview, also and mainly including such ideas, shall be given here.

## 4.1 Boltzmann's Equation and $H$-Theorem

Boltzmann's work was probably one of the first scientific approaches to irreversibility (1866) [5]. It was basically meant to explain and quantify the observation that a gas, which is at first located in one corner of a volume, will always spread over the whole volume, whereas a gas uniformly distributed over the full volume is never found to suddenly shrink to one corner. This seems to contradict Hamiltonian dynamics according to which any process that is possible forward in time, should also be possible backward in time.

Instead of describing a system in real space (configuration space), Boltzmann tried to describe systems in $\mu$-space, the six-dimensional space spanned by the positions $\mathbf{q}$ and the velocities $\mathbf{v}$ of one particle being a point-like object in a three-dimensional configuration space. Now, to describe the state of the whole system consisting of very many, $N$, particles, Boltzmann did not introduce $N$ points in this

$\mu$-space, he rather used a continuous function $f(\mathbf{q}, \mathbf{v}, t)$ meant as a sort of particle density in $\mu$-space. His basic idea was to divide $\mu$-space into cells on an intermediate length scale, one cell of size $dxdydzdv_xdv_ydv_z$ being big enough to contain very many particles, but small compared to a length scale on which the number of particles within one cell would substantially change from one cell to the next. If such an intermediate scale could be introduced, $f(\mathbf{q}, \mathbf{v}, t)$ would simply be the number of particles in the cell around $(\mathbf{q}, \mathbf{v})$. Thus, if $f$ was large at some point, this would simply mean that there are many particles at the corresponding point in configuration space, moving in the same direction with the same velocity.

This description is definitely coarser than the full microscopic description, since information about the exact position of particles within one cell is discarded, which will turn out to be an important point, and it excludes a class of states, namely all those for which the number of particles per cell cannot be given by a smooth continuous function.

Having introduced such a description, Boltzmann tried to give an evolution equation for this function $f$, which is today known as the Boltzmann equation. Since a full derivation of the Boltzmann equation is beyond the scope of this text (the interested reader will find it in [6]) we only give a qualitative account of the basic ideas and describe in some detail the assumption on which this theory relies.

For a change of $f$ at some point in $\mu$-space and at some time $t$ two different mechanisms have to be taken into account: a change of $f$ due to particles that do not interact and a change due to particles that do interact (scatter). The part corresponding to particles that do not collide with each other does not cause any major problems and results only in some sort of sheer of the function $f$ that changes positions but leaves velocities invariant.

More problematic is the part due to particles that do interact with each other. First, only the case of interactions that are short ranged compared to the mean free path are considered. Due to this restriction the dynamics can be treated on the level of scattering processes, just relating incoming to outgoing angles, rather than computing full trajectories. Furthermore, as this is the most important assumption in this context, it is assumed that the particles in cells that collide with each other are uncorrelated within those cells before they scatter. This is called the "assumption of molecular chaos." To understand this assumption in more detail, we consider an infinitesimal time step of the evolution of some special function $f$ depicted in Fig. 4.1. We restrict ourselves here to a two-dimensional "gas"; nevertheless, the $\mu$-space is already four-dimensional, thus we have to visualize $f$ by projections. The special function $f$ corresponds to a situation with all particles concentrated in one cell in configuration space but moving in opposite directions with the same velocity. By some "center of mass" coordinate transformation any collision process may by mapped on this one. After a period $dt$ $f$ will look as shown in Fig. 4.2. Due to momentum and energy conservation $f$ will only be non-zero on a circle. However, where exactly on this circle the particles end up depends on their exact positions within the cells before the collisions. If, e.g., the particle configuration at $t_0$ had been such that all particles had collided head-on, there could only be particles in the marked cells in Fig. 4.2 (right). This corresponds to a strong correlation of

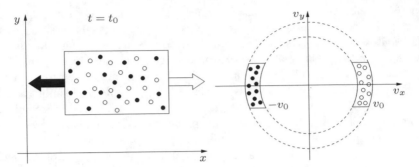

**Fig. 4.1** Two-dimensional gas in $\mu$-space at $t = t_0$ before any collision. Projection onto position space (*left*). Projection onto velocity space (*right*). *White particles* are flying with $v_0$ to the right-hand side, *black ones* with $-v_0$ to the left

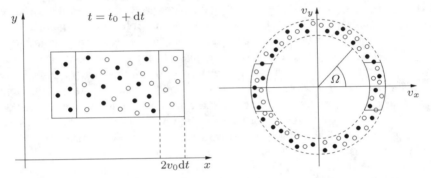

**Fig. 4.2** Two-dimensional gas in $\mu$-space at $t = t_0 + dt$. Projection onto position space (*left*). Because of momentum conservation all particles are now concentrated on a ring in velocity space (*right*). In the boxes at $v_0$ and $-v_0$ there are only particles which did not collide or which collided head-on

the particles before the collision. If the particles had been uniformly distributed and completely uncorrelated, the distribution of particles onto the circle at $t_0 + dt$ would simply be given by the differential scattering cross section $\sigma(\Omega)$ corresponding to scattering into the respective angle $\Omega$. This is exactly what Boltzmann assumed. By simply plugging in the differential cross section for $f$ after the collision process he could derive an autonomous evolution equation for $f$ which no longer contains the exact positions of the particles.

This assumption, which seems intuitively appealing, has been criticized by other scientists for the following reason: even if the positions of particles within their cells were uncorrelated before the collision, they will no longer be uncorrelated after the collision. To understand this we look at the configuration of the particles before the collision more closely, see Fig. 4.3. Some particles are going to collide head-on (pair of particles with number 4), some are going to collide on a tilted axis (e.g., pair of particles with number 1), and some are not going to collide at all during $dt$ (e.g., pair of particles with number 3). If we ask which of those particles will still move in the

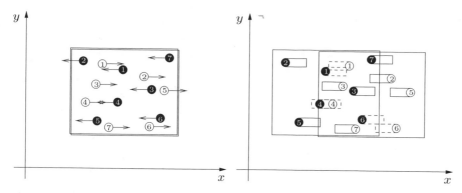

**Fig. 4.3** Configuration (hypothetical) in two cells before (*left*) and after the collision (*right*). Configurations as those depicted for particles "1" and "4" cannot result. Thus, after the collision, particle configurations in corresponding cells are correlated

initial direction after the collision, i.e., occupy the marked cells in Fig. 4.2 (right), it is the particles that collided head-on (particles 4) and the ones that did not collide at all (particles 2, 3, 5, 7). That means that within a cylinder of length $2v_0 dt$ on the left side of the particles moving to the right (white ones), there is either no particle (c.g., particle 5) to the left or there is exactly one in the cylinder (e.g., particle 4). Thus there are, definitely, correlations between the positions of particles in the cells marked in Fig. 4.2 (right) after the collision process. The same is true for all cells on opposite positions on the ring.

To meet this criticism Boltzmann argued that there might be correlations of particles after collisions but not before collisions, since particles would collide so many times with other particles before they collided with themselves again that in these intermediate collisions all correlations were erased. This is why the assumption of molecular chaos is also called the "Stoßzahlansatz" ("large number of collisions approach").

One modern view on this problem follows the lines of "typicality," cf. Sect. 6.1: There are initial microstates for which $f(\mathbf{q}, \mathbf{v}, t)$ dynamically obeys the Boltzmann equation and there are microstates for which $f(\mathbf{q}, \mathbf{v}, t)$ does not. However, the relative weight of the amount of initial microstates for which $f(\mathbf{q}, \mathbf{v}, t)$ does not follow the Boltzmann equation decreases to zero in a certain limit of infinitely "dense" gases consisting of infinitely small particles (cf. [7] and references therein).

Exploiting the above-described Boltzmann equation, Boltzmann was able to prove that the function

$$H(t) = \int f(\mathbf{v}, t) \log f(\mathbf{v}, t) \, d\mathbf{v} \tag{4.1}$$

(already very reminiscent of the entropy proper) which he formulated as a functional of $f$ can only decrease in time, regardless of the concrete scattering potential. Thus, in a way, irreversibility was introduced. This was the beginning of modern statistical mechanics.

Based on the Boltzmann equation it is possible to derive the Maxwell–Boltzmann distribution, which describes the distribution of velocities of particles in a gas, to derive the equation of state for an ideal gas, identifying the mean energy of a particle with $\frac{3}{2}k_B T$, and even set up transport equations.

Despite the enormous success of these ideas, Boltzmann later abandoned these approaches and turned to ideas centered around ergodicity. In this way he responded to the fact that he could not get rid of the assumption of molecular chaos, which appeared unacceptable to him.

## 4.2  Ergodicity

The basis of this approach, also pursued by Boltzmann, is the assumption that any possible macroscopic measurement takes a time which is almost infinitely long compared to the timescale of molecular motion. Thus, the outcome of such a measurement can be seen as the time average over many hypothetical instantaneous measurements. Hence, if it were true that a trajectory ventured through all regions of the accessible volume in phase space, no matter where it started, the measured behavior of a system would be as if it were at any point at the same time, regardless of its starting point. This way irreversibility could be introduced, entropy being somehow connected to the volume that the trajectory ventured through during the observation time.

In order to state this idea in a clearer form, the so-called "ergodic hypothesis" had been formulated:

> The trajectory of a representative point of the system in phase space eventually passes through every point on the energy surface (accessible volume).

If this statement were taken for granted, it could be shown that the amount of time that the trajectory spends in a given volume is proportional to that volume [8]. This leads to another formulation of the ergodic hypothesis stating that *the time average equals the ensemble average*, the latter in this case being an average over all system states within the energy shell.

Unfortunately, the ergodic hypothesis in this form is necessarily wrong for any system, since the trajectory is a one-dimensional line, whereas the so-called energy surface is typically a very high dimensional volume, hence the trajectory cannot pass through all points of the energy surface in any finite time [8]. To circumvent this limitation, the quasi-ergodic hypothesis was introduced, which states that the representative point passes arbitrarily close to any given point in the accessible volume in phase space.

Birkhoff and von Neumann actually demonstrated that there are systems which are quasi-ergodic in this sense and that their representing points actually spend equal time in equal phase space cells [9, 10]. This proof, however, cannot be generalized to the class of thermodynamic systems as a whole, and it remains unclear how exactly

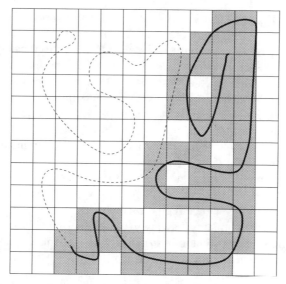

**Fig. 4.4** Ergodicity approach. Entropy is defined by the volume in phase space that is occupied by the cells that the trajectory has already ventured through (*gray region*). This obviously depends on the observation time (*solid line*) and the cell size

an entropy should be introduced. It has been suggested to divide phase space into finite size cells (*coarse graining*) and simply count the cells that the trajectory passes through in a given time (see Fig. 4.4), or to count the cells weighted with the time the representing point spent within the cell.

Thus there is a fair amount of arbitrariness. Obviously a lot has to be introduced artificially, such as averaging (counting) time, cell size.

## 4.3  Ensemble Approach

The term "ensemble" was introduced by Gibbs in about 1902 [11]. The idea is that, in general, a macroscopic observation will be consistent with a very large number of microscopic configurations. All these, represented by their corresponding points in phase space, form the "ensemble." The ensemble, therefore, is basically represented by a density in phase space which is normalized and non-zero everywhere where the system could possibly be found.

To describe the evolution of a system, one now considers the evolution of this density rather than the evolution of a single representing point. The most important theorem for the analysis of the evolution of such a density is Liouville's theorem. This theorem states that the volume of any region in phase space is invariant under Hamiltonian evolution. This theorem has two important consequences: first, if the system is described by a density which is uniform throughout all the accessible energy surface, it will be in a stationary state because this distribution cannot change in time. Thus, such a state that somehow fills the entire accessible region can be seen as an equilibrium state. Therefore one could be tempted to connect the volume, in which such a density is non-zero, with the entropy. Unfortunately, the second consequence is that such a volume cannot change in time. This means that if a system does not start off in an equilibrium state, it can never reach one.

In order to save this concept, Ehrenfest and others introduced *coarse graining* also into this idea [12]. They claimed that entropy should not be connected to the volume in which the density is non-zero, but rather to the number of cells, in which the density is non-zero somewhere. If a smooth-shaped region, in which an initial density is non-zero, would then be mapped by the Hamiltonian evolution onto a "sponge"-like structure featuring the same volume but stretched over the whole energy shell (see Fig. 4.5), such an entropy could be said to grow up to the limit, where the structures of the sponge region become small compared to the cell size. Such a behavior is called "mixing," owing to a metaphor by Gibbs, who compared the whole scenario to the procedure of dripping a drop of ink into a glass of water and then stirring it until a mixture results [11].

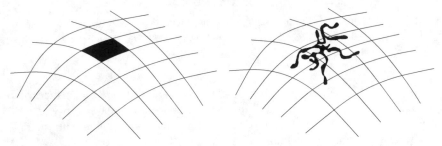

**Fig. 4.5** Ensemble approach. The volume of a region in phase space cannot grow during a Hamiltonian evolution, due to Liouville's law. Nevertheless a simple initial density (*left*) can be transformed to a complicated structure (*right*) that may eventually be found in any cell, if some graining is introduced

However, even if the phase space played a unique role here, since Liouville's theorem is only true in phase space or in any canonical transformation of it, the cell size has to be introduced artificially, and, of course, mixing has to be shown for any system under consideration. This has been done for some systems, but again, there is no generalization to thermodynamic systems at all. Another objection raised against this idea is concerned with the fact that entropy seems to be here due to the observer's inability to find out in exactly which microstate the system is. It has been argued that this would introduce an unacceptable amount of subjectivity into this field of physics.

## 4.4 Macroscopic Cell Approach

The idea of the "macroscopic cells" is also due to Ehrenfest who called them "stars" [8]. Such a star is a region in phase space that only consists of points that are con-

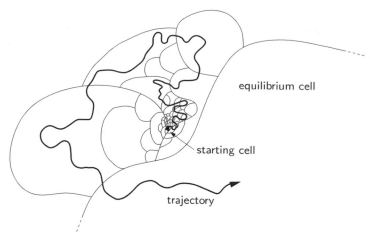

**Fig. 4.6** Macroscopic cell approach. The phase space is divided into cells, according to the macrostate of the system. One of these cells is much larger than all the others, the equilibrium cell. Any trajectory that is not subject to further restrictions is likely to end up in the biggest cell

sistent with one macroscopic description of the system. For example, if we want to describe a gas by the volume it occupies, $V$, and its total internal energy, $U$, all points in phase space corresponding to a gas in this macrostate will form the macroscopic cell labeled by those specific macroscopic variables, $V$ and $U$. This way, phase space is not grained into equal-sized Cartesian cells like in the former approaches, but into strangely shaped macroscopic cells that may be extremely different in size from each other (see Fig. 4.6).

This difference in size is crucial here, for it is assumed that the "equilibrium cell," i.e., the cell in which the gas occupies the biggest possible volume, $V$, would be by far the largest one, i.e., be large enough to almost fill the entire phase space. Technically this also needs to be proven for any thermodynamic system individually, but it seems much more plausible than the assumption of ergodicity or mixing. From a conceptual point of view this approach is very closely related to the "typicality" approach as described in Sect. 6.1.

This plausibility is connected to the so-called "law of large numbers." It is usually established by considering some abstract space (basically identified with $\mu$-space), grained into a large number of equally sized cells and divided into two halves, each one containing an equal number of cells (see Fig. 4.7). If now the set of all possible distributions of a large number of points into those cells is examined, it turns out that the vast majority of such distributions feature the same amount of points in both halves.

The larger the number of cells and the number of points, the more drastic is this result. Transferring this result to phase space, it can be argued that almost all points in phase space, corresponding to distributions of points in $\mu$-space, belong to one macroscopic state, specified by one macroscopic variable that just measures the amount of points, say, in the left half [13].

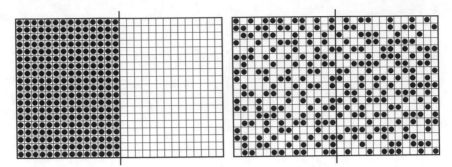

**Fig. 4.7** Macroscopic cell approach, "the law of large numbers." The vast majority of distributions of points in these cells feature the same amount of points in both halves (*right*). There is just one configuration where all points are in the left half (*left*)

Having established such a structure of phase space, one does not need the strict ergodicity hypothesis any longer, for if the trajectory wanders around in phase space without any further restrictions, it will most likely eventually spend almost all time in the biggest cell, even if it started in a small one (see Fig. 4.6).

In such an approach, entropy is connected to the size of the cell the representing point is wandering through. However, here new problems are to be faced: to decide whether or not a given microstate belongs to a macroscopic cell one has to assign a volume to the configuration of gas particles. This is more subtle than it might seem at first sight. One method is to coarse grain configuration space into standard Cartesian cells and count the cells that are occupied by at least one particle. However, this only yields satisfactory results for a certain ratio of the cell size to the diluteness of the gas particles. Other approaches proceed by taking the convex cover or other measures defined for a set of points, thus there is a fair amount of arbitrariness.

Another problem with this idea arises if one tries to examine situations, in which the internal energy of a system is not rigidly fixed, but the system can exchange energy with another, in some sense, bigger system called a heat bath. In this case, one finds empirically that the probability of finding the system in a state of energy $E$ is proportional to $\exp(-E/k_B T)$ the Boltzmann distribution (see Sect. 3.3.3). There are attempts to explain this behavior by enlarging the phase space to contain both the system and the bath. Qualitatively the argument then is that the cells containing less energy in the considered system will have more energy in the bath due to overall energy conservation, and are, thus, bigger than cells corresponding to higher energies in the considered system. Therefore, so the argument goes, it is more likely to find the considered system at lower energies. However, in order to derive a Boltzmann distribution more properties are needed to make this concept work: an exponential growth of the cell sizes of the bath (see Chap. 12) and, most importantly, ergodicity of the full system. Otherwise cell sizes will not map onto probabilities. So again, in this sense this approach only holds under the assumption of ergodicity, the very condition one tried to get rid of.

## 4.5 The Problem of Adiabatic State Change

One issue that has attracted much less attention in the past than the second law is the fact that entropy should be shown to be invariant during so-called adiabatic processes [14]. As known from countless practical experiments, the state of a system controlled by the change of an external parameter, like volume or magnetic moment, proceeds in such the way that entropy is left unchanged, if the system is thermally insulated and the parameter change happens slowly enough. As obvious as this fact may seem from an experimental perspective, it is surprising from the theoretical side. From all the possible ways in which, e.g., energy can change, exactly that one needs to be singled out that leaves a quantity as complicated as entropy unchanged! This is true for all sorts of processes that fulfill the conditions mentioned above. From a microscopical point of view this is an enormously large class of processes. For the phenomenological theory of thermodynamics it is a very important property since, otherwise, the identification of pressure with the negative derivative of energy with respect to volume under constant entropy, a basic statement of the first law, would be without justification [see Sect. 3.1.3 and especially (3.18)].

The most popular answer to this question on the basis of classical mechanics is the "law of the invariance of the phase space volume" [14]. This law is introduced by investigating a short time step of the evolution of a system while an external parameter $a$ is changed. The considered time step is short enough to neglect the motion of any representing point of the system in phase space during that step. However, with the parameter change the "energy landscape" of the system in phase space also changes by an infinitesimal amount. This means all points that belonged to one energy shell of energy $E$ and thickness $\Delta E$ before the step may belong, in general, to different energy shells after the step (see Fig. 4.8).

Now the ensemble average is computed, i.e., the mean energy change $\overline{\Delta E}$ corresponding to all the points on the energy shell before the time step. Hereby it turns out by using differential geometry that the volume below the energy surface $\Omega_{a+\Delta a}(E+\overline{\Delta E})$ corresponding to the energy $E+\overline{\Delta E}$ defined by the "new" Hamiltonian encloses a volume in phase space that is exactly as large as the volume enclosed by the energy surface $\Omega_a(E)$ corresponding to the energy $E$ defined by the "old" Hamiltonian,

$$\Omega_{a+\Delta a}(E + \overline{\Delta E}) = \Omega_a(E).  \tag{4.2}$$

For the exact definition of such phase space volumes, see Sect. 3.3.2.

The idea now is that in an adiabatic process the actual representing point of the system moves much faster than the Hamiltonian changes, and that it moves in an ergodic way. If this were the case, the system would undergo many such infinitesimal changes, while the energy changes corresponding to the points on the energy shell would remain practically the same. Within a time interval in which the representing point passes all points on the energy shell and in which the external parameter only changes by a very small amount, the system would "average itself." If the change of the external parameter were performed sufficiently slowly,

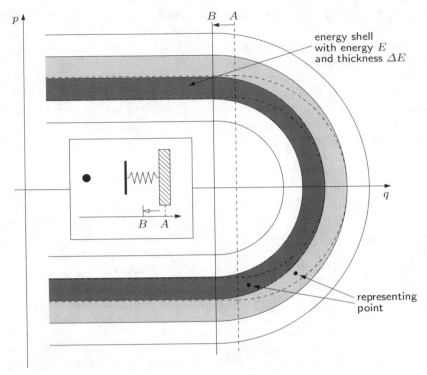

**Fig. 4.8** Invariance of entropy. Phase space of a particle interacting with a harmonic wall (see inset). If we were going to change an external parameter by moving the wall from position $A$ to $B$, the Hamiltonian would change and therefore also the phase space. If the particle was initially within the region between the *dashed lines* of the phase space, it could be in the two different *gray regions* after the change. Two representing points are shown belonging to the same energy surface before the process, which are on different surfaces afterward

the representing point would thus travel through phase space in such a way that the energy surfaces the point can be actually found on will always enclose the same volume. If one now considered the evolution of many representing points, initially on equidistant energy surfaces, it may be argued that the volumes in between two adjacent energy surfaces corresponding to two representing points cannot change, since the volumes enclosed by those energy surfaces do not change. The conclusion of this consideration is that the volume of the energy shell of a representing point does not change in a sufficiently slow process.

This reasoning faces two major shortcomings: first, the whole concept relies on ergodicity, a behavior that, in general, can only be postulated (see Sect. 4.2). Furthermore, the Boltzmann definition connects entropy with the volume of the energy shell [see (3.47)]. This volume depends linearly on the thickness of the shell, $\Delta E$ and may be written as $\Delta E\, G(E)$, where $G(E)$ is the classical state density, which may be computed from the volume enclosed by some energy surface $\Omega(E)$ by $G(E) = \partial \Omega(E)/\partial E$ (see Sect. 3.3.2). The thickness $\Delta E$ is controversial, if only entropy changes are considered. However, its precise value is irrelevant, as long as it is not changed during the process. This is why entropy is usually defined simply on the basis of $G(E)$ as $S = k_B \ln G(E)$. The latter, however, may lack invariance, as will be shown below.

Consider, e.g., a particle bound to a two-dimensional rectangular surface which may even be quasi-ergodic (see Fig. 4.9, upper part). The state density $G(E)$ of such a system is constant, i.e., does not vary with $E$ (see Fig. 4.9 lower part). If one edge of the surface is moved out, the state density changes but remains constant with respect to $E$, shifted only to a higher value. Thus, although the energy changes, the system cannot have the same entropy after the process, according to the above definition. The volumes enclosed by energy surfaces corresponding to representing points at the upper and lower energies of the energy shell may nevertheless be left invariant (see Fig. 4.9). The above consideration involves a change of $\Delta E$, which is problematic, since it is usually left out of the entropy definition. This problem does not show in standard calculations simply because, contrary to our example, one has for typical systems like ideal gases $G(E) \approx \Omega(E)$. However, in principle, the problem remains unsolved (cf. Sect. 14.1).

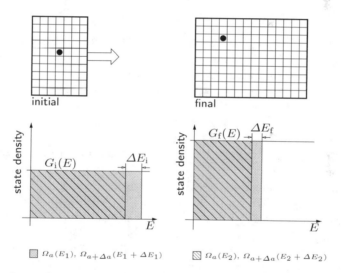

**Fig. 4.9** Invariance of entropy. Special example of a particle in a two-dimensional box (*upper part*). The respective state densities and the problem of different $\Delta E$ before and after the change of the external parameter (volume of a box) from $a$ to $a + \Delta a$ are shown

## 4.6 Shannon Entropy, Jaynes' Principle

In Sect. 3.3.3 Jaynes' principle, i.e., maximization of Shannon entropy (3.49), has been introduced basically as a recipe of how to calculate thermodynamic behavior from knowledge about microscopic structures. Nevertheless, since maximization of entropy plays a major role in this concept, it is sometimes brought forth as a justification for the second law. As explained in Sect. 3.3.3, this consideration is neither based on classical mechanics nor on quantum theory, indeed its applicability is not even meant to be restricted to physics at all. The basic idea is that entropy represents a measure of lack of information, thus the whole concept is introduced as a rational way of dealing with incomplete knowledge, wherever it may occur [15, 16]. Thus, this approach somehow radicalizes the idea that has already been underlying some of the previously described concepts, namely that thermodynamic behavior is due to the observer's inability to measure precisely and in enough detail. This would mean that the origin of the second law was no longer searched for in the physical world but would take place more or less in the observer's brain.

And this, so the most common objection against this point of view, causes principal problems. From the fact that entropy and thus a basic thermodynamic quantity is based on the subjective lack of knowledge of the observer it must follow that, if an observer gains more information about a system, its entropy decreases and thus, e.g., its temperature decreases. This means that a change in the observer's mind could induce a macroscopic physical change of a system. This is considered unacceptable, at least as long as it is unclear whether or not there are principal limits to the observer's possibility to overcome his lack of knowledge.

All other approaches mentioned above define entropy on the basis of the dynamics of a microscopic picture; thus entropy is eventually defined as a function of time, the evolution of which is controlled by some underlying equation of motion. A reduction proper is not even attempted in Jaynes' approach, microscopic dynamics does not play any role there.

At first sight it appears to be an advantage of this approach that properties of systems under canonical (rather than microcanonical) conditions follow naturally from the basic idea. However, for this to be true in general, one has to accept that keeping an intensive quantity fixed leads to a fixed average value of the conjugate extensive quantity. For this claim no further justification is given.

Furthermore, if this principle is applied to the exchange of any other extensive quantity, the resulting states are not necessarily stationary anymore, which is inconsistent, for the resulting states should be equilibrium states.

A last objection against this theory is that the limits of its applicability are not stated clearly. Technically, this concept might be applied, e.g., to low-dimensional "few body problems," for which the laws of thermodynamics are obviously not valid.

## 4.7 Time-Averaged Density Matrix Approach

This concept is explicitly quantum mechanical and refers to the von Neumann entropy (see Sect. 2.2.4 or [10])

$$S = -k_B \text{Tr} \{\hat{\rho} \ln \hat{\rho}\} \ . \tag{4.3}$$

Since any possible state $\hat{\rho}$ of a quantum system can be written as a weighted mixture of states that form an orthogonal set, the von Neumann entropy reduces to Shannon's entropy with those orthogonal states taken as the distinguishable states and their weights as the corresponding probabilities. The von Neumann entropy is invariant under unitary transformations. This property has two consequences: it is independent of the chosen basis and it is time independent, just like the Gibbs entropy in the ensemble approach.

Since one needs an entropy that can possibly change in time, it has been suggested to calculate $S$ using a time-averaged density matrix rather than the actual instantaneous one.

The elements of the time-dependent density matrix read

$$\langle i|\hat{\rho}(t)|j\rangle = e^{i(E_j - E_i)t} \langle i|\hat{\rho}(0)|j\rangle , \qquad (4.4)$$

where $|i\rangle$, $|j\rangle$ are energy eigenstates and $E_i$, $E_j$ the respective eigenenergies. Since all off-diagonal elements are oscillating, they will vanish if the density matrix is time averaged [17]. Moreover, it can be shown that the von Neumann entropy indeed rises, if the off-diagonal elements vanish. There are investigations indicating that the respective quantum dynamics are in a sense ergodic, at least for a class of systems [18].

The problem of this idea is the averaging time. If systems get big, typically the energy-level spacing becomes very small, and the averaging time that is necessary to actually see entropy rises significantly, may very well exceed typical relaxation times of thermodynamic systems. In the limit of degenerate energy eigenstates this averaging time becomes infinitely long. Thus, technically speaking, a system with a precisely given energy (microcanonical conditions), and therefore only occupying degenerate energy eigenstates, would never exhibit an increasing entropy, regardless of the averaging time.

## 4.8 Open System Approach and Master Equation

The open system approach is based on the fact that any description of a real system will necessarily be incomplete. This means that the subsystem selected for investigation will eventually be subject to the interaction with some unobserved and, to a large extent, uncontrollable "environment." Isolated models should thus be taken as an idealization of rather limited validity for classical as well as for quantum systems. Embeddings of the target system prevail, which do not only limit the accessible information but also drastically change the dynamics of the parts. Note, however, that the partition between system and environment does not follow from the underlying complete description; it has to be put in "by hand."

There is no space for a complete introduction into the theory of open quantum systems here. Let us, thus, just repeat some central aspects and comment on common results. To learn more we refer the interested reader to the excellent and comprehensive books by Breuer and Petruccione [19] or Weiss [20] discussing most aspects of open quantum systems.

The total Hamiltonian of the full system is defined by

$$\hat{H} = \hat{H}_0 + \lambda\hat{V}, \tag{4.5}$$

where $\hat{H}_0$ describes an uncoupled local Hamiltonian of system and environment, and $\hat{V}$ their interaction with strength $\lambda$. It is possible to consider different partitioning schemes here as well, not only system–environment partitions. Important is the possibility to split up the Hamiltonian into two parts, one part which is fully under control, i.e., we know the spectrum of this part, and the rest which is small in some perturbative sense. Here, the full system is considered to be a closed quantum system, i.e., constituting the whole "quantum universe." Thus, the dynamics of the combined system state in the interaction picture is given by the Liouville–von Neumann equation ($\hbar = 1$)

$$\frac{\partial}{\partial t}\hat{\rho}(t) = -i\lambda[\hat{V}(t), \hat{\rho}(t)] = \lambda\,\hat{\mathcal{L}}(t)\,\hat{\rho}(t), \tag{4.6}$$

where $\hat{\rho}(t)$ is the density operator for the state of the full system. The time argument of the interaction results from the transformation into the interaction picture

$$\hat{V}(t) = e^{i\hat{H}_0 t}\,\hat{V}\,e^{-i\hat{H}_0 t}. \tag{4.7}$$

The right-hand side of the dynamical equation (4.6) can also be written by using the Liouvillian $\hat{\mathcal{L}}(t)$, a superoperator acting on density operators of the Liouville superspace.

The discrimination between relevant and irrelevant parts does not follow from the underlying complete description, it is typically motivated by a certain partition between system proper and environment. Formally, the relevant part can be projected out from the full state of the system, by using the projection superoperator

$$\hat{\rho}_{\mathrm{rel}}(t) = \hat{\mathcal{P}}\hat{\rho}(t). \tag{4.8}$$

Accordingly, a superoperator projecting on the irrelevant part may be introduced by

$$\hat{\rho}_{\mathrm{irrel}}(t) = \hat{\mathcal{Q}}\,\hat{\rho}(t) = (\hat{\mathcal{I}} - \hat{\mathcal{P}})\,\hat{\rho}(t), \tag{4.9}$$

with $\hat{\mathcal{I}}$ being the unit operator in Liouville space. Both superoperators are maps in the state space of the combined system featuring standard properties

$$\hat{\mathcal{P}} + \hat{\mathcal{Q}} = \hat{\mathcal{I}}, \tag{4.10}$$

$$\hat{\mathcal{P}}\hat{\mathcal{Q}} = \hat{\mathcal{Q}}\hat{\mathcal{P}} = 0, \tag{4.11}$$

$$\hat{\mathcal{P}}^2 = \hat{\mathcal{P}}, \tag{4.12}$$

$$\hat{\mathcal{Q}}^2 = \hat{\mathcal{Q}}. \tag{4.13}$$

Using these projection superoperators and considering initial states which are identically reproduced by the projection, i.e., $\hat{\mathcal{P}}\hat{\rho}(0) = \hat{\rho}(0)$ one finally finds the exact dynamical equation for the relevant part, called the Nakajima–Zwanzig (NZ) equation [21, 22]

$$\frac{\partial}{\partial t}\hat{\mathcal{P}}\hat{\rho}(t) = \int_0^t ds \, \hat{\mathcal{K}}(t, s)\hat{\mathcal{P}}\hat{\rho}(s), \tag{4.14}$$

with the memory kernel $\hat{\mathcal{K}}(t, s)$ which is a superoperator as well. For a full derivation of this equation, see [19]. In case of $\hat{\mathcal{P}}\hat{\rho}(0) \neq \hat{\rho}(0)$, i.e. for initial correlations, the above given NZ equation has to be extended by a further inhomogeneity which may potentially alter the resulting dynamics significantly. However, in order to simplify the discussion below we have skipped this term. For more details on this, see Chap. 11.

So far this integro-differential equation is as complicated to solve as the full, above-given Schrödinger equation. Hence, the memory kernel is expanded in the coupling strength $\lambda$ finding

$$\hat{\mathcal{K}}(t, s) = \sum_{i=1}^{\infty} \lambda^i \hat{\mathcal{K}}_i(t, s) = \lambda^2 \hat{\mathcal{P}}\hat{\mathcal{L}}(t)\hat{\mathcal{L}}(s)\hat{\mathcal{P}} + O(\lambda^4), \tag{4.15}$$

where all odd terms vanish due to the properties of the projection operators (for details, see [19]). In case of a weak interaction, a truncation of the expansion after second-order called Born approximation seems to be plausible. This approximation leads to the second-order equation for the relevant part

$$\frac{\partial}{\partial t}\hat{\mathcal{P}}\hat{\rho}(t) = \lambda^2 \int_0^t ds \, \hat{\mathcal{P}}\hat{\mathcal{L}}(t)\hat{\mathcal{L}}(s)\hat{\mathcal{P}}\hat{\rho}(s). \tag{4.16}$$

Basically, the Born approximation is a perturbation expansion in the coupling strength between system and environment, thus an approximation for weakly interacting bipartite systems. Since higher oder terms are connected to multiple integrations of time, these are also connected to later times within the dynamics of the relevant part. This makes the above approximation also some kind of approximation within time which gets worse for larger times of the dynamics. But if the system has already reached a stationary equilibrium state on a timescale where higher orders are typically very small, the truncation at second order can be rather good for the whole dynamics. In Chap. 19, however, we will consider an example where the second-order approximation is correct for short times and completely wrong on a longer timescale. Equation (4.16) remains to be an integro-differential equation, containing all the history of the dynamics and is therefore in principle suitable to treat non-Markovian situations, i.e., distinct memory effects.

Within a system–environment partitioning scheme a standard projection operator is defined by the action

$$\hat{\rho}_{\rm rel}(t) = \hat{P}\hat{\rho} = {\rm Tr}_{\rm E}\{\hat{\rho}(t)\} \otimes \hat{\rho}_{\rm B} = \hat{\rho}_{\rm S}(t) \otimes \hat{\rho}_{\rm B}, \qquad (4.17)$$

with an arbitrary time-independent state $\hat{\rho}_{\rm B}$ of the environment mostly chosen to be a thermal one. Using both this standard projection operator and the definition of the Liouvillian, (4.16) turns into

$$\frac{\partial}{\partial t}\hat{\rho}_{\rm S}(t) = -\lambda^2 \int_{t_0}^{t} ds \, {\rm Tr}_{\rm E}\left\{[\hat{V}(t), [\hat{V}(s), \hat{\rho}_{\rm S}(s) \otimes \hat{\rho}_{\rm B}]]\right\}. \qquad (4.18)$$

This type of equation is also obtained from an iterative approach to the Liouville–von Neumann equation (4.6) for weakly interacting systems under the assumption that the total state of the system remains factorizable for all times, i.e., $\hat{\rho}(t) = \hat{\rho}_{\rm S}(t) \otimes \hat{\rho}_{\rm B}$ (see [23]). However, this does not mean that the state of the system is really separable for all times, since even weak interactions may lead to serious quantum correlations for longer times (cf. [24]). Furthermore, it is hard to see how local entropy could rise, without developing entanglement between the system and its environment, local entropy being an entanglement measure in this case. This issue is also addressed in some detail in Sect. 11.1. However, under the non-entanglement assumption environments (linearly coupled to the system proper) may even be treated classically [25].

Since such an equation is still very hard to solve one intends to get back to a time-local form. This, in turn, cannot be achieved without specializations, approximations, and/or assumptions. Different authors propose slightly different schemes, here. All of them proceed in the interaction picture and require the interaction between the subsystems to be small. A widely applied and famous procedure is using the Markov assumption to remove the time convolution of the kernel by replacing the state of the system at time $s$ on the right-hand side of (4.18) by the state at time $t$. With this approximation one obtains the Redfield master equation (see [26–28])

$$\frac{\partial}{\partial t}\hat{\rho}_{\rm S}(t) = -\lambda^2 \int_{0}^{\infty} ds \, {\rm Tr}_{\rm E}\left\{[\hat{V}(t), [\hat{V}(t-s), \hat{\rho}_{\rm S}(t) \otimes \hat{\rho}_{\rm B}]]\right\}. \qquad (4.19)$$

This Markovian equation does not resolve the dynamics beyond the correlation time, and thus, assumes that environmental correlations decay much faster than the state of the system changes. Equation (4.19) is also obtained from another approach (see, e.g., [29]), constructing a Dyson series and truncating at some order due to the smallness of the interaction. This might be in conflict with evaluating the result for arbitrarily long times, or even times long enough to see the effect of the environment on the system. It can be shown that at least one stationary state of the above given Markovian quantum master equation (4.19) is the canonical equilibrium state

$$\lim_{t \to \infty} \hat{\rho}_S(t) \propto e^{-\beta \hat{H}_S} ,$$
(4.20)

where $\hat{H}_S$ is the local Hamiltonian of the system in $\hat{H}_0$ given in (4.5) and $\beta$ is the inverse temperature of the chosen thermal state $\hat{\rho}_B$ of the environment (for a proof, see [30]).

Starting from (4.19) and using a concrete microscopic model for the environment, e.g., a set of decoupled harmonic oscillators, with an eigenfrequency density suitable to fulfill the Markov assumption (i.e., the fast decay of correlations within the environment), and assuming a special type of interaction (rotating wave approximation), an autonomous master equation can be derived. Before writing down this closed differential equation for the relevant part, let us illuminate the problem from a different perspective.

Instead of the above shown microscopic derivation, it is possible to propose the most general form of a differential equation describing the dynamics of the relevant part in a closed manner, i.e., without a concrete dynamical model for the irrelevant part. If the Markovian quantum master equation for the relevant part is given by the autonomous differential equation

$$\frac{\partial}{\partial t} \hat{\rho}_S(t) = \hat{\mathcal{L}} \, \hat{\rho}_S(t) ,$$
(4.21)

according to the time-independent generator $\hat{\mathcal{L}}$, the dynamical map from the state space of density operators into itself is defined as

$$\hat{\mathcal{V}}(t) = e^{\hat{\mathcal{L}} t} .$$
(4.22)

Here, of course, the dynamics has to conserve the properties of the density operator, and thus, $\hat{\mathcal{V}}(t)$ must be a convex linear, completely positive and trace-preserving map satisfying the semigroup property (cf. [19]). Due to these requirements Lindblad [31, 32] was able to derive the most general form of the generator $\hat{\mathcal{L}}$.

To write down the most general form of the generator we need a suitable basis for the Liouville space which has $n^2$ dimensions if $n$ is the dimension of the respective Hilbert space. A suitable basis to span this superspace is defined by the set of $n^2$ operators $\{\hat{F}_i\}$. We require that these operators are orthogonal due to the Hilbert–Schmidt scalar product

$$\mathrm{Tr} \left\{ \hat{F}_i \, \hat{F}_j^\dagger \right\} = \delta_{ij} ,$$
(4.23)

and choose one to be the unit operator within this state space. In terms of these operators the generator is defined by

$$\hat{\mathcal{L}} \hat{\rho}_{S} = -i[\hat{H}_{S}, \hat{\rho}_{S}] + \sum_{i,j=1}^{n^2-1} \gamma_{ij} \left( \hat{F}_i \, \hat{\rho}_{S} \hat{F}_j^\dagger - \frac{1}{2}[\hat{F}_j^\dagger \hat{F}_i \, , \hat{\rho}_{S}]_+ \right) \qquad (4.24)$$

excluding the unit operator from the sum. This form is called the first standard form of the generator. Since the coefficient matrix $\gamma_{ij}$ has to be positive due to the properties of the dynamical map (4.22), it is easily brought into Lindblad form by diagonalizing the matrix $\gamma_{ij}$ finding

$$\hat{\mathcal{L}} \hat{\rho}_{S} = -i[\hat{H}_{S}, \hat{\rho}_{S}] + \sum_{i=1}^{n^2-1} \gamma_i \left( \hat{L}_i \hat{\rho}_{S} \hat{L}_i^\dagger - \frac{1}{2}[\hat{L}_i^\dagger \hat{L}_i \, , \hat{\rho}_{S}]_+ \right), \qquad (4.25)$$

with the Lindblad operators $\hat{L}_i$ which are linear combinations of the basis operators $\hat{F}_i$. The first term describes the coherent evolution due to the Hamiltonian of the system, whereas the following terms describe decoherence and damping effects following from the coupling to the other degrees of freedom not explicitly modeled here. The rates $\gamma_i$ depend on the coupling strength $\lambda$ and eventually on a temperature of the associated environment. The generator (4.25) is the most general form which preserves all properties of the density operator. The Lindblad form can also be obtained from a microscopic model, say (4.20), using several approximations. Finally, one needs the Born, Markov, Redfield, and rotating wave approximations or secular approximation to derive the Markovian–Lindblad quantum master equation (4.25).

Besides the Nakajima–Zwanzig (NZ) technique there is another different approach to the reduced relevant dynamics, which does not contain a time convolution within the memory kernel. This approach is therefore called time convolutionless method or TCL (see [33–35]). The idea behind this approach is to remove the influence of the future time evolution on the history of the dynamics by using the backward propagator in time, and thus, obtain a time-local differential equation. This, finally, produces the TCL master equation (here again for factorizing initial conditions)

$$\frac{\partial}{\partial t} \hat{\mathcal{P}} \hat{\rho} = \hat{\mathcal{K}}(t) \hat{\mathcal{P}} \hat{\rho}(t), \qquad (4.26)$$

with the time-dependent TCL generator which is a very complicated superoperator. This time-local equation is again exact, however, as difficult to treat as the full problem. Again a perturbation expansion of the generator

$$\hat{\mathcal{K}}(t) = \sum_{i=1}^{\infty} \lambda^i \hat{\mathcal{K}}_i(t) \qquad (4.27)$$

leads to the second-order equation

$$\frac{\partial}{\partial t}\hat{\mathcal{P}}\hat{\rho} = \lambda^2 \int_0^t ds\ \hat{\mathcal{P}}\hat{\mathcal{L}}(t)\hat{\mathcal{L}}(s)\hat{\mathcal{P}}\hat{\rho}(t), \tag{4.28}$$

which is similar to the NZ equation, but local in time. For a complete derivation, a discussion of differences between NZ and TCL and some examples, see [19].

Thus, all these techniques come with the remarkable advantage of being able to describe not only equilibrium states but also the way to equilibrium. However, although there are extensions, these approaches typically rely on many crucial approximations. Furthermore, state changes of the environment are not included, which introduces a certain asymmetry between system and environment which becomes a problem in case of finite small environments. It is the environment already being in an equilibrium state that induces equilibrium in the considered system.

# References

1. J.L. Lebowitz, Phys. Today **46**(9), 32 (1993)
2. H. Primas, Mind and Matter **1**(1), 81 (2003)
3. G.P. Beretta, A.F. Ghoniem, G.N. Hatsopoulos (eds.), *Meeting the Entropy Challenge* (American Institute of Physics, 2008)
4. W.R. Theis, *Grundzüge der Quantentheorie* (Teubner, Stuttgart, 1985)
5. L. Boltzmann, *Lectures on Gas Theory* (University of California Press, Los Angeles, 1964)
6. G. Adam, O. Hittmair, *Wärmetheorie*, 4th edn. (Vieweg, Braunschweig, Wiesbaden, 1992)
7. C. Cercignani, *The Boltzmann Equation and its Applications* (Springer, 1988)
8. P. Davies, *The Physics of time Asymmetry* (University of California Press, Berkeley, 1974)
9. G. Birkhoff, Proc. Natl. Acad. Sci. USA **17**, 656 (1931)
10. J.V. Neumann, *Mathematischen Grundlagen der Quantenmechanik*. Die Grundlehren der Mathematischen Wissenschaften (Springer, Berlin, Heidelberg, New York, 1968)
11. J.W. Gibbs, *Elementary Principles in Statistical Mechanics* (Dover Publications, New York, 1960)
12. P. Ehrenfest, T. Ehrenfest, *The Conceptual Foundations of the Statistical Approach in Mechanics* (Cornell University Press, Ithaca, 1959)
13. R. Penrose, *The Emperor's New Mind* (Oxford University Press, New York, 1990)
14. A. Münster, *Statistische Thermodynamik* (Springer, Berlin, Heidelberg, 1956)
15. E. Jaynes, Phys. Rev. **106**, 620 (1957)
16. E. Jaynes, Phys. Rev. **108**, 171 (1957)
17. E. Fick, G. Sauermann, *The Quantum Statistics of Dynamic Processes* (Springer, Berlin, Heidelberg, 1990)
18. J.V. Neumann, Z. Phys. **57**, 30 (1929)
19. H.P. Breuer, F. Petruccione, *The Theory of Open Quantum Systems* (Oxford University Press, Oxford, 2002)
20. U. Weiss, *Quantum Dissipative Systems, Series in Modern Condensed Matter Physics*, vol. 10, 2nd edn. (World Scientific, Singapore, New Jersey, London, Hong Kong, 1999)
21. S. Nakajima, Prog. Theor. Phys. **20**, 948 (1958)
22. R. Zwanzig, J. Chem. Phys. **33**, 1338 (1960)
23. M.O. Scully, M.S. Zubairy, *Quantum Optics* (Cambridge University Press, Cambridge, 1997)
24. J. Gemmer, G. Mahler, Europhys. Lett. **59**, 159 (2002)
25. P. Gong, J. Brumer, Phys. Rev. Lett. **90**(5), 050402 (2003)
26. A. Redfield, IBM J. Res. Dev. **1**, 19 (1957)

27. G. Mahler, V.A. Weberruß, *Quantum Networks*, 2nd edn. (Springer, Berlin, Heidelberg, 1998)
28. K. Blum, *Density Matrix Theory and Applications*, 2nd edn. (Plenum Press, New York, London, 1996)
29. D.F. Walls, G.J. Milburn, *Quantum Optics*, 2nd edn. (Springer, Berlin, Heidelberg, 1995)
30. R. Kubo, M. Toda, N. Hashitsume, *Statistical Physics II: Nonequilibrium Statistical Mechanics*, 2nd edn. No. 31 in Solid-State Sciences (Springer, Berlin, Heidelberg, New-York, 1991)
31. G. Lindblad, Commun. Math. Phys. **48**, 119 (1976)
32. G. Lindblad, *Non-equilibrium Entropy and Irreversibility*, *Mathematical Physics Studies*, vol. 5 (Reidel, Dordrecht, 1983)
33. F. Shibata, Y. Takahashi, N. Hashitsume, J. Stat. Phys. **17**, 171 (1977)
34. S. Chaturvedi, F. Shibata, Z. Phys. B **35**, 297 (1979)
35. F. Shibata, T. Arimitsu, J. Phys. Soc. Jpn. **49**, 891 (1980)

# Part II
# Equilibrium

# Chapter 5
# The Program for the Foundation of Thermodynamics

*Nobody knows, what entropy really is.*

— J. Von Neumann [1]

**Abstract** In this chapter a "checklist" of properties of thermodynamic quantities will be given, which is meant as a set of rules by which thermodynamic behavior (as we observe it) is completely defined. Thus, if all those rules can be shown to result from an underlying theory, here quantum mechanics, thermodynamics might be considered as emerging from this underlying theory; if the approach failed to demonstrate any of them, the task would not be accomplished fully.

For a foundation of thermodynamics it is tempting to give an abstract but, nevertheless, intuitively appealing, ontological definition of entropy, such as entropy is disorder, entropy is some volume in phase space, entropy is the lack of knowledge. However, then one is left with the task of showing that the quantity so defined indeed behaves in a way that is empirically observed (see Chap. 4, Sect. 3.1).

Thus, here we introduce thermodynamics on the basis of some axiomatic structure which only spells out how thermodynamic quantities are expected to behave without giving explanations of their "intrinsic nature." Usually some of the axioms deal with the notorious irreversibility of thermodynamic systems, others with the possibility of formulating the internal energy $U$ as a state function in terms of the entropy $S$ and other extensive variables. Starting from first principles it is evident that much of the behavior of thermodynamic quantities (including their mere existence) needs explanation.

The axiomatic structure or "checklist" given here should neither be considered unique nor irreducible. If, e.g., ergodicity (and mixing) were assumed, the validity of the second law (Checklist 2) and the equality of intensive variables (Checklist 4) would follow; thus, one could replace those two items by demanding ergodicity. This, however, would not be suitable for the approach at hand. The choice of properties given here is expected to be complete and appropriate for starting from the theory of quantum mechanics.

Gemmer, J. et al.: *The Program for the Foundation of Thermodynamics*. Lect. Notes Phys. **784**, 65–68 (2009)
DOI 10.1007/978-3-540-70510-9_5

## 5.1 Basic Checklist: Equilibrium Thermodynamics

1. **Definition of Thermodynamic Quantities:** All thermodynamic quantities (entropy, temperature, energy, pressure, etc.) should be precisely defined as functions of the variables of an underlying theory, such that this underlying theory describes the dynamics of those variables. If introduced in this way, the thermodynamic quantities would "always" be well defined, i.e., even outside equilibrium, or for systems that are not thermodynamic (see Sect. 5.2). However, only for thermodynamic systems and for processes close to equilibrium the functions are required to behave as formulated below. (In the work at hand such definitions of thermodynamic quantities are given in Chaps. 6, 13, and 14.)

2. **Second Law of Thermodynamics** (Maximum Principle for Entropy): This axiom establishes the irreversibility of thermodynamic evolutions and processes. It postulates the existence of a certain stationary (up to fluctuations, cf. below) equilibrium state, into which a thermodynamic system will evolve eventually. Under given constraints this equilibrium state is stable with respect to perturbations. It should be shown that the system in quest reaches such a state for which the fluctuations of all well-defined thermodynamic quantities are negligible. This state has to be controllable by macroscopic constraints. Since those constraints can be imposed in different ways, i.e., by keeping different sets of intensive and extensive variables fixed or controlled, at least two cases have to be distinguished:

   a. *Microcanonical Conditions* (energy $U$ kept fixed): In this case the entropy should only increase during the evolution, and the final state should only depend on the energy distribution of the initial state. (This behavior is established in Sect. 10.1.)

   b. *Canonical Conditions* (temperature $T$ kept fixed): Since under these conditions the equilibrium state of the system is controlled by the contact with a heat bath, the only specifying parameter is its temperature; the equilibrium state should only depend on this temperature, regardless of its initial state. (This behavior is established in Sect. 10.2.)

3. **Gibbsian Fundamental Form** (Possibility of macroscopic control): From this law, eventually, connections between measurable, macroscopic intensive and extensive quantities are inferred. Thus, it guarantees that for a certain class of processes that involve a change of those macroscopic variables, a detailed microscopic picture is dispensable and can be replaced by a simpler, macroscopic picture.

   a. *State Function:* It should be shown that if the extensive variables, say, volume $V$ and (equilibrium) entropy $S$, take on certain values, the internal energy $U$ necessarily has a corresponding value, regardless of the path by which the state has been reached.

   b. *Temperature as a Conjugate Variable:* It should be shown that standard close-to-equilibrium processes (heating, cooling, etc.), in which all extensive variables are kept fixed except for energy and entropy, proceed according to

$$\left(\frac{\partial U}{\partial S}\right)_{V=\text{const.}} = T \, . \tag{5.1}$$

Of course, here and below the above-mentioned definitions of temperature $T$, entropy $S$, etc. (that are in principle based on the microstate) have to be used. (This behavior is established in Sect. 13.3.)

c. *Pressure as a Conjugate Variable:* It should be shown that there are processes (isentropic), in which the extensive variable volume $V$ changes, while all others, including especially entropy, remain constant. The analysis of such a process then has to yield,

$$\left(\frac{\partial U}{\partial V}\right)_{S=\text{const.}} = -P \tag{5.2}$$

where $P$ is the pressure. (This behavior is established in Sect. 14.1.)

d. *Other Conjugate Variables:* It should be shown that processes may and will occur such that while some additional extensive variable changes, all others remain constant. The derivative of $U$ should yield the respective conjugate-intensive variable.

4. **Classes of Thermodynamic Variables:**

a. *Extensive Variables:* It should be shown that thermodynamic variables that are claimed to be extensive, in particular the entropy $S$, are indeed extensive quantities. (This is shown in Chap. 12.)

b. *Intensive Variables:* It should be shown that two systems allowed to exchange some extensive quantity will end up in an equilibrium state having the same conjugate intensive variable. (This behavior is, e.g., established in Sect. 13.2.)

Those properties of thermodynamic quantities and the various relations allow for an application of the standard techniques and methods of thermodynamics. Thus, if they are shown to result as claimed from quantum mechanics, the field of thermodynamics can, in this sense, be considered reducible to quantum mechanics.

Such a reconstruction, theoretically satisfying as it might be, will eventually have to be judged by the results it produces. Thus, in order to make it a physically meaningful theory rather than just an abstract mathematical consideration, the limits of its applicability have to be examined just as much as its connection to the standard classical theory. This is the subject of the supplementary checklist.

## 5.2 Supplementary Checklist: Quantum Mechanical Versus Classical Aspects

1. **Thermodynamic Systems:** It is necessary to explain and clarify the relationship between the emerging theory and its underlying theory. If the emerging properties would inevitably result from the underlying theory, one could discard the latter completely. In the present context this cannot be the case: the underlying theory is supposed to be quantum mechanics, and it is obvious that not all systems that obey quantum mechanics can be described thermodynamically, while Schrödinger-type quantum theory is believed to underlie all sorts of non-relativistic systems. Thus, a fairly precise definition of the class of systems that can be expected to behave thermodynamically should be given. This definition should not result in a tautology like "all systems that show the properties mentioned above are thermodynamic systems," but should result in a criterion that can be checked with acceptable efforts.

2. **Quantum Versus Classical Entropy**: Despite its problematic foundation standard "classical" thermodynamics works pretty well for almost all practical purposes. If this is not just incidental, it should be possible to show that entropy as a function of, say, volume and energy should be the same, no matter whether it is calculated based on a standard classical definition or the quantum mechanical one that can be shown to have the above properties. Here, this would eventually amount to showing that quantum state density and the volume of energy shells in classical phase space are proportional for large classes of systems (that have a classical analog). Such a relation would prove a kind of "correspondence principle."

## Reference

1. M. Tribus, E.C. McIrvine, Sci. Am. **225**(9), 179 (1971)

# Chapter 6
# Outline of the Present Approach

*One person's mystery is another person's explanation.*
— N. D. Mermin [1]

**Abstract** As already indicated in previous chapters we intend to explain the validity of the laws of thermodynamics on the basis of quantum dynamics as given by the Schrödinger equation. In this chapter we provide the background of this approach which does not rely on ergodicity, mixing, etc., but on the concept of "typicality." The importance of a system–environment scheme for a concise definition of entropy and the role of entanglement in this context is discussed. Furthermore we investigate the nature of thermodynamical uncertainties (often referred to as "fluctuations"). Within this quantum version of the typicality approach these uncertainties appear as quantum uncertainties and are thus of fundamental nature.

## 6.1 Typicality and Dynamics

Despite the fact that some call it a "tired old question" the search for the origin and the true nature of the second law of thermodynamics is ongoing [2]. The idea of any (thermodynamic) system always approaching a "state" (equilibrium) in which it behaves as if it occupied an immense multitude of microstates at the same time (ensemble), even if it started from a perfectly concrete microstate, is still quite puzzling. How can the system end up in a multitude of states at all, and even worse, always in the same well-defined set, no matter where it started from? How can this behavior that obviously breaks time symmetry be explained on the basis of a time-reversal-invariant underlying theory? The traditional answers are of course well-known and may be found in Sect. 4.

Although conceptually already being present in the (later) work of Boltzmann and Ehrenfest, the idea of typicality seems to have attracted less attention than the above-mentioned approaches. The term "typicality" has (to the authors best knowledge) been coined by Lebowitz et al., who have done pioneering work on this concept (cf. [3–6] and references therein). Especially within the quantum context typicality has recently been picked up and worked on also by others [7, 8]. What is special about the typicality approach?

Within the typicality approach there is neither time averaging like in the ergodicity approach nor does the microstate evolve into an ensemble as in the approaches based on mixing. The idea simply is that very many microstates from some well-

Gemmer, J. et al.: *Outline of the Present Approach.* Lect. Notes Phys. **784**, 69–76 (2009)
DOI 10.1007/978-3-540-70510-9_6

defined region in the microstate space of the system (phase space, Hilbert space, etc.) may yield, for some observables (or rather, function defined on this microstate space), very similar outcomes. If this is the case one may draw states from the above region at random, consider the above observable, and for a majority of states, the corresponding outcomes will be almost the same and furthermore in accord with the idea of the system occupying all states within the respective region "at the same time." Thus, with respect to the above observable most microstates are almost indistinguishable from the ensemble.

Let us try to put this concept in a more rigorous form thereby defining the term "typicality" as it will be used throughout the remainder of this book. Consider a system that may occupy various microstates. Call those microstates $\mathbf{X}$. These could be, e.g., points in phase space for a classical system, points (for the case of pure states) in Hilbert space for a quantum system, the set of occupied sites in a "lattice hopping model". Now consider some set of microstates $\mathbf{X}$ and call the region formed by them the "accessible region" (AR). In principle it could be an arbitrary set, but most conveniently it will be defined by a common feature of all microstates belonging to it. If this common feature is, e.g., a value of an observable that is conserved under given dynamics, e.g., energy, a microstate cannot leave the AR, irrespective of the details of the dynamics. (This consideration also motivates the term "accessible region".) Consider, furthermore, a scalar function $f(\mathbf{X})$ of those microstates. That could be, e.g., any observable in the classical case, some expectation value or some local entropy, purity or anything alike in the quantum case. Now we define the restricted average of this function $f$ with respect to the AR as follows:

$$E_{AR}[f] \equiv \int_{AR} f(\mathbf{X}) dV_{\mathbf{X}}. \tag{6.1}$$

(In the case of a discrete microstate the integral should be replaced by a corresponding sum.) This definition only makes sense with respect to a certain "metric" or definition of the volume element in the space of the microstates. In the classical case, e.g., the outcome of (6.1) depends on whether one chooses the volume element as, e.g., $dV_{\mathbf{X}} \equiv dx dp$ (as one usually does for good reasons) or as $dV_{\mathbf{X}} \equiv dx^2 dp^2$ or anything alike. In principle the choice of this metric is arbitrary. However, if one aims at inferring statements on dynamics from the quantities considered here, a certain choice of the volume element is to be preferred. This will be considered in more detail below. We, furthermore, define the variance of $f$ with respect to the AR as follows:

$$V_{AR}[f] \equiv E_{AR}[(f - E_{AR}[f])^2] = E_{AR}[f^2] - E_{AR}^2[f]. \tag{6.2}$$

Equipped with those definitions we may now define typicality. The observable $f(\mathbf{X})$ will be almost the same for all $\mathbf{X}$ from the AR if the variance is small. This means that $E_{AR}[f]$ is the typical $f$ with respect to the AR if and only if

$$V_{AR}^{\frac{1}{2}}[f] \ll f_{\max} - f_{\min}, \tag{6.3}$$

where $f_{max}$, $f_{min}$ are the extremal values of $f$ within the AR. Such a property is in general called "typicality."

The generalization of typicality from scalars to vectors or tensors $\hat{M}(\mathbf{X})$ is rather straightforward. Define the average and (possibly) the typical tensor simply by the tensor formed by the averages of all its components. Denote that average tensor as $\hat{M}_0 \equiv \hat{E}_{AR}[\hat{M}]$. Define, furthermore, a distance measure $D$ in the space of those tensors, i.e., let $D(\hat{P}, \hat{Q})$ be the distance between $\hat{P}$ and $\hat{Q}$. Then $\hat{M}_0$ is typical with respect to the AR if

$$E_{AR}[D(\hat{M}(\mathbf{X}), \hat{M}_0] \ll D_{max}(\hat{M}_0), \tag{6.4}$$

where $D_{max}(\hat{M}_0)$ is the maximum distance of any tensor from $\hat{M}_0$ within the AR. To detect typicality it is in this case sufficient to consider averages rather than variances, since distances are by construction non-negative. If some average distance is small, almost all distances must be small. One could also define typicality for scalars on the basis of distances, but that would obviously also lead to mean deviations as considered in (6.2).

So far these considerations on typicality have been restricted to statements on relative frequencies of certain "observables" in certain regions in microstate space. However, the second law is eventually not a statement on relative frequencies of features of microstates which are drawn at random, but on dynamics. So how can the previous considerations be linked with dynamics?

To those ends assume that the pertinent dynamics preserves the microstate volume, for example, in the continuous case through an equation of motion of the form

$$\dot{\mathbf{X}} = \mathbf{G}(\mathbf{X}) \quad \text{with} \quad \text{div}_{\mathbf{X}}\mathbf{G} = 0, \tag{6.5}$$

whether or not (6.5) applies, depends, of course, on the above-mentioned representation of the microstate space. In the case of classical mechanics, e.g., it applies for the standard position–momentum representation of phase space and canonical transforms thereof. (A representation of Hilbert space to which such a description applies will be introduced in Sect. 7.1.) We furthermore partition the AR into two sets of states $\alpha$ and $\beta$ according to some (scalar) observable $f(\mathbf{X})$: Let $\mathbf{X}$ belong to $\alpha$ if

$$|f(\mathbf{X}) - E_{AR}[f]| \leq \epsilon \tag{6.6}$$

and to $\beta$ otherwise. Define the "size" $\Omega(\gamma)$ of a set $\gamma$ by the volume it occupies in microstate space, i.e.,

$$\Omega(\gamma) \equiv \int_\gamma dV_{\mathbf{X}}. \tag{6.7}$$

Now choose an $\epsilon$ such that $\epsilon \ll f_{max} - f_{min}$. In case there is typicality of $f$ with respect to the AR one will nevertheless have $\Omega(\alpha) \gg \Omega(\beta)$, i.e., there are many more states featuring an $f$ very close to the average than states that do not. Let us, in this case, call $\alpha$ the "equilibrium set" and $\beta$ the "non-equilibrium set." Now denote by $\Omega_t(\beta \rightarrow \alpha)$ the size of the set of states that initially ($t = 0$) belonged to $\beta$, but evolved under (6.5) into $\alpha$ at time $t$, i.e., the amount of states that evolved from the non-equilibrium set to the equilibrium set. Let, respectively, $\Omega_t(\alpha \rightarrow \beta)$ stand for the amount of states that evolved from the equilibrium set to the non-equilibrium set. If the dynamics preserves the microstate volume those two amounts are equal $\Omega_t(\beta \rightarrow \alpha) = \Omega_t(\alpha \rightarrow \beta)$. However, due to the equilibrium set being much larger than the non-equilibrium set we may state

$$\frac{\Omega_t(\alpha \rightarrow \beta)}{\Omega(\alpha)} \ll \frac{\Omega_t(\beta \rightarrow \alpha)}{\Omega(\beta)} . \tag{6.8}$$

Or to state in words, if there is typicality the relative frequency of states that evolve from the non-equilibrium set to the equilibrium set is much larger than the relative frequency of states that evolve from the equilibrium set to the non-equilibrium set. In a sense this may be viewed to explain the origin of the second law. Note that up to here nothing has been said about entropy, as a matter of fact we did not even define it so far. Thus, the second law is to be understood here simply as a statement claiming that some observables show a strong dynamical tendency toward constant values that are then called equilibrium values. To explain this central result in a non-formal way we non-literally quote here a little parable that H. J. Schmidt[1] told in this context:

> Consider America and Osnabrück (a little German town, ca. 160 000 inhabitants). Assume both have reached their maximum number of inhabitants, i.e., if people from Osnabrück want to move to America the same amount of Americans has to move to Osnabrück. (It is understood there are no other places to go to except for America and Osnabrück). Now if there is some migration going on at all, one is much more likely to meet somebody in Osnabrück who is about to leave for America, than to meet somebody in America heading for Osnabrück.

Here, obviously, people are microstates. Living in America corresponds to being in equilibrium while living in Osnabrück amounts to being non-equilibrated. This little story also illustrates the very general and overall character of the statement. There may very well be people who remain in Osnabrück forever, as well as people moving back and forth between America and Osnabrück all the time. While the general statement on relative frequencies holds, the individual migration history depends on the person considered. Or, to come back to physics, the above statement (6.8) is a statement on the statistics of all possible initial states featuring some non-equilibrium property, compared to the statistics of all possible initial states featuring the respective equilibrium property. Even if there is typicality, special initial states may not equilibrate or even evolve from equilibrium to non-equilibrium; however, in a mathematical sense, those are few. To verbally illustrate once more, we want

---

[1] At this point the authors like to thank H. J. Schmidt for his interest in this work, the fruitful discussions, valuable comments, and extensions. His contribution is faithfully acknowledged.

to quote Penrose here with a comment made in the context of the closely related macroscopic cell approach, Sect. 4.4 [9]:

> We would seem now to have an explanation for the second law! For we may suppose that our phase space point does not move about in any particularly contrived way, and if it starts off in a tiny phase space volume, [...], then, as time progresses, it will indeed be overwhelmingly likely to move into successively larger and larger phase space volumes, [...].

In the context of strongly interacting many-particle quantum systems (Hubbard model, etc.), there have recently been considerable (numerical) efforts to investigate whether or not some observables or correlation functions dynamically equilibrate [10, 11]. It should again be noted here that considerations based on typicality cannot replace such investigations of concrete dynamics in concrete models. Nevertheless the typicality approach can provide a picture of the scenario which is generally to be expected.

### 6.1.1 On the Analysis of Quantum Typicality

This paragraph is meant as a short "roadmap" for our concrete approach to typicality in quantum systems. Thus, roughly speaking, we essentially analyze in the following (Sect. 7–Sect. 10) if and in which sense the typicality paradigm as mathematically defined by (6.3) and (6.4) holds for quantum systems. (Comparable considerations may also be found in [6–8, 12–14].) In this analysis the microstate space will be concretely formed by all pure states, i.e., will essentially be the Hilbert space, represented in a pertinent way, cf. Sect. 7. It will be demonstrated that typicality applies to a wide range of observables, or precisely, quantum mechanical expectation values, if the systems feature high-dimensional Hilbert spaces and the observables-bound spectra. Furthermore we find that the typical values are in accord with the Boltzmann principle of "equal a priori probabilities."

However, it turns out that typicality of the full state, i.e., not only of some observables, cannot be found in unpartitioned system scenarios. There is also no increase of the von Neumann entropy in those scenarios, cf. Sect. 8. If one considers, in contrast, bipartite total quantum systems, i.e., total systems consisting of a (considered) system and an environment, one finds that there may even be typicality of states. "State" in this case refers to the reduced, local density matrix of the system. Again, it turns out that the typical state (toward which local states then tend to evolve) is in accord with the Boltzmann principle, which implies that the typical state is a maximum local entropy state. Thus, in this case evolutions from low toward high entropies may be expected. This holds for microcanonical as well as for canonical conditions, cf. Sect. 9 and Sect. 10. In Sect. 11 we comment in some detail on the concept of system–environment factorizability and the increase of local entropy.

## 6.2  Compound Systems, Entropy, and Entanglement

Before we start with the investigations outlined in the last paragraph of Sect. 6.1 we would like to comment on some related, more conceptual issues in advance.

In all theoretical approaches to thermodynamics entropy plays a central role. As already mentioned the entropy we are going to consider is the von Neumann entropy (Sect. 2.2.4; [15]), which is defined for a system on the basis of its density operator.

As already explained in Sect. 2.4 this entropy is invariant with respect to unitary transformations. Since the von Neumann equation (2.49) gives rise to a unitary evolution of the density operator $\hat{\rho}$, this entropy can never change for the total system and is thus not a candidate for the thermodynamic entropy with all its properties (cf. Chap. 4 and Chap. 5). Nevertheless the present approach also provides an explanation for the tendency toward equilibrium of many observables in pertinent closed systems. However, from Chap. 9 on we will be dealing with bipartite systems, i.e., systems the Hilbert spaces of which may be represented as direct products of the Hilbert spaces of their two "local" systems (cf. Sect. 2.2.5). In this case a valid and complete description of, say, a subsystem is given by the reduced density operator $\hat{\rho}^{(1)}$, rather than by the density operator of the full system $\hat{\rho}$. Contrary to the entropy of the total system, the entropy of subsystem 1 defined according to (2.42) on the basis of $\hat{\rho}_1$ can very well change under unitary transformations of the compound system. It would not change if the unitary transformation generated by the von Neumann equation factorized as $\hat{U}(t) = \hat{U}^{(1)}(t) \otimes \hat{U}^{(2)}(t)$ which, in turn, would be the case if the Hamiltonian of the full system did not contain any interactions between the subsystems. However, if the Hamiltonian contains such interactions, regardless of how small they might be, the subsystem entropy is no longer a conserved quantity and may change in time [16]. And, according to the outline in the last paragraph of Sect. 6.1, it can be expected to increase substantially for sufficiently long times.

As already mentioned, in the following it will be assumed that the state of the total system is always a pure state. (We want to remark here that this course of action is taken more for mathematical simplicity than for conceptual reasons.) Since for such a pure state the local entropy is an entanglement measure, an evolution toward increasing local entropy is also an evolution toward increasing entanglement. Thus in contrary to approaches in which the local relaxation dynamics is inferred on the basis of the idea of system and environment remaining factorizable (cf., Sect. 4.8, [17]), here relaxation and growing correlations are impartial. A more detailed argument in favor of this general view, in spite of it being possibly somewhat counterintuitive, will be given in Chap. 11. Thus a typicality of high entropy reduced local states, as announced in the outline in the last paragraph of Sect. 6.1, is equivalent to stating that the Hilbert space of pertinent compound systems is essentially filled with entangled states. Results in that direction are also presented in [12, 13, 18, 19]. This may also appear to contradict the idea of entanglement as being something exotic, rare, or hard to produce.

It has often been argued that the influence of the environment should not play any crucial role, since entropy rises also (or especially) in the case of an isolated system. However, in the context of thermodynamics this isolation only means that the system is not allowed to exchange any extensive quantities like energy or particles, with the environment. It does not mean that there is no interaction at all between the system and its environment. In particular, this is not to be confounded with a microscopically closed system. For example, gas particles within a thermally insulating container, nevertheless, interact with the particles that make up the container walls, otherwise the gas would not even stay inside the container. Quantum mechanically such an interaction, even if it does not allow for an exchange of energy with the environment, will, nevertheless, typically give rise to entanglement [16].

## 6.3  Fundamental and Subjective Lack of Knowledge

It is often stated that entropy should somehow be a measure for the lack of knowledge. However, then the question arises whether the observer, by overcoming his deficiency to calculate or observe more precisely, i.e., by reducing his subjective lack of knowledge, could indeed influence the entropy and the resulting thermodynamic behavior of real physical systems (cf. also remark in Sect. 3.1.2).

Within classical mechanics lack of knowledge may always be considered subjective: in principle, any observable could be known with "unlimited" precision. This is fundamentally different in quantum mechanics. From the uncertainty principle we know that there are always observables that are undetermined. Nevertheless, in single-system scenarios (no compound systems), at least one observable can, in principle, be known exactly at any time, if the initial state is a pure state; hence, the fundamental lack of knowledge does not grow. However, for compound systems there are pure states for which all observables referring to a specific subsystem are unknown, even if some compound observable of the full system is exactly predictable, just like the position of a particle is necessarily unknown to anybody, if its momentum is exactly predictable. Thus, in the latter case, the fundamental lack of local knowledge is considerably larger than in the former case. Those states are the entangled states mentioned in Sect. 6.2 [20, 21]. Compound systems might evolve from states that contain exact knowledge about some observable of each subsystem (pure product states) into the above-mentioned states, featuring this fundamental lack of knowledge about any local observable [16].

So, in the quantum domain we have two possible sources of ignorance: one being due to our inability to identify the initial state and calculate the evolution exactly, the other being intrinsic to the momentary state and thus present even for a demon with unlimited abilites. Or, to rephrase, even if we knew the microstate with precision, there will most likely be uncertainties (in the context of thermodynamics often called "fluctuations") of some (single system) or all (compound system) observables. These are fundamental quantum uncertainties.

Here we want to show that in typical thermodynamic situations the fundamental lack of knowledge by far dominates over the subjective lack of knowledge in the following sense. Almost all the possible evolutions (of which we are typically unable to predict the actual one) will eventually lead to states that are characterized by a maximum fundamental lack of knowledge about the considered subsystem; this lack is only limited by macroscopic constraints.

# References

1. A. Peres, Found. Phys. **33**(10), 1543 (2003)
2. G.P. Beretta, A.F. Ghoniem, G.N. Hatsopoulos (eds.), *Meeting the Entropy Challenge* (American Institute of Physics, 2008)
3. J.L. Lebowitz, Phys. Today **46**(9), 32 (1993)
4. J.L. Lebowitz, cond-mat/9605183 (1996)
5. J.L. Lebowitz, cond-mat/0709.0724 (2007)
6. S. Goldstein, J.L. Lebowitz, R. Tumulka, N. Zanghi, Phys. Rev. Lett. **96**, 050403 (2006)
7. P. Reimann, Phys. Rev. Lett. **99**(16), 160404 (2007)
8. S. Popescu, A.J. Short, A. Winter, Nat. Phys. **2**, 754 (2006)
9. R. Penrose, *The Emperor's New Mind* (Oxford University Press, New York, 1990)
10. T. Barthel, U. Schollwöck, Phys. Rev. Lett. **100**, 100601 (2008)
11. M. Rigol, A. Muramatsu, M. Olshanii, Phys. Rev. A **74**, 053616 (2006)
12. S. Lloyd, H. Pagels, Ann Phys. (N.Y.) **188**, 186 (1988)
13. E. Lubkin, J. Math. Phys. **19**, 1028 (1978)
14. P. Bocchieri, A. Loinger, Phys. Rev. **114**, 948 (1959)
15. J.V. Neumann, *Mathematischen Grundlagen der Quantenmechanik*. Die Grundlehren der Mathematischen Wissenschaften (Springer, Berlin, Heidelberg, New York, 1968)
16. J. Gemmer, G. Mahler, Eur. Phys. J. D **17**, 385 (2001)
17. M.O. Scully, M.S. Zubairy, *Quantum Optics* (Cambridge University Press, Cambridge, 1997)
18. J. Gemmer, G. Mahler, Eur. Phys. J. B **31**, 249 (2003)
19. J. Gemmer, A. Otte, G. Mahler, Phys. Rev. Lett. **86**, 1927 (2001)
20. L.B. Ballentine, *Quantum Mechanics*, 2nd edn. (World Scientific, Singapore, 1998)
21. K. Blum, *Density Matrix Theory and Applications*, 2nd edn. (Plenum Press, New York, London, 1996)

# Chapter 7
# Dynamics and Averages in Hilbert Space

*The use of statistical methods, essentially ignoring the*
*dynamics, is a very powerful tool; but like a very powerful*
*drug, it must be used with proper safeguards and under adult*
*supervision.*

— H. Grad [1]

**Abstract** In the previous chapter the concept of typicality has been introduced for a rather abstract space of the states of a system. Also the properties of the dynamics which are imperative for the applicability of the concept have been denoted quite formally. Here we introduce the Hilbert space of pure quantum states as our concrete state space and formulate a representation of Hilbert space. Within this representation the dynamics as described by the Schrödinger equation indeed meets the above requirements. Furthermore the averages and variances that occur in the context of typicality are mathematically concretized for this quantum case and thus accordingly called Hilbert space average and Hilbert space variance, respectively.

## 7.1 Representation of Hilbert Space

Contrary to the real configuration space, Hilbert space (see Sect. 2.2), the space on which quantum mechanical state vectors are defined, is neither three dimensional nor real, which makes it almost inaccessible to intuition. Thus there is no obvious way for having quantum mechanical states parametrized, i.e., specified by a set of numbers, quantum mechanical states. Obviously one has to choose a basis $\{|i\rangle\}$ such that

$$|\psi\rangle = \sum_i \psi_i |i\rangle \, . \tag{7.1}$$

It is, however, unclear which basis one should prefer and how one should represent the set of complex numbers $\{\psi_i\}$. This could be done in terms of real and imaginary parts, absolute values and phases, or in many other ways. Eventually one will always have a set of real numbers that somehow specify the state. To decide now, how big a region in Hilbert space really is, and this is a very important question within our approach (see Sect. 6.1), the only way is to calculate the size of the region that the corresponding specifying parameters occupy. Therefore one eventually has to orga-

Gemmer, J. et al.: *Dynamics and Averages in Hilbert Space.* Lect. Notes Phys. **784**, 77–83 (2009)
DOI 10.1007/978-3-540-70510-9_7     © Springer-Verlag Berlin Heidelberg 2009

nize the parameter space as a real Cartesian space of some dimension. The problem now is that the size of this region will depend on the parametrization chosen. Thus, if one wants to compare such regions in terms of size, one has to explain why one does this on the basis of the special chosen parametrization.

This question does not only arise in the context of quantum mechanics or Hilbert spaces, but it needs to be answered for classical phase space considerations also. It is not mandatory to parametrize the microstates of classical systems in terms of their positions and momenta. This parametrization (or canonical transforms thereof) is simply routinely chosen to guarantee the validity of Liouville's law. Since the invariance of the microstate volume with respect to dynamics is imperative for the typicality approach (see Sect. 6.1) we primarily search for a parametrization which provides this invariance. To those ends we first introduce a parametrization and demonstrate the invariance property afterward.

Consider a representation of a state in terms of the real $\eta_i$ and imaginary $\xi_i$ parts of its amplitudes with respect to some orthonormal basis $\{|i\rangle\}$,

$$|\psi\rangle = \sum_i (\eta_i + i\xi_i)|i\rangle . \tag{7.2}$$

If the $\eta_i$ and $\xi_i$ are organized in a Cartesian parameter space, a representation of Hilbert space with a real, regular, Cartesian metric is defined. All vectors that represent physical states, i.e., that are normalized, lie on a hypersphere of unit radius,

$$\langle\psi|\psi\rangle = \sum_i (\eta_i - i\xi_i)(\eta_i + i\xi_i) = \sum_i (\eta_i^2 + \xi_i^2) = 1 ; \tag{7.3}$$

this property is obviously independent of the choice of the basis $\{|i\rangle\}$.


## 7.2 Dynamics in Hilbert Space

Within this representation of the state in Hilbert space the Hamiltonian itself can be represented as a Hermitian matrix with the elements

$$H_{ij} = g_{ij} + ih_{ij}, \tag{7.4}$$

with $h_{ii} = 0$, $h_{ij} = -h_{ji}$, and $g_{ij} = g_{ji}$ to be a Hermitian matrix. The Schrödinger equation of the vector components is then given by

$$\dot{\eta}_i + i\dot{\xi}_i = \sum_k (g_{ik} + ih_{ik})(\eta_k + i\xi_k) . \tag{7.5}$$

Considering both real and imaginary parts of these complex equations separately, the following system of differential equations results:

$$\dot{\eta}_i = h_{ik}\eta_k + g_{ik}\xi_k,\tag{7.6}$$

$$\dot{\xi}_i = -g_{ik}\eta_k + h_{ik}\xi_k,\tag{7.7}$$

where we use the Einstein convention to sum over double indices. This system of differential equations may be written in vector form, according to the vector of numbers $\mathbf{q} = \{\eta_i, \xi_i\}$ and the Hamiltonian matrix

$$\dot{\mathbf{q}} = \mathsf{H}\mathbf{q} := \mathbf{G}(\mathbf{q}),\tag{7.8}$$

where we have introduced the vector $\mathbf{G}$ representing the right-hand side of the differential equation. To show that the dynamics given by (7.8) conserves the volume we have to investigate the divergence of the right-hand side, i.e.,

$$\operatorname{div}\mathbf{G}(\mathbf{q}) = \frac{\partial \mathbf{G}_k}{\partial \mathbf{q}_k},\tag{7.9}$$

using again the Einstein convention. Plugging in (7.6) and (7.7) the divergence reads

$$\operatorname{div}\mathbf{G}(\mathbf{q}) = h_{ik}\frac{\partial \eta_k}{\partial \eta_i} + g_{ik}\frac{\partial \xi_k}{\partial \eta_i} - g_{ik}\frac{\partial \eta_k}{\partial \xi_i} + h_{ik}\frac{\partial \xi_k}{\partial \xi_i}.\tag{7.10}$$

Since real and imaginary parts are independent of each other the two terms in the middle vanish. The other derivative is just a delta function, finally finding for the divergence

$$\operatorname{div}\mathbf{G}(\mathbf{q}) = 2h_{ii} = 0,\tag{7.11}$$

which is zero according to the hermiticity of the Hamiltonian (as required for an application of the typicality approach, cf. (6.5)). That implies that the quantum dynamics indeed conserves the volume which directly follows from the vanishing divergence (see [2–4]). We remark here that in this case an even more rigid statement holds for the dynamics of sets in Hilbert space: Consider the (squared) distance between any two points (states) in this representation of Hilbert space $d^2(\psi_1, \psi_2) \equiv \sum_i (\eta_{i,2} - \eta_{i,1})^2 + (\xi_{i,2} - \xi_{i,1})^2 = |||\psi_2\rangle - |\psi_1\rangle|^2$. It is easy to show that $d$ is a constant of motion. Thus, not only the volume of a set of states is invariant but also its shape. It does not undergo deformation. (From this fact it has often been inferred that there is no "quantum chaos" [5].)

There is a further interesting general feature of the quantum dynamics: In this parametrization of Hilbert space the effective Hilbert space velocity, of some trajectory, $v$, can be defined by

$$v = \sqrt{\sum_i \left[\left(\frac{d\eta_i}{dt}\right)^2 + \left(\frac{d\xi_i}{dt}\right)^2\right]}.\tag{7.12}$$

The square of this velocity may be written as

$$v^2 = \sum_i \frac{\mathrm{d}}{\mathrm{d}t}(\eta_i + \mathrm{i}\,\xi_i)\frac{\mathrm{d}}{\mathrm{d}t}(\eta_i - \mathrm{i}\,\xi_i) = \left|\frac{\mathrm{d}}{\mathrm{d}t}|\psi\rangle\right|^2 . \tag{7.13}$$

This squared absolute value of $v$ can now be calculated from the Schrödinger equation in the following form:

$$\mathrm{i}\frac{\mathrm{d}}{\mathrm{d}t}|\psi\rangle = (\hat{H} - E_0)|\psi\rangle , \tag{7.14}$$

where $E_0$ is an arbitrary, real, constant zero-point adjustment of the energy that results just in an overall phase factor of $\exp(\mathrm{i}E_0 t)$ for any solution. It has, however, an influence on the Hilbert space velocity, for it could, e.g., make a stationary state be represented by a moving point in Hilbert space. Thus, one wants to choose $E_0$ such as to make the Hilbert space velocity as small as possible, since any motion that is due to $E_0$ just reflects a changing overall phase which has no physical significance.

Inserting (7.14) into (7.13) we find for the square of the Hilbert space velocity

$$\begin{aligned}v^2 &= \langle\psi|(\hat{H} - E_0)(\hat{H} - E_0)|\psi\rangle\\ &= \langle\psi|\hat{H}^2|\psi\rangle - 2E_0\langle\psi|\hat{H}|\psi\rangle + E_0^2\langle\psi|\psi\rangle .\end{aligned} \tag{7.15}$$

Obviously $v$ is constant, since all terms in (7.15) that could possibly depend on time are expectation values of powers of $\hat{H}$ and thus constants of motion.

Searching the minimum of $v^2$ with respect to $E_0$ now yields

$$E_0 = \langle\psi|\hat{H}|\psi\rangle = \langle\psi(0)|\hat{H}|\psi(0)\rangle , \tag{7.16}$$

which inserted into (7.15) gives the Hilbert space velocity

$$\begin{aligned}v &= \sqrt{\langle\psi|\hat{H}^2|\psi\rangle - (\langle\psi|\hat{H}|\psi\rangle)^2}\\ &= \sqrt{\langle\psi(0)|\hat{H}^2|\psi(0)\rangle - (\langle\psi(0)|\hat{H}|\psi(0)\rangle)^2} ,\end{aligned} \tag{7.17}$$

which is just the energy uncertainty, or the variance of the energy probability distribution of the corresponding state. Accordingly, stationary states, i.e., energy eigenstates, are represented by non-moving points in Hilbert space. If the states belonging to some AR all feature roughly the same energy probability distribution (which will apply to many cases considered below, cf. Sects. 8.1, 9.3.1 and 9.3.2) and thus very similar energy variances, then they will all venture through Hilbert space with approximately the same velocity.

## 7.3 Hilbert Space Average and Variance

This and the remaining sections of this chapter are meant to introduce the mathematical ideas behind the methods used later in Chap. 8 on a rather abstract level. (These methods are explained in full detail in the Appendix.) Though we strongly recommend that the reader should go through these sections first, they may nevertheless be skipped by the reader who is primarily interested in results. In this case the reader may proceed to Chap. 9.

In the following we will often be interested in the average of a certain quantity over a subregion of the complete Hilbert space called the accessible region (AR), cf. Sect. 6.1. Here, the quantity itself depends on the complete state of the system. This state of the full system is constrained to the accessible region within the high-dimensional Hilbert space, which results from the conservation of some quantities, as will be seen later. Before we can compute these mean values, we need to know how to evaluate such an average of a quantity in Hilbert space in general.

Let $f$ be a function of the complete state $|\psi\rangle$ of the system in the AR. To calculate the Hilbert space average $[\![f]\!]$ of $f$ over the AR we use the parametrization for a state $|\psi\rangle$ introduced in the last section, the real and imaginary parts $\{\eta_i, \xi_i\}$ of the complex amplitudes $\psi_i$. The Hilbert space is now represented by a $2n_{tot}$-dimensional Cartesian space, in which the Hilbert space average (HA) over the AR is defined as

$$[\![f]\!] = \frac{\int_{AR} f(\{\eta_i, \xi_i\}) \prod_{n=1}^{n_{tot}} d\eta_n d\xi_n}{\int_{AR} \prod_{n=1}^{n_{tot}} d\eta_n d\xi_n} , \qquad (7.18)$$

where the integral in the denominator is just the volume of the AR we are integrating over. Such a constraint average can always be also represented by an unconstrained average involving a suitable "weight function" $G$

$$[\![f]\!] = \int f(\{\eta_i, \xi_i\}) \, G(\{\eta_i, \xi_i\}) \prod_{n=1}^{n_{tot}} d\eta_n d\xi_n . \qquad (7.19)$$

Here, the integration is over the whole infinite Cartesian space spanned by all $\{\eta_i, \xi_i\}$. To make (7.18) and (7.19) indeed equivalent, $G$ has to be zero everywhere outside the AR and must take on equal values anywhere inside the AR. Furthermore of course $\int G \prod_{n=1}^{n_{tot}} d\eta_n d\xi_n = 1$ is required. Obviously this HA meets the following standard properties of averages:

$$[\![c\,f]\!] = c[\![f]\!] \quad \text{with} \quad c \in \mathbb{C}, \qquad (7.20)$$
$$[\![f + f']\!] = [\![f]\!] + [\![f']\!],$$
$$[\![f^*]\!] = [\![f]\!]^* .$$

The restrictions which define the AR arise from peculiarities of the dynamics, e.g., constants of motion. For reasons given below, Sect. 8.1, we will exclusively focus on restrictions which yield ARs of a special type. This type of AR may be described as follows: Partition the total set of coordinates $\{\eta_i, \xi_i\}$ into subsets labeled by $J$, i.e., $\{\eta_j^J, \xi_j^J\}$. Denote the number of coordinates within the set $J$ by $2n^J$. With these definitions our type of ARs or rather the weight functions $G$ may be defined in the following way:

$$G \equiv \prod_J g_J(\{\eta_j^J, \xi_j^J\}), \tag{7.21}$$

with the $g_J$ defined by

$$g_J(\{\eta_j^J, \xi_j^J\}) \equiv \frac{\delta\left(\sqrt{\sum_{j=1}^{n_{\text{tot}}}\left((\eta_j^J)^2+(\xi_j^J)^2\right)}-R_J\right)}{\int \delta\left(\sqrt{\sum_{j=1}^{n_{\text{tot}}}\left((\eta_j^J)^2+(\xi_j^J)^2\right)}-R_J\right)\prod_j^{N_J} d\eta_j^J d\xi_j^J}. \tag{7.22}$$

Obviously $G$ is by construction normalized. Furthermore it represents a confinement onto hyperspheres within the subsets $J$. The radius of each hypersphere within a subset is given by $R_J$. Thus, in order to fulfill the normalization of the full state, i.e., $\langle \psi | \psi \rangle = 1$ we have to require $\sum_J R_J^2 = 1$. It is also important to notice that this $G$ has product form with respect to coordinates from different subsets. As a direct consequence there are no correlations between observables or quantities which are defined on the basis of the coordinates from different subsets. Or, to state in equations: Assume the total $f$ to be a product in the sense of $f \equiv \prod_J f_J(\{\eta_j^J, \xi_j^J\})$, then the HA takes on the form

$$[\![f]\!] = \left[\!\!\left[\prod_J f_J\right]\!\!\right] = \prod_J [\![f_J]\!]. \tag{7.23}$$

To actually evaluate $[\![f_J]\!]$ it then suffices to consider only the subset $J$, i.e.,

$$[\![f_J]\!] = \int g_J(\{\eta_j^J, \xi_j^J\}) f_J(\{\eta_j^J, \xi_j^J\}) \prod_j^{N_J} d\eta_j^J d\xi_j^J. \tag{7.24}$$

This is simply the average of $f_J$ over a single hypersphere of dimension $N_J$ and radius $R_J$. How such averages can concretely be calculated is described in Appendix A. Thus, since the quantities $f$ we are going to consider are always sums of products of the above type, we can do all our HAs by exploiting (7.20), (7.23), and (7.24)

According to Sect. 6.2 the crucial quantity to establish typicality is not so much the average of some observable, but the variance. If the variance is small there is typicality. However, from a formal, mathematical point of view the variance may of course be introduced on the basis of averages. In our case we may define the Hilbert space variance $\Delta_{\text{H}}^2(f)$ simply as

$$\Delta_{\text{H}}^2(f) := [\![f^2]\!] - [\![f]\!]^2. \tag{7.25}$$

The only term that needs particular consideration is the first term on the right-hand side. But since squaring of a sum of products results also in a sum of products, this term can as well be evaluated using the techniques described above.

# References

1. H. Grad, *Delaware Seminar in the Foundation of Physics* (Springer, Berlin, Heidelberg, New York, 1967), *Studies in the Foundations, Methodology and Philosophy of Science*, vol. 1, chap. "Levels of Description in Statistical Mechanics and Thermodynamics", pp. 49–76
2. V.I. Arnold, *Mathematical Methods of Classical Mechanics*, 2nd edn. (Springer, Heidelberg, 1989)
3. A. Münster, *Statistische Thermodynamik* (Springer, Berlin, Heidelberg, 1956)
4. A. Münster, *Statistical Thermodynamics*, vol. 1, 1st edn. (Springer, Berlin, Wien, New York, 1969)
5. F. Haake, *Quantum Signatures of Chaos*, 2nd edn. (Springer, Berlin, Heidelberg, New York, 2004)

# Chapter 8
# Typicality of Observables and States

*... the positions and velocities of every particle of a classical plasma or a perfect gas cannot be observed, nor can they be in an atom nor in a molecule; the theoretical requirement now is to find the gross macroscopic consequences of states that cannot be observed in detail.*

— A. Cook [1]

**Abstract** The implementation of the typicality approach to quantum systems as introduced in Chaps. 6 and 7 is further elaborated on. Thus, a concrete class of accessible regions as formally introduced in Sect. 6 is given. To start with, expectation values of observables are investigated for typicality. We essentially find that typicality can be expected for observables which are defined on high-dimensional Hilbert spaces, but feature bound spectra within the accessible state space. The typical values of the observable are in accord with Boltzmann's principle of "equal a priori probabilities." Furthermore, typicality of a state rather than for an observable is introduced. We find that, while there may very well be such typicality of some observables there is no typicality of states for non-composite systems. This points toward the investigation of composite systems.

## 8.1 A Class of Accessible Regions

As already mentioned, we intend to show here that many relevant properties or features of a quantum state (expectation values of relevant observables, entropies, etc.) do not depend on the precise "position" of the state in Hilbert space, but only on a certain (possibly large) region in which the state can be found. But how can we know in which region a state may be found? Due to physical invariants the Hilbert space may be "coarse grained" into accessible regions (ARs) onto which a state is confined. That means, under a given unitary evolution the state can never leave an AR to enter another one, it can only venture through the AR it started in. Thus if we know the AR of some initial state we know its AR at any later time even if we are unable to calculate its exact dynamics.

In order to define such ARs, we divide the Hilbert space in the following into subspaces labeled by $\alpha$. Each subspace $\alpha$ is associated with a projection operator

Gemmer, J. et al.: *Typicality of Observables and States*. Lect. Notes Phys. **784**, 85–93 (2009)
DOI 10.1007/978-3-540-70510-9_8      © Springer-Verlag Berlin Heidelberg 2009

$$\hat{\Pi}_\alpha = \sum_i |\alpha, i\rangle\langle\alpha, i| \tag{8.1}$$

according to the orthonormal states $|\alpha, i\rangle$ belonging to the subspace $\alpha$. Furthermore we require orthogonality and completeness of the projectors in the sense of

$$\sum_\alpha \hat{\Pi}_\alpha = \hat{1}, \qquad \hat{\Pi}_\alpha \hat{\Pi}_\beta = \delta_{\alpha\beta}\hat{\Pi}_\alpha . \tag{8.2}$$

The dimension of the subspace $\alpha$ is thus given by

$$N_\alpha = \mathrm{Tr}\left\{\hat{\Pi}_\alpha\right\} . \tag{8.3}$$

If we represent an operator as a matrix in an orthonormal basis according to those subspace projection operators, each block within this matrix corresponds to one (diagonal) or two (off-diagonal) such subspaces, as shown in Fig. 8.1. If, e.g., some operator only acts on one given subspace it only has non-zero entries in the corresponding diagonal block. Applying the projection operator (8.1) to an arbitrary state $|\psi\rangle$ of the considered system we define the non-normalized vector

$$|\psi_\alpha\rangle \equiv \hat{\Pi}_\alpha|\psi\rangle . \tag{8.4}$$

According to the definition of the projection operator the probability to find the system within the subspace $\alpha$ is defined by the expectation value of the projection operator and is given by

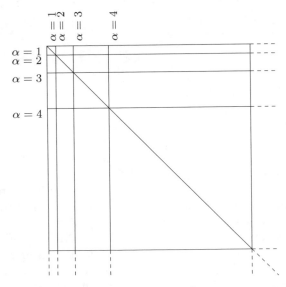

**Fig. 8.1** Arrangement of states according to subspaces $\alpha$. Operators represented by matrices, e.g., the Hamiltonian, feature a block form

$$\langle \psi | \hat{\Pi}_\alpha | \psi \rangle = \langle \psi_\alpha | \psi_\alpha \rangle = W_\alpha . \tag{8.5}$$

We can now define the AR by the set of projectors $\hat{\Pi}_\alpha$ and the set of corresponding probabilities $W_\alpha$. If and only if some given $|\psi\rangle$ is in accord with (8.5), it belongs to accessible region $\alpha$:

$$\text{AR} := \{ |\psi\rangle : \langle \psi | \hat{\Pi}_\alpha | \psi \rangle = \langle \psi_\alpha | \psi_\alpha \rangle = W_\alpha \} . \tag{8.6}$$

Note that this definition of the AR exactly corresponds to the confinement of a state onto hyperspheres as discussed in Sect. 7.3.

As explained above this choice of the AR has primarily dynamical reasons: If the $\hat{\Pi}_\alpha$ correspond to invariants of the system such that

$$[\hat{\Pi}_\alpha, \hat{H}] = 0 \tag{8.7}$$

($\hat{H}$ being the Hamiltonian of the system, which then, in the representation of Fig. 8.1 assumes block-diagonal form), then the (pure) state of the system $|\psi(t)\rangle$ can never leave the AR it started in, simply because the $W_\alpha$ are constant. What are such possible invariant subspaces $\hat{\Pi}_\alpha$? Of course subspaces spanned by some exact eigenstates of $\hat{H}$ always fulfill (8.7) and thus are invariant projective subspaces. However, if there is a small perturbation of $\hat{H}$ of strength $\epsilon$, say, one can nevertheless expect subspaces spanned by the eigenstates of the unperturbed system corresponding to an energy interval $\Delta E > \epsilon$ to be approximately invariant. That means, even the perturbed system will not substantially leave the AR defined on the basis of such subspaces. This construction of adequate $\hat{\Pi}_\alpha$s is of great practical importance, since most likely, for a system in quest, the exact eigenstates are unknown and thus now concrete computation may be based upon them. More accessible projectors $\hat{\Pi}_\alpha$ may arise from natural invariants of a system. If, e.g, the number of particles in a system is conserved, a subspace spanned by all states featuring the same amount of particles is a valid $\hat{\Pi}_\alpha$, even if its energy eigenstates are not known at all.

All further considerations will be restricted to this type of AR. In the following it will be investigated whether or not there is typicality with respect to those ARs, i.e., whether or not crucial properties of a state primarily depend on AR the state belongs to, but not so much where exactly within this AR it is located.

## 8.2 Hilbert Space Average of Observables

Let us start by considering the expectation values of some Hermitian operator $\hat{A}$, i.e., $\langle \psi | \hat{A} | \psi \rangle$, with $|\psi\rangle$s belonging to some specific AR. If there were typicality with respect to the expectation value of this $\hat{A}$ and the chosen AR, that would mean that the various possible $\langle \psi | \hat{A} | \psi \rangle$s would not differ much from each other. Of primary interest may be the question whether or not this is the case. It is, however, technically simpler to answer the following question first: if there was a typical $\langle \psi | \hat{A} | \psi \rangle$, what would it be? Or to rephrase, what is the average over all possible $\langle \psi | \hat{A} | \psi \rangle$s from the AR? Such an average is a Hilbert space average (HA) over a given AR and as already mentioned we denote it by $[\![\langle\psi|\hat{A}|\psi\rangle]\!]$. (Thus the double brackets always refer to some AR.) Its abstract formulation has already been discussed in Sect. 7.3. (For additional literature, cf. [2–6].)

To compute the HA of a general operator $\hat{A}$ we decompose it into subspaces defined by the projection operators (8.1) finding

$$\hat{A} = \sum_{\alpha\beta} \hat{A}_{\alpha\beta}, \quad \hat{A}_{\alpha\beta} \equiv \hat{\Pi}_\alpha \hat{A} \hat{\Pi}_\beta. \tag{8.8}$$

(In the representation sketched in Fig. 8.1 those $\hat{A}_{\alpha\beta}$ assume the mentioned block form). According to this decomposition the HA yields

$$[\![\langle\psi|\hat{A}|\psi\rangle]\!] = \sum_{\alpha\beta} [\![\langle\psi|\hat{\Pi}_\alpha \hat{A}\hat{\Pi}_\beta|\psi\rangle]\!]$$
$$= \sum_{\alpha\beta} [\![\langle\psi_\alpha|\hat{A}|\psi_\beta\rangle]\!]. \tag{8.9}$$

Using a concrete representation for the states in Hilbert space as introduced in Sect. 7.1, the average reduces to

$$[\![\langle\psi_\alpha|\hat{A}|\psi_\beta\rangle]\!] = \sum_{ij} [\![A_{\alpha i,\beta j} \psi_{\alpha i}^* \psi_{\beta j}]\!]$$
$$= \sum_{ij} A_{\alpha i,\beta j} [\![\psi_{\alpha i}^* \psi_{\beta j}]\!]. \tag{8.10}$$

Here we have used the notations

$$\psi_{\alpha i} \equiv \langle\alpha, i|\psi\rangle, \quad \psi_{\beta j} \equiv \langle\beta, j|\psi\rangle, \quad A_{\alpha i,\beta j} \equiv \langle\alpha, i|\hat{A}|\beta, j\rangle. \tag{8.11}$$

As discussed in Appendices A and B.1, such averages can only be non-zero for $\alpha = \beta$ and $i = j$. Thus the average yields

$$[\![\langle\psi_\alpha|\hat{A}|\psi_\alpha\rangle]\!] = \sum_i A_{\alpha i,\alpha i} [\![|\psi_{\alpha i}|^2]\!]_{AR}$$
$$= \frac{W_\alpha}{N_\alpha} \mathrm{Tr}\{\hat{A}_{\alpha\alpha}\}, \tag{8.12}$$

where we have used the special average (B.3). For more details on the derivation of this result, see Appendix B.1.

It is instructive to re-derive this result (8.12) from a slightly different perspective. To those ends we rewrite (8.9) in a somewhat different fashion:

$$[\![\langle\psi|\hat{A}|\psi\rangle]\!] = \sum_{\alpha\beta}[\![\langle\psi_\alpha|\hat{A}|\psi_\beta\rangle]\!]$$

$$= \sum_{\alpha\beta}[\![\mathrm{Tr}\left\{\hat{A}|\psi_\beta\rangle\langle\psi_\alpha|\right\}]\!]. \tag{8.13}$$

In the above expression the operators $|\psi_\beta\rangle\langle\psi_\alpha|$ are multiplied by $\hat{A}$, then the the trace is taken, then the Hilbert space average is taken, then the double sum over all subspaces is taken. From the definitions of these operations follows that, to some extent, the order in which they are performed may be changed without changing the result. From this consideration the following relations may be inferred:

$$[\![\langle\psi|\hat{A}|\psi\rangle]\!] = \mathrm{Tr}\left\{\hat{A}\hat{\Omega}\right\}, \quad \hat{\Omega} \equiv \sum_{\alpha\beta}[\![|\psi_\beta\rangle\langle\psi_\alpha|]\!] = [\![|\psi\rangle\langle\psi|]\!]. \tag{8.14}$$

We mention in passing that same technique is used to find slightly more complicated HAs in Sect. 19.4.

Now the HA of the observable $\hat{A}$ is given as the expectation value of $\hat{A}$ corresponding to some fictitious state $\hat{\Omega}$. We may compute $\hat{\Omega}$ in more detail

$$\hat{\Omega} = \sum_{\alpha\beta ij}[\![\psi_{\alpha i}^*\psi_{\beta j}]\!]|\beta, j\rangle\langle\alpha, i|. \tag{8.15}$$

With the same results from Appendix B.1 that we already used to derive (8.12) we find

$$\hat{\Omega} = \sum_{\alpha i}\frac{W_\alpha}{N_\alpha}|\alpha, i\rangle\langle\alpha, i| = \sum_{\alpha}\frac{W_\alpha}{N_\alpha}\hat{\Pi}_\alpha. \tag{8.16}$$

The central comment that should be made on this result is that it is in some sense in agreement with Boltzmann's postulate of a priori equally distributed probabilities. If one, following the reasoning of Boltzmann, distributes the probabilities $W_\alpha$ equally among the states that span the subspaces $\alpha$ one gets the very same state $\hat{\Omega}$. Or, to rephrase, $\hat{\Omega}$ represents the corresponding Boltzmann ensemble. Thus, if there is a typical outcome for $\langle\psi|\hat{A}|\psi\rangle$ it will be the same that one obtains from applying standard Boltzmann ensemble theory.

## 8.3 Hilbert Space Variance of Observables

Now, we eventually have to decide whether or not there is a typical outcome for $\langle\psi|\hat{A}|\psi\rangle$ within the considered AR. Or, to rephrase, whether or not the distribution of $\langle\psi|\hat{A}|\psi\rangle$s, which one gets from drawing states from the AR at random is broad or small. As a measure for the width of this distribution we calculate its Hilbert space variance (HV) as already explained in Sect. 7.3. The HV of the expectation value of an observable is defined as

$$\Delta_H^2(\langle\psi|\hat{A}|\psi\rangle) := [\![\langle\psi|\hat{A}|\psi\rangle^2]\!] - [\![\langle\psi|\hat{A}|\psi\rangle]\!]^2 . \tag{8.17}$$

The second term is just the square of the above-derived average. Therefore we primarily concentrate on the first term. Since the calculation is in principle similar to the above one and rather lengthy we omit it here, but the interested reader may find it in Appendix B.2. The first term of the variance (8.17), thus, reads

$$[\![\langle\hat{A}\rangle^2]\!] = \sum_{\alpha\beta} \frac{W_\alpha W_\beta}{N_\alpha(N_\beta + \delta_{\alpha\beta})} \left(\mathrm{Tr}\left\{\hat{A}_{\alpha\beta}\hat{A}_{\alpha\beta}^\dagger\right\} + \mathrm{Tr}\left\{\hat{A}_{\alpha\alpha}\right\}\mathrm{Tr}\left\{\hat{A}_{\beta\beta}\right\}\right). \tag{8.18}$$

With this term and the squared HA, cf. (8.12), the HV of an expectation value of an observable yields

$$\Delta_H^2(\langle\hat{A}\rangle) = \sum_{\alpha\beta} \frac{W_\alpha W_\beta}{N_\alpha(N_\beta + \delta_{\alpha\beta})} \left(\mathrm{Tr}\left\{\hat{A}_{\alpha\beta}\hat{A}_{\alpha\beta}^\dagger\right\} + \mathrm{Tr}\left\{\hat{A}_{\alpha\alpha}\right\}\mathrm{Tr}\left\{\hat{A}_{\beta\beta}\right\}\right)$$

$$- \sum_{\alpha\beta} \frac{W_\alpha W_\beta}{N_\alpha N_\beta}\mathrm{Tr}\left\{\hat{A}_{\alpha\alpha}\right\}\mathrm{Tr}\left\{\hat{A}_{\beta\beta}\right\}$$

$$= \sum_{\alpha\beta} \frac{W_\alpha W_\beta}{N_\alpha(N_\beta + \delta_{\alpha\beta})} \left(\mathrm{Tr}\left\{\hat{A}_{\alpha\beta}\hat{A}_{\alpha\beta}^\dagger\right\} - \delta_{\alpha\beta}\frac{\mathrm{Tr}\left\{\hat{A}_{\alpha\alpha}\right\}\mathrm{Tr}\left\{\hat{A}_{\beta\beta}\right\}}{N_\alpha}\right). \tag{8.19}$$

This is a rigid result and may, in principle, be evaluated for any given AR and any operator $\hat{A}$. Whenever it is small, the average result (8.16) can be considered as being typical. For what scenarios can such typicality be expected? Specializing without substantial loss of generality to observables with $\mathrm{Tr}\left\{\hat{A}\right\} = 0$ the variance of the spectrum of $\hat{A}$ reads

$$\Delta_S^2(\hat{A}) = \frac{1}{N}\mathrm{Tr}\left\{\hat{A}^2\right\} = \frac{1}{N}\sum_{\alpha\beta}\mathrm{Tr}\left\{\hat{A}_{\alpha\beta}\hat{A}_{\alpha\beta}^\dagger\right\} , \quad N \equiv \sum_\alpha N_\alpha . \tag{8.20}$$

An HV is by construction positive, so are both terms that appear in the difference in (8.19), omitting the second of those will only make the outcome larger. Thus an upper bound to the HV written in a suggestive way reads

$$\Delta_H^2(\langle\hat{A}\rangle) \leq \frac{1}{N^2}\sum_{\alpha\beta}\left(\frac{W_\alpha W_\beta}{n_\alpha n_\beta}\right)\mathrm{Tr}\left\{\hat{A}_{\alpha\beta}\hat{A}_{\alpha\beta}^\dagger\right\} , \tag{8.21}$$

with $n_{\alpha(\beta)} \equiv N_{\alpha(\beta)}/N$. Thus, whenever the system occupies with significant probability only subspaces that are large enough to represent a substantial fraction of the full dimension of the system, the HV is roughly by a factor $N$ smaller than the variance of the spectrum of $\hat{A}$. Thus, a result for $\langle\hat{A}\rangle$ as calculated from $\hat{\Omega}$

(cf. (8.15)) classifies as typical for a wide range of accessible regions if $\hat{A}$ has a bounded spectrum and is defined on a large-dimensional Hilbert space. This finding is in accord with the results by Reimann [7].

In which cases can this be expected? Consider as an instructive example the case of no restriction, i.e., the AR being all Hilbert space. One may be interested in a "local" variable $A$ which should really be written as $A \otimes I$, where $I$ denotes the unit operator acting on the (for this inquiry "irrelevant") rest of the system. If this rest of the system is enlarged $N$ increases drastically while $\Delta_S^2(\hat{A})$ remains constant. Thus the corresponding HV will decrease according to (8.21). This means whenever a bounded local variable of interest is "embedded" in a large surrounding featuring a high dimensionality, a small HV can be expected. Such a scenario is naturally implemented if a considered system is coupled to a large "environmental" system. One may then even observe a set of local variables which determine the reduced (local) state of the considered system completely, finding that they all relax to equilibrium due to all their HVs being small. Such a scenario will be considered in substantially more detail in Chaps. 9 and 10. However, since the above reasoning does not require weak coupling, it also applies in principle to scenarios in which the system–environment partition in the traditional sense is absent. If one, e.g., considers a many-particle system of some solid state type, one may be interested in the number of particles that can be expected in some spatial region of the system. The variance of the corresponding number operator surely remains unchanged if the whole system is increased (at constant particle density) but the dimension on which the number operator is defined increases exponentially. Thus a strongly typical "occupation number" will result from this scenario.

The same overall picture can be considered more or less appropriate even if there are restrictions to different subspaces $\hat{\Pi}_\alpha$. Thus, in very many scenarios and for many observables one finds small HVs and in all those cases the typicality argument applies.

## 8.4 Hilbert Space Averages of Distances Between States

Thus, due to typicality, measuring only one or a few observables $\hat{A}$, one may not be able to distinguish some $|\psi\rangle$ from the AR from $\hat{\Omega}$. However, measuring more and more observables one will eventually be able to determine the full, true quantum state $|\psi\rangle$ of the system. Thus, one may ask whether a true, pure quantum state from the AR is typically close or similar to the state that represents the Boltzmann ensemble, i.e., $\hat{\Omega}$. To quantify this question we need a distance measure. We will employ the Bures metric (2.19) which has already been introduced in Sect. 2.2.3 to measure the distance between two density operators. Here we want to quantify the distance $D$ between some pure state $\hat{\rho} \equiv |\psi\rangle\langle\psi|$ from the AR and the ensemble state $\hat{\Omega}$

$$D^2(\hat{\rho}, \hat{\Omega}) \equiv \text{Tr}\left\{(\hat{\rho} - \hat{\Omega})^2\right\}. \tag{8.22}$$

To repeat, the measure $D$ is always positive, vanishes for $\hat{\rho} = \hat{\Omega}$, reaches a maximum of $\sqrt{2}$ for $\hat{\Omega}$ pure and orthonormal to $\hat{\rho}$ ("maximum distance"), etc. To decide whether or not the $\hat{\rho}$s ($|\psi\rangle$s) from the AR are typically close to $\hat{\Omega}$ we can thus simply compute the HA of $D^2$. If the result was small compared to, say, 1, there would be a "typicality of states." To do so we consider

$$
\begin{aligned}
[\![D^2(\hat{\rho}, \hat{\Omega})]\!] &= \left[\!\left[\mathrm{Tr}\left\{(\hat{\rho} - \hat{\Omega})^2\right\}\right]\!\right] \\
&= \left[\!\left[\mathrm{Tr}\left\{\hat{\rho}^2 - 2\hat{\rho}\hat{\Omega} + \hat{\Omega}^2\right\}\right]\!\right],
\end{aligned}
\tag{8.23}
$$

where we have combined the mixed terms due to the properties of the trace operation. Exchanging the average and the trace operation (which can be done due to the same reasons as given before (8.14)), we find

$$
[\![D^2(\hat{\rho}, \hat{\Omega})]\!] = \mathrm{Tr}\{[\![\hat{\rho}^2]\!] - 2[\![\hat{\rho}\hat{\Omega}]\!] + \hat{\Omega}^2\}.
\tag{8.24}
$$

Since $\hat{\Omega}$ in the second term on the right-hand side is constant it can be taken out of the average. Furthermore, using $\hat{\Omega} = [\![\hat{\rho}]\!]$, we finally find

$$
\begin{aligned}
[\![D^2(\hat{\rho}, \hat{\Omega})]\!] &= \mathrm{Tr}\left\{[\![\hat{\rho}^2]\!] - \hat{\Omega}^2\right\} = [\![\mathrm{Tr}\left\{\hat{\rho}^2\right\}]\!] - \mathrm{Tr}\left\{\hat{\Omega}^2\right\} \\
&= [\![P(\hat{\rho})]\!] - P(\hat{\Omega}),
\end{aligned}
\tag{8.25}
$$

where $P$ denotes the purity of a state as introduced in Sect. 2.2.3. Since in this example $\hat{\rho}$ is always a pure state its purity is always 1. If the purity of $\hat{\Omega}$ is low, which will be the case in most realistic, thermodynamical scenarios, the above mean distance $[\![D^2(\hat{\rho}, \hat{\Omega})]\!]$ will be close to 1 and thus not small at all.

This finding essentially implies that there is no typicality of states, whatsoever, in such a "single-system scenario." While there may be many observables for which almost all pure states from the AR produce almost the same outcome as $\hat{\Omega}$ (typicality of some observables) there will always be other observables that produce outcomes that are significantly different from the ones corresponding to $\hat{\Omega}$. This simply reflects that there is no typicality of states.

This may drastically change if one considers compound systems, i.e., systems with a principle structure as outlined in Sect. 2.2.5. If, in such a scenario, we concentrate on the state of only one subsystem, which is mathematically formulated in terms of a reduced density matrix, there may very well be typicality of this reduced state. This means that then essentially all reduced states $\hat{\rho}^{(S)}(\psi)$ (cf. (2.39)) corresponding to pure states $|\psi\rangle$ from some AR (defined on the full compound system) may be very close to the HA over all those reduced states. In this case the corresponding $[\![D^2(\hat{\rho}, \hat{\Omega})]\!]$ will indeed be very small compared to 1. As a consequence in this case there will also be typicality for all possible observables that can be defined on the considered subsystem. This scenario and its consequences will be discussed in some detail in Chap. 10.

# References

1. A. Cook, *The Observational Foundations of Physics* (Cambridge University Press, Cambridge, 1994)
2. J. Gemmer, G. Mahler, Europhys. Lett. **59**, 159 (2002)
3. J. Gemmer, *A Quantum Approach to Thermodynamics*. Dissertation, Universität Stuttgart (2003)
4. J. Gemmer, G. Mahler, Eur. Phys. J. B **31**, 249 (2003)
5. J. Gemmer, G. Mahler, Eur. Phys. J. D **17**, 385 (2001)
6. J. Gemmer, A. Otte, G. Mahler, Phys. Rev. Lett. **86**, 1927 (2001)
7. P. Reimann, Phys. Rev. Lett. **99**(16), 160404 (2007)

# Chapter 9
# System and Environment

*The results in decoherence theory strongly suggest that*
*interactions with the environment are crucial in the emergence*
*of quasi-classical and thermodynamic behavior.*
                                    — M. Hemmo and O. Shenker [1]

**Abstract** Since the considerations in Chap. 8 suggest that typicality of states or increasing von Neumann entropy may only be found in composite systems, we here formulate a system–environment scenario in some detail. Technically an adequate notation is introduced. We comment on the fact that any thermodynamic scenario comprises some sort of environment. This statement holds if the notion of "environment" is enlarged to include not only systems with which extensive quantities (heat, particles, etc.) may be exchanged but any interacting quantum system. If the system is, e.g., a gas, the environment is or includes at least the vessel that contains the gas. We explain in which sense weak system–environment coupling plays a crucial role in our scenario. We, furthermore, discuss the implications of this weak coupling on a pertinent accessible region. Eventually couplings are classified according to whether or not they allow for energy exchange between system and environment.

## 9.1 Partition of the System and Pertinent Notation

We assume that the full Hamiltonian may be partitioned as

$$\hat{H} = \hat{H}_{\mathrm{S}} + \hat{H}_{\mathrm{E}} + \hat{I}_{\mathrm{SE}}, \tag{9.1}$$

where $\hat{H}_{\mathrm{S}}$ and $\hat{H}_{\mathrm{E}}$ are the local Hamiltonians of the system S and the environment E, respectively, which act on two different parts of a product Hilbert space in the sense discussed in Sect. 2.2.5 (see also Fig. 9.1). The Hamiltonian part $\hat{I}_{\mathrm{SE}}$ describes the interaction between system and environment. Note that in the framework of this approach the environment must not necessarily be a standard reservoir, say a set of harmonic oscillators, etc., as required in the context of traditional open quantum systems. No decay of bath correlation functions or anything alike is required for the applicability of the typicality approach. Thus the environment may be any quantum system, possibly as small as a molecule or so.

System and environment could hypothetically be mutually non-influencing physical objects, this is mathematically reflected by the fact that the two local

Gemmer, J. et al.: *System and Environment.* Lect. Notes Phys. **784**, 95–105 (2009)
DOI 10.1007/978-3-540-70510-9_9                    © Springer-Verlag Berlin Heidelberg 2009

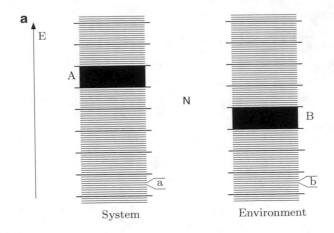

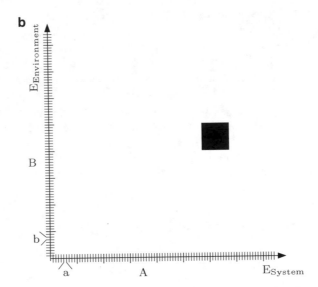

**Fig. 9.1** (**a**) Sketch to illustrate the notation. The eigenstates of system and environment are coarse grained with respect to energy. The "grains" are labeled by capitals, $A$, $B$, respectively. The energy eigenstates within the grains are labeled by $a$, $b$. Thus a pair $A$, $B$ (marked in black) indicates a set of states representing a given (gross) distribution of energy onto system and environment (**b**) A two-dimensional representation of the basis states of the full system. The axes correspond to system energy and environmental energy, respectively. The *black spot* indicates the same set of states that is also marked in Fig. 9.1

Hamiltonians always commute,

$$[\hat{H}_S, \hat{H}_E] = 0. \tag{9.2}$$

In the following we are mostly going to take the interaction $\hat{I}_{SE}$ between system and environment as some sort of weak coupling. Such a weak coupling description might require a reorganization of the partition, as will be outlined in Sect. 9.2. Depending on its concrete structure $\hat{I}_{SE}$ may or may not allow for energy transfer between system and environment. In case it does not allow for energy transfer it nevertheless may give rise to some sort of dephasing. The interaction $\hat{I}_{SE}$ thus specifies the macroscopic (or even microscopic) constraints, such as microcanonical, canonical, as will be seen later.

In the following we essentially write the Hamiltonian of the whole system in the product energy eigenbasis of the decoupled system and environment. Let us introduce our nomenclature for such a bipartite system in some more detail. In order to adjust our notation to the type of accessible region (AR) considered in the previous section (cf. (8.6), Fig. 8.1) the energy spectra of system and environment are "coarse grained" as depicted in Fig. 9.1. These compartments could in principle have different widths (with respect to energy) but in the following we are going to choose them to be all of some width $\Delta E$. The mean energy of the compartments in the system and the environment is denoted by $E_A^S$ and $E_B^E$, respectively. Thus the indices $A$ and $B$ specify compartments in the spectrum of the system and the environment, respectively. To label the energy eigenstates of the decoupled subsystems (system, environment) within the compartments, we introduce a subindex counting the states in each subspace $A$ or $B$,

$$a = 1, \ldots, N_A \quad \text{and} \quad b = 1, \ldots, N_B . \tag{9.3}$$

Thus $|A, a\rangle$ denotes the $a$th energy eigenstate within the compartment $A$ of the system, whereas $|B, b\rangle$ denotes the $b$th state of subspace $B$ of the environment (cf. Fig. 9.1). For later reference we already introduce here the notation for the projectors corresponding to these compartments

$$\hat{\Pi}_A \equiv \sum_a |A, a\rangle\langle A, a|, \quad \hat{\Pi}_B \equiv \sum_b |B, b\rangle\langle B, b| . \tag{9.4}$$

A pure state of the full system will be denoted as a superposition of product eigenstates (of the decoupled system) $|A, a\rangle \otimes |B, b\rangle$,

$$|\psi\rangle = \sum_{A,B} \sum_{a,b} \psi_{ab}^{AB} |A, a\rangle \otimes |B, b\rangle . \tag{9.5}$$

In general it is not possible to write the state of one subsystem as a pure state. This is due to the entanglement between system and environment (cf. Sect. 2.2.5), which will most likely emerge during the time evolution, even if we start with a separable state in the beginning, cf. Chap. 11. The state of a single subsystem is thus a mixed state and must be described by a density matrix. From the density operator of the whole system $\hat{\rho} = |\psi\rangle\langle\psi|$, the reduced density operator of the subsystem, S, is found by tracing over the environment (see Sect. 2.2.5)

$$\hat{\rho}^{(S)} := \sum_{A,A',B} \sum_{a,a',b} \psi_{ab}^{AB} \left(\psi_{a'b}^{A'B}\right)^* |A,a\rangle\langle A',a'| . \tag{9.6}$$

Analogously, it is possible to evaluate the density operator of the environment, but mostly we are not interested in this state.

The diagonal elements of the density operator are the probabilities of finding the system in the respective eigenstate. Since they are frequently needed, we introduce these quantities now. The joint probability of finding the system at energy $E_A^S$ and the environment at the energy $E_B^E$ is given by

$$W_{AB} = W(E_A^S, E_B^E) := \sum_{a,b} |\psi_{ab}^{AB}|^2 . \tag{9.7}$$

Of course the individual probabilities $W_A$, $W_B$ to find the subsystems at $E_A^S$, $E_B^E$, respectively, are

$$W_A = \sum_B W_{AB}, \qquad W_B = \sum_A W_{AB} . \tag{9.8}$$

Another important quantity is the probability of finding the complete system (approximately, up to $\Delta E$) at the energy $E$. This probability is a summation of all possible joint probabilities $W_{AB}$ under the subsidiary condition of overall energy conservation,

$$W(E) = \sum_{A,B} W_{AB} M_{AB,E}, \quad \text{with} \quad M_{AB,E} \equiv \delta(E_A^S + E_B^E - E), \tag{9.9}$$

where $\delta$ assumes the value 1 if its argument vanishes and 0 otherwise.

## 9.2  Comment on Bipartite Systems and Physical Scenarios

This section is meant as a little example how the previously mentioned partitioning scheme (system–environment) can be applied to the standard thermodynamical system of, e.g., an (ideal) gas g in a container c. It is intended to clarify the physical background, but it is not imperative for the understanding of the following sections.

Let us analyze the above-mentioned example in some detail under this point of view. Writing down the Hamiltonian for such a gas–container system, the sum over the kinetic energies of all gas particles $\mu$ (mass $m$, momentum operator $\hat{\mathbf{p}}_\mu^g$, assume for the moment for simplicity that the gas particles are somehow distinguishable) is the only part of this Hamiltonian that acts on the gas subspace alone,

$$\hat{H}_S = \sum_\mu \frac{\hbar^2}{2m_g} (\hat{\mathbf{p}}_\mu^g)^2 . \tag{9.10}$$

$\hat{H}_E$, the Hamiltonian of the container provides the environment that has to be present to make the gas particles a thermodynamic system. It reads

$$\hat{H}_E = \sum_\mu \frac{\hbar^2}{2m_c} (\hat{\mathbf{p}}_\mu^c)^2 + \frac{1}{2} \sum_{\mu,\nu} \hat{V}^c(\mathbf{q}_\mu^c, \mathbf{q}_\nu^c), \qquad (9.11)$$

where $\hat{V}^c(\mathbf{q}_\mu^c, \mathbf{q}_\nu^c)$ are the interactions that bind the container particles (mass $m$) at positions $\mathbf{q}_\mu^c$ and $\mathbf{q}_\nu^c$ to each other to form a solid, and acts exclusively in the container subspace. Thus, as required, $\hat{H}_S$ and $\hat{H}_E$ commute.

Now, $\hat{I}$ contains the interactions of all gas particles with all container particles and reads

$$\hat{I} = \sum_{\mu,\nu} \hat{V}^{gc}(\mathbf{q}_\mu^g, \mathbf{q}_\nu^c). \qquad (9.12)$$

This part contains the repelling interactions between the gas particles and the container particles and establishes the container as a boundary for the gas particles from which they cannot escape. Starting from first principles, the Hamiltonian has to be written in this way, especially the last part is indispensable (see Fig. 9.2).

Unfortunately, any stationary state of $\hat{H}_S$, and it is such a state we want to see the system evolve into, is unbounded and thus not confined to any volume that might be given by the container. This is due to the fact that such a state is a momentum eigenstate and therefore not localized in position space. This means the expectation value of $\hat{I}$, for an energy eigenstate of the uncoupled problem, $\hat{H}_S + \hat{H}_E$ would definitely not be small, and thus the system would miss a fundamental prerequisite, the weak coupling, for a thermodynamic system accessible from our method.

The standard way to overcome the above-mentioned deficiency is to define an effective potential for the gas particles generated by the container, in which all gas particles are trapped. Fortunately, an effective local Hamiltonian and an effective interaction can be defined so that the weak coupling limit is fulfilled by

$$\hat{H}^{g'} := \hat{H}_S + \hat{V}^g, \quad \hat{I}' := \hat{I} - \hat{V}^g, \quad \hat{V}^g = \hat{V}^g(\{\mathbf{q}_\mu^g\}). \qquad (9.13)$$

Here, $\hat{V}^g(\{\mathbf{q}_\mu^g\})$ is some potential for gas particles alone and depends on all position vectors $\{\mathbf{q}_\mu^g\}$. $\hat{V}^g$ will be chosen to minimize the coupling energy. Substituting the real parts by the effective parts of the Hamiltonian obviously leaves the full Hamiltonian unchanged, but now there is a chance that the partition will fit into the above scheme (see Sect. 9.1). A good candidate for $\hat{V}^g$ will be some sort of effective "box" potential for each gas particle, comprising the mean effect of all container particles. This makes $\hat{I}'$, the deviation of the true "particle-by-particle" wall interaction from the "effective box" wall interaction, likely to be small. The eigenstates of the gas system alone are then simply the well-known bound eigenstates of particles in a box of corresponding size (see Fig. 9.3).

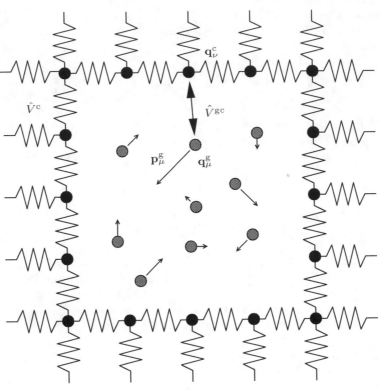

**Fig. 9.2** Bipartite system: gas in a container, represented by an interacting net of particles (*black dots*)

The effective potential $\hat{V}^g(\{\mathbf{q}^g_\mu\})$ is a sort of a weighted summation over all inter-actions of the gas particle with all particles of the container (see Fig. 7.3). According to the pre-condition of these considerations, this potential will indeed be a weak interaction. Thus, the Hamiltonian is reorganized so that a partition into system and environment with a weak interaction is possible.

In general, however, the effective interaction $\hat{I}'$ cannot be made zero and repre-sents a coupling, i.e., a term that cannot be written as a sum of terms that act on the different subspaces separately. Such a coupling, however small it might be, can, and in general will, produce entanglement, thus causing local entropy to increase. (For a specific example of this partition scheme, see [2].)

## 9.3  Weak Coupling and Corresponding Accessible Regions

Generically, weak coupling between system S and environment E is assumed in standard thermodynamic considerations. Already the concept of energy being either in the system or the environment, i.e., the idea that the total energy of the full system

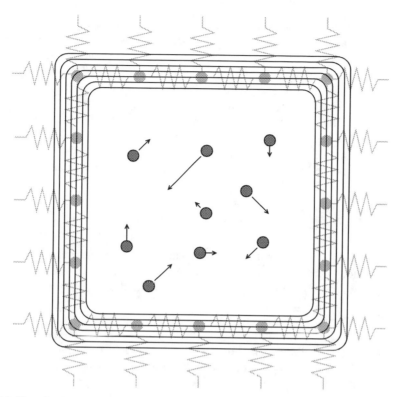

**Fig. 9.3** Bipartite system: gas in a container, represented by an effective single particle potential (indicated by equipotential lines)

may be computed as a sum of the energy contained in the system and the energy contained in the environment, makes the assumption of a weak coupling indispensable. Furthermore, it can be shown that the concept of intensive and extensive variables relies on a weak coupling limit (cf. [3] and also [4]).

In our case, since we eventually want to define our AR in terms of products of energy eigenstates, the coupling strength (perturbation strength) has a crucial influence on the AR as has already been outlined in Sect. 8.1. For an infinitesimally weak interaction a state with given (sharp) energies, $U_S$ for the isolated system and $U_E$ for the isolated environment, can only evolve into states featuring the very same pair of energies $U_S$, $U_E$ if energy exchange is forbidden. It can only evolve into states featuring the very same sum of energies $U_S + U_E$ if energy exchange is allowed. An interaction with a finite strength $\delta\epsilon$, however, may "absorb" or "release" a corresponding amount of energy. Thus during an evolution the $U_S$, $U_E$ or $U_S + U_E$, respectively, are no longer exactly conserved, but only confined to some energy interval $\Delta E > \delta\epsilon$. This essentially sets the scale for the "energy coarse graining" outlined in Sect. 9.1. Hence an appropriate AR of the type discussed in Sect. 8.1 will eventually involve invariant subspaces defined in terms of $\hat{\Pi}_A$, $\hat{\Pi}_B$. The concrete definition of this AR will be discussed in detail in the two following sections.

However, we want to mention here that choosing $\Delta E$ arbitrarily large will always produce a permissible AR in the sense of a pure state being unable to leave the AR during its evolution. Nevertheless this could be an inappropriate AR for it may be "too large," i.e., it may contain enormous regions into which that state can never venture. The whole typicality argument only holds if the ARs are chosen such that the state could, under given constraints, hypothetically reach any point within the AR. Thus one should reasonably choose $\Delta E$ one order of magnitude or so larger than $\delta\epsilon$ but not more. Eventually the whole approach, of course, only yields a reliable result, if the result does not severely depend on how large $\Delta E$ is chosen, within this limits. We are going to address this question below whenever it occurs.

### 9.3.1 Microcanonical Conditions

It has often been claimed that a system under so-called microcanonical conditions would not interact with its environment. This, however, is typically not true (cf. [5, 6]). A thermally isolated gas in a container, e.g., definitely interacts with the container, otherwise the gas could not even have a well-defined volume, as explained above (Sect. 9.2). If a system is thermally isolated, it is not necessarily isolated in the microscopic sense, i.e., not interacting with any other system. The only constraint is that the interaction with the environment should not give rise to energy exchange. As will be seen later, this does not mean that such an interaction has no effect on the considered system, a fact that might seem counterintuitive from a classical point of view but is, nevertheless, true in the quantum regime. This constraint, however, leads to an immense reduction of the region in Hilbert space, AR, which the wave vector is confined to.

We start with the so-called microcanonical contact, which is similar to the pure dephasing models from open system theory. An interaction $\hat{I}_{SE}$ is present, but it is neither allowed to (substantially) exchange energy between the system and its environment, nor should it contain much energy in itself, thus we take it to be of some strength $\delta\epsilon$ as already done in the last section. For such a scenario we now have to define an appropriate AR. Following the argument in the last section we choose some $\Delta E$ with $\Delta E > \delta\epsilon$. A substantial exchange of energy is impossible if we define the invariant subspaces of the full system in the sense of Sect. 8.1 by

$$\hat{\Pi}_\alpha = \hat{\Pi}_A \hat{\Pi}_B, \quad \alpha = (A, B). \tag{9.14}$$

The dimensions of subspaces $\alpha$ are obviously given by the product of the dimensions of $A$ and $B$, $N_\alpha = N_A N_B$. The AR is then defined by a set $W_\alpha = W_{AB}$ where $W_{AB}$ is the (initial and conserved) probability to find the system S in the energy intervals $A$ and the environment E in the energy interval $B$, cf. (9.7). Thus we may, using the notation introduced in Sect. 8.1, state the AR as follows:

$$AR := \{\langle\psi|\hat{\Pi}_A\hat{\Pi}_B|\psi\rangle = W_{AB}\}. \tag{9.15}$$

So far we have defined an AR which is in accord with a microcanonical contact. But what structural properties must $\hat{I}_{SE}$ feature in order to give rise to such an AR? It essentially only has to leave $W_A$ and $W_B$ invariant under the evolution. This only specifies some commutator relations of $\hat{I}_{SE}$:

$$[\hat{\Pi}_A, \hat{I}_{SE}] = [\hat{\Pi}_B, \hat{I}_{SE}] = 0. \tag{9.16}$$

Obviously there are a lot of, at least mathematically possible, non-zero interactions that do fulfill these conditions.

In Fig. 9.4 we illustrate the AR as resulting from microcanonical conditions. The average reduced state and its typicality corresponding to microcanonical conditions, i.e., the above-described AR, will be addressed in Sect. 10.2.

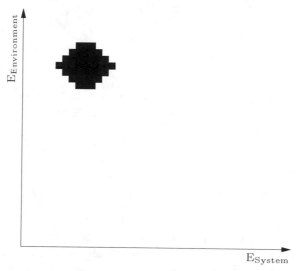

**Fig. 9.4** Under microcanonical conditions only a certain set of states is accessible to the evolution of the full system. The set of states depends on the initial state and may, e.g., be indicated by the *black region*. The representation of states follows the scheme introduced in Fig. 9.1

### 9.3.2 Energy Exchange Conditions

So far, we only considered a contact scenario, for which no energy transfer between system and environment was allowed. However, many systems do exchange energy with their environment, and therefore it is necessary to allow also for this possibility in our considerations.

Finally, for environments with a special increase of the state density with energy, i.e., an exponential increase, the system will reach the canonical equilibrium state. This special scenario is called canonical contact. However, first let us consider here

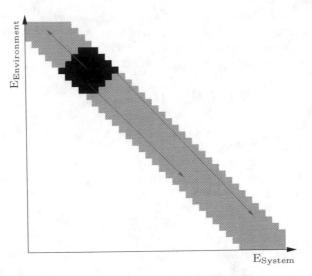

**Fig. 9.5** Under energy exchange conditions only a certain set of states is accessible to the evolution of the full system. The set of states depends on the initial state which may, e.g., belong to the *black* region. The corresponding accessible set of states, however, is indicated in *gray*. The *arrows* represent some "transitions" which are allowed under energy exchange conditions. The representation of states follows the scheme introduced in Fig. 9.1

the more general situation of an energy exchange contact condition, without any assumptions about the spectrum of the environment.

Our approach to the "energy exchange conditions" will be based on similar techniques as before. The possibility of a partition according to Sect. 9.1 is still assumed. We concretely consider an interaction which allows for energy exchange between system and environment and which is again of overall strength $\epsilon$, cf. Sect. 8.1. Thus now we assume subspaces spanned by the eigenstates of the full decoupled system from an energy interval of width $\Delta E$ (which is again chosen as $\Delta E > \epsilon$, just like above) as invariant. Hence, choosing to label the invariant subspaces by their mean energy $E$, rather than by $\alpha$ as done in Sect. 8.1, we may characterize the new AR by the projectors

$$\hat{\Pi}_E = \sum_{AB} \hat{\Pi}_A \hat{\Pi}_B \, M_{E,AB}. \tag{9.17}$$

Here $M_{AB,E}$ stands again for the delta-like function which warrants that $E_A^S + E_B^E \approx E$, as defined in (9.9). The corresponding set of probabilities $W(E)$ is again to be computed from the initial state by

$$W(E) := \sum_{A,B} W_{AB} \, M_{AB,E} = \sum_{A,B} \sum_{a,b} M_{AB,E} |\psi_{ab}^{AB}(0)|^2. \tag{9.18}$$

Those sets, $\hat{\Pi}_E$, $W(E)$, specify the AR if energy exchange is allowed. Using again the notation introduced in Sect. 8.1 we may state

$$\text{AR} := \{\langle\psi|\hat{\Pi}_E|\psi\rangle = W_E\}. \tag{9.19}$$

Other than in the case of microcanonical conditions this AR does not imply any other property of $\hat{1}_{SE}$ except for its weakness, i.e., no additional commutator relations, etc. The only constraint that this AR implements is the approximate conservation of the sum of the energies contained in the subsystems, i.e., considered system and environment. Since the total energy is conserved anyway, the sum of the local energies is approximately conserved whenever the interaction is weak.

In Fig. 9.5 the energy exchange coupling is depicted. Starting again in some region in Hilbert space, the figure shows further subspaces the system could evolve into. That means that all these states belong to the AR here, which are much more than for the microcanonical case.

The average reduced state and its typicality corresponding to energy exchange conditions, i.e., the above-described AR, will be addressed in Sect. 10.3.

# References

1. M. Hemmo, O. Shenker, Stud. Hist. Phil. Mod. Phys. **32**, 555 (2001)
2. J. Gemmer, G. Mahler, Europhys. Lett. **59**, 159 (2002)
3. B. Diu, C. Guthmann, D. Lederer, B. Roulet, *Elements de Physique Statistique* (Hermann Editeurs des Sciences et des Arts, Paris, 1989)
4. A.E. Allahverdyan, T.M. Nieuwenhuizen, Phys. Rev. Lett. **85**, 1799 (2000)
5. P. Bergmann, J. Lebowitz, Phys. Rev. **99**, 578 (1955)
6. J.M. Blatt, Progr. Theor. Phys. **22**(6), 745 (1959)

# Chapter 10
# The Typical Reduced State of the System

*Indeed my favourite key to understanding quantum mechanics
is that a subsystem cannot be isolated by tracing from an
enveloping pure state without generating impurity: the
probability associated with measurement develops because the
observer must implicitly trace himself away from the observed
system.*

— E. Lubkin [1]

**Abstract** In Chap. 8 we found that there is no typicality of the full state of the
non-composite system. In this chapter we investigate whether or not there is typicality of the reduced state of some considered system which interacts with some
sort of environment. This question is answered for couplings that may exchange
energy between system and environment and couplings that cannot. It is essentially
found that, roughly speaking, large environments yield typical reduced states of the
system. Those typical states turn out to be, e.g., the Boltzmann equilibrium state
under microcanonical or the Gibbs equilibrium state under canonical conditions.
Thus the composite system scenario may give rise to standard thermodynamical
relaxation of the whole considered system, including increase of local entropy.

## 10.1 Reduced State and Typicality

In Sect. 8.4 we addressed the question whether or not there can be typicality of the
full state of a single system, which implies that all possible expectation values would
have to take on the same values for almost all states from some accessible region
(AR). We found that to be impossible in the context of single systems, but it has
already been announced in Sect. 8.4 that such a typicality can occur for the reduced
state (as introduced in Sect. 2.2.5) of one subsystem in a compound, e.g., bipartite
system. In this case all reduced states for some subsystem corresponding to all full
states from some AR have to be essentially the same. They may, furthermore, all
take the form of some standard ensemble.

Such a scenario can then clarify the reason of the immense success of all sorts
of computations based on equilibrium ensembles despite the fact that no pure
microstate can ever evolve into such an ensemble (cf. Sect. 4). If there is typicality
of the reduced state the full system may very well always be in some evolving pure

Gemmer, J. et al.: *The Typical Reduced State of the System.* Lect. Notes Phys. **784**, 107–117
(2009)
DOI 10.1007/978-3-540-70510-9_10

state, nevertheless all possible measurements on the subsystem will yield the same outcomes as if its state was the equilibrium ensemble.

Hence in this section we are going to analyze in which cases such a typicality is indeed present for bipartite systems and the AR arising from microcanonical or energy exchange conditions, as described in Sect. 9.3. Of course we will also, or in fact first, analyze whether the possibly typical, mean state takes the form of a standard ensemble.

For simplicity of notation we are going to use now some symbols that have already been introduced before for slightly different objects. Other than in (8.22) $\hat{\rho}$ will in the following denote the reduced density operator of the considered system S with the environment E already being traced out (as explained in Sect. 2.2.5), i.e.,

$$\hat{\rho} \equiv \text{Tr}_E\{|\psi\rangle\langle\psi|\}, \tag{10.1}$$

with $|\psi\rangle$ being the pure state of the full system as before (for an explicit formulation of the reduced density operator, see (9.6)). In the sense of (8.22) and (8.14) we had $[\![\hat{\rho}]\!] = \hat{\Omega}$. However, now we define the Hilbert space average (HA) of the reduced state $\hat{\rho}$ as given in (10.1) by

$$\hat{\omega} \equiv [\![\hat{\rho}]\!] . \tag{10.2}$$

After these introductory remarks we proceed as follows:

1. Find the average reduced state $\hat{\omega}$ corresponding to the pertinent AR.
2. Check whether $\hat{\omega}$ is in accord with a standard Boltzmann ensemble.
3. Check whether the mean distance $[\![D^2(\hat{\rho}, \hat{\omega})]\!]$ between the average state and the actual reduced state is indeed small.

We start by considering matrix elements of the reduced density operator $\hat{\rho}$. If the considered system itself consists of subspaces labeled by $A$ ($C$) with states $a$ ($c$) there is a suitable basis for it in terms of the states $|A, a\rangle$, as already used in Sect. 9.1. From (10.1) it is straightforward to show that a matrix element of $\hat{\rho}$ in this basis, $\rho^{Aa,Cc} \equiv \langle C, c|\hat{\rho}|A, a\rangle$, can be expressed in terms of the full state $|\psi\rangle$ as

$$\rho^{Aa,Cc}(\psi) = \langle\psi||A, a\rangle\langle C, c| \otimes \hat{1}|\psi\rangle, \tag{10.3}$$

with $\hat{1}$ being the unit operator of the environment. This way the HA of the reduced state may be accessed on the basis of HAs of expectation values of operators for which already some results have been derived in Chap. 8. However, since the operators $|A, a\rangle\langle C, c| \otimes \hat{1}$ are not Hermitian and our previous results are on HAs of observables, i.e., Hermitian operators, we prefer to consider the operators

$$\hat{X}^{Aa,Cc} := \frac{1}{2}(|A, a\rangle\langle C, c| + |C, c\rangle\langle A, a|) \otimes \hat{1}, \tag{10.4}$$

$$\hat{Y}^{Aa,Cc} := \frac{i}{2}(|A, a\rangle\langle C, c| - |C, c\rangle\langle A, a|) \otimes \hat{1}, \tag{10.5}$$

which are obviously Hermitian. Thus, the operator used in (10.3) reads

$$|A, a\rangle\langle C, c| \otimes \hat{1} = \left(\hat{X}^{Aa,Cc} - i\hat{Y}^{Aa,Cc}\right) \otimes \hat{1} . \tag{10.6}$$

The expectation values of the operators given in (10.4) and (10.5) refer to the real and imaginary parts of the density matrix elements:

$$\langle\hat{X}^{Aa,Cc}\rangle \equiv \langle\psi|\hat{X}^{Aa,Cc}|\psi\rangle = \text{Re}\{\rho^{Aa,Cc}\} , \tag{10.7}$$

$$\langle\hat{Y}^{Aa,Cc}\rangle \equiv \langle\psi|\hat{Y}^{Aa,Cc}|\psi\rangle = \text{Im}\{\rho^{Aa,Cc}\} . \tag{10.8}$$

The elements of the reduced density operator are thus given in terms of $\langle\hat{X}^{Aa,Cc}\rangle$, $\langle\hat{Y}^{Aa,Cc}\rangle$ as

$$\rho^{Aa,Cc} = \langle\hat{X}^{Aa,Cc}\rangle - i\langle\hat{Y}^{Aa,Cc}\rangle . \tag{10.9}$$

Hence their HAs simply read

$$[\![\rho^{Aa,Cc}]\!] = [\![\langle\hat{X}^{Aa,Cc}\rangle]\!] - i[\![\langle\hat{Y}^{Aa,Cc}\rangle]\!] . \tag{10.10}$$

Since on the right-hand side of (10.10) only HAs of Hermitian operators appear we may now use the results from Sect. 8 to compute the mean matrix elements of $\hat{\rho}$. The mean reduced state $\hat{\omega}$ is simply the operator formed by all averaged matrix elements of $\hat{\rho}$, i.e.,

$$\hat{\omega} = \sum_{AC}\sum_{ac}[\![\rho^{Aa,Cc}]\!]|A, a\rangle\langle C, c| . \tag{10.11}$$

Below we will use this form to analyze whether $\hat{\omega}$ is in accord with standard equilibrium ensembles.

But as mentioned above another central aim is to show that this state is indeed the most probable reduced state of the system S. As discussed in Sect. 8.4 this can be shown by using the average over the distance between the actual states $\hat{\rho}$ and the mean state $\hat{\omega}$. It is straightforward to show that the considerations on such distances from Sect. 8.4 directly carry over to reduced states such that, in complete analogy to (8.25), we may write

$$[\![D^2(\hat{\rho}, \hat{\omega})]\!] = [\![\text{Tr}\{\hat{\rho}^2\}]\!] - \text{Tr}\{\hat{\omega}^2\} . \tag{10.12}$$

This, as turns out after some technical computations, may be written in terms of Hilbert space variances of Hermitian operators as

$$[\![D^2(\hat{\rho}, \hat{\omega})]\!] = \sum_{AC}\sum_{ac}\left(\Delta_{\text{H}}^2(\langle\hat{X}^{Aa,Cc}\rangle) + \Delta_{\text{H}}^2(\langle\hat{Y}^{Aa,Cc}\rangle)\right) . \tag{10.13}$$

For details concerning the derivation of this result see Appendix C.1, especially (C.6). This is of course not necessarily small. However, we will use this result in the following to show that there are several big classes of systems with different contact conditions for which (10.12) is indeed very small.

## 10.2 Microcanonical Conditions

### 10.2.1 Microcanonical Average State

Let us start, for simplicity, with a microcanonical contact scenario, i.e., no energy exchange between system and environment. Thus the pertinent AR is given by (9.15). To compute the average of the reduced density matrix elements according to (10.10) we have to account for the average of the operators $\hat{X}^{Aa,Cc}$ and $\hat{Y}^{Aa,Cc}$ first. The average of the expectation value of a Hermitian operator is given by (8.12), thus, we need the "suboperators"

$$
\begin{aligned}
\hat{X}_{\alpha\alpha}^{Aa,Cc} &= \hat{X}_{A'B,A'B}^{Aa,Cc} \\
&= \frac{1}{2}\hat{\Pi}_{A'}\hat{\Pi}_{B}\Big((|A,a\rangle\langle C,c| + |C,c\rangle\langle A,a|)\otimes\hat{1}\Big)\hat{\Pi}_{A'}\hat{\Pi}_{B} \\
&= \frac{1}{2}\delta_{A'A}\,\delta_{A'C}(|A,a\rangle\langle C,c| + |C,c\rangle\langle A,a|)\otimes\hat{\Pi}_{B}\,.
\end{aligned}
\tag{10.14}
$$

Taking the trace of this operator as required by (8.12) yields

$$
\mathrm{Tr}\left\{\hat{X}_{A'B,A'B}^{Aa,Cc}\right\} = \delta_{A'A}\,\delta_{A'C}\,\delta_{AC}\,\delta_{ac}N_B\,.
\tag{10.15}
$$

Equipped with this expression we may now compute the HA of $\hat{X}^{Aa,Cc}$ with respect to the AR corresponding to microcanonical conditions, finding again from (8.12)

$$
\begin{aligned}
[\![\langle\hat{X}^{Aa,Cc}\rangle]\!] &= \sum_{A'B}\frac{W_{A'B}}{N_{A'}N_B}\,\mathrm{Tr}\left\{\hat{X}_{A'B,A'B}^{Aa,Cc}\right\} \\
&= \sum_{B}\frac{W_{AB}}{N_A}\,\delta_{AC}\,\delta_{ac}\,.
\end{aligned}
\tag{10.16}
$$

Further explicit calculations of averages and variances can be found in Appendix C. Here we only quote the result that the HA of $\langle\hat{Y}^{Aa,Cc}\rangle$ vanishes (see Appendix C.2). Thus using (10.10), the microcanonical HA of the reduced density operator is eventually found to be

$$
[\![\rho^{Aa,Cc}]\!] = \sum_{B}\frac{W_{AB}}{N_A}\,\delta_{AC}\,\delta_{ac}\,.
\tag{10.17}
$$

With (9.8)

$$[\![\rho^{Aa,Cc}]\!] = \frac{W_A}{N_A}\delta_{AC}\delta_{ac}.\tag{10.18}$$

From (10.18) follows the average of the full density operator for system S: all off-diagonal elements vanish whereas we get equal a priori probabilities on the diagonal elements, i.e., a Boltzmann state

$$\hat{\omega} = [\![\hat{\rho}]\!] = \sum_A \frac{W_A}{N_A}\hat{\Pi}_A.\tag{10.19}$$

Since it is in accord with Boltzmann's a priori principle its von Neumann entropy $S(\hat{\omega})$ is the maximum entropy compatible with some given set of $W_A$. For the case of one specific $W_A = 1$ and all others vanishing, $\hat{\omega}$ reduces to the standard equilibrium state corresponding to the microcanonical ensemble. And its von Neumann entropy is of course the standard Boltzmann entropy, cf. (3.38). The crucial question is now whether or not $\hat{\omega}$ is a typical state. Only if it is, will it appear as an effective, attractive local equilibrium state.

## 10.2.2  Typicality of the Microcanonical Average State

To show that the state given in (10.19) is indeed the typical state for the AR we have to compute the mean distance of this state to all actual reduced states of the system as discussed in Sects. 8.4, 6.2, and 10.1. To evaluate the mean squared distance according to (10.13) we need the Hilbert space variances (HVs) of the expectation value of the operator $\hat{X}^{Aa,Cc}$ and $\hat{Y}^{Aa,Cc}$. These variances are computed in Appendix C.3 and read

$$\Delta_{\mathrm{H}}^2(\langle\hat{X}^{Aa,Cc}\rangle) = \frac{1}{2}\sum_B \frac{W_{AB}W_{CB}}{N_A(N_CN_B + \delta_{AC})} - \delta_{AC}\delta_{ac}\sum_B \frac{W_{AB}^2}{N_A^2(N_AN_B + 1)},\tag{10.20}$$

$$\Delta_{\mathrm{H}}^2(\langle\hat{Y}^{Aa,Cc}\rangle) = \frac{1}{2}\sum_B \frac{W_{AB}W_{CB}}{N_A(N_CN_B + \delta_{AC})}.\tag{10.21}$$

From these results the mean distance of the states within AR from the Hilbert space average over all states reads

$$\llbracket D^2(\hat{\rho}, \hat{\omega}) \rrbracket = \sum_{AC} \sum_{ac} \left( \sum_B \frac{W_{AB} W_{CB}}{N_A(N_C N_B + \delta_{AC})} \right.$$

$$\left. - \delta_{AC} \delta_{ac} \sum_B \frac{W_{AB}^2}{N_A^2(N_A N_B + 1)} \right). \tag{10.22}$$

Carrying out the summation over $a$ and $c$ which just yields the sizes $N_A$ and $N_C$ of the respective subspaces, we find

$$\llbracket D^2(\hat{\rho}, \hat{\omega}) \rrbracket = \sum_{AC} \sum_B \left( \frac{W_{AB} W_{CB} N_C}{(N_C N_B + \delta_{AC})} - \delta_{AC} \frac{W_{AB}^2}{N_A(N_A N_B + 1)} \right). \tag{10.23}$$

Considering the special case for large environments, we may neglect the 1 in the denominator, and factorizing initial conditions $W_{AB} = W_A W_B$ we find

$$\llbracket D^2(\hat{\rho}, \hat{\omega}) \rrbracket \approx \sum_{AC} \sum_B \left( \frac{W_A W_C W_B^2}{N_B} - \delta_{AC} \frac{W_A^2 W_B^2}{N_A^2 N_B} \right). \tag{10.24}$$

Realizing that $\sum_A W_A = 1$, this simplifies further to

$$\llbracket D^2(\hat{\rho}, \hat{\omega}) \rrbracket \approx \sum_B \frac{W_B^2}{N_B} \left( 1 - \sum_A \frac{W_A^2}{N_A^2} \right). \tag{10.25}$$

Obviously this average distance will be very small compared to 1 (which implies typicality), if either the distribution of probabilities onto the environmental subspaces is broad and/or, even more important, if the occupied environmental subspaces are large. Consider, to state this more rigorously, an upscaling of the system by $N_A \to x N_A$ and of the environment by $N_B \to y N_B$. In such a scaling the upper bound to the average distance, i.e., $\llbracket D^2 \rrbracket^+ > \llbracket D^2(\hat{\rho}, \hat{\omega}) \rrbracket$, based on (10.25) scales as

$$\llbracket D^2 \rrbracket^+ \to \frac{1}{y} \llbracket D^2 \rrbracket^+. \tag{10.26}$$

Thus whenever the system is coupled microcanonically to a large environment there is typicality of the system's local reduced state. In this case the average state (10.19) is indeed a typical state and thus plays the role of an equilibrium state. This means $\hat{\omega}$ appears as an effective attractor state and entropy evolutions that increase up to the maximum von Neumann entropy $S(\hat{\omega})$ (cf. (2.26)) can be expected. This, as already mentioned, corresponds to the famous Boltzmann entropy formula as written on his tombstone (Zentralfriedhof, Vienna). This finding is the central result of [2]. For a numerical illustration of this principle with a concrete example, see Sect. 16.1.

## 10.3 Energy Exchange Conditions

### 10.3.1 Energy Exchange Average State

In the following we allow for energy exchange between system S and environment E. The pertinent AR for this scenario has already been given in (9.19). The Hilbert space average of the reduced density matrix elements with respect to the above pertinent AR is computed in a straightforward way, details may be found in Appendix C.4. One finds

$$\llbracket \rho^{Aa,Cc} \rrbracket = \sum_E \frac{W_E}{N_E} M_{E,AB}\, \delta_{AC}\, \delta_{ac}\, N_B , \qquad (10.27)$$

with the notation introduced in Sect. 9.3.2. $M_{E,AB}$ is the $\delta$-function (9.9). Thus, we get again a diagonal state. In order to understand the structure of this mean reduced state, consider a hypothetical (mixed) state which is in accord with the Boltzmann a priori principle and in accord with the AR. This means a state for which the probability of the total system to be found in some invariant subspace, $W_E$, is uniformly distributed onto all states belonging to this subspace. (This subspace is of course represented by $\hat{\Pi}_E$). Now, for such a state the probability $W_A$ of the local system S to occupy a subspace $A$ is given by

$$W_A = N_A \sum_E \frac{W_E}{N_E} N_B M_{E,AB} . \qquad (10.28)$$

In Sect. 10.4, it will be shown that for a certain type of environmental spectrum this probability distribution reduces to a Boltzmann distribution. Using this quantity (10.28) the average state as given in terms of matrix elements in (10.27) may be rewritten in the form

$$\hat{\omega} = \sum_A \frac{W_A}{N_A} \hat{\Pi}_A . \qquad (10.29)$$

Formally this looks exactly like the typical state in the microcanonical case, (10.19). But, of course, here $W_A$ is defined on the basis of (10.28) and is, with respect to dynamics, not a conserved quantity. However, just like the microcanonical equilibrium state the average energy exchange state (10.29) has the probability $W_A$ equally distributed onto the states that span the subspace $A$. Thus $W_A$ is in accord with the Boltzmann principle on the full system, and the occupation probabilities of local system states are in accord with the Boltzmann principle under the restriction of given $W_A$. In this sense it is, just like the microcanonical equilibrium state, a maximum entropy state. Eventually, with respect to the interpretation of (10.29), it is important to mention that, as will be shown in Sect. 10.4, for a common type of environmental spectra $\hat{\omega}$ takes the form of a standard Gibbs equilibrium state.

Another point should be mentioned: the energy probability distribution $W_A$ (10.28) is, in general, not independent of the initial state, since different energy probability distributions of the local initial state may result in different overall energy probability distributions $W_E$, and those clearly enter (10.28) and thus even (10.29). Normally the canonical contact of standard thermodynamics leads to an equilibrium state, which does not depend on the initial state. Here, we have considered a more general contact scenario from which the canonical contact seems to be a special subclass, as we will demonstrate in Sect. 10.4.

### 10.3.2  Typicality of the Energy Exchange Average State

To show that the above-derived equilibrium state is really the typical state within the accessible region we proceed again as already done in the microcanonical situation: The distances between the averaged reduced state and the actual reduced states from the AR are investigated with the aid of (10.13). The rather lengthy calculation can be found in Appendix C.5. Eventually, the mean squared distance reads

$$
[\![ D^2(\hat{\rho}, \hat{\omega}) ]\!] = \sum_{EE'} \sum_{ABC} \frac{W_E W_{E'} N_B N_A N_C}{N_E (N_{E'} + \delta_{EE'})} M_{E,AB} M_{E',CB}
$$
$$
- \sum_{E} \sum_{AB} \frac{W_E^2 N_B^2 N_A}{N_E^2 (N_E + 1)} M_{E,AB} . \tag{10.30}
$$

For large environments one gets $N_{E'} + \delta_{EE'} \approx N_{E'}$. Exploiting this and skipping the last term because it is presumably much smaller than the first term (and we are eventually interested in an upper bound anyway) we find

$$
[\![ D^2(\hat{\rho}, \hat{\omega}) ]\!] \leq \sum_{ABC} N_A N_B N_C \sum_E \frac{W_E}{N_E} M_{E,AB} \sum_{E'} \frac{W_{E'}}{N_{E'}} M_{E',CB} . \tag{10.31}
$$

The $M_{E,AB}$ are the $\delta$-functions according to (9.9). The dimensions in the denominator of (10.31) are products of system and environment dimensions defined in (8.3) with the projector (9.17). Thus, scaling up the system by $N_A \to x N_A$ and the environment by $N_B \to y N_B$ an upper bound to the mean distance, $[\![ D^2 ]\!]^+$, scales according to

$$
[\![ D^2 ]\!]^+ \to \frac{1}{y} [\![ D^2 ]\!]^+ , \tag{10.32}
$$

i.e., decreases for larger environments. Based on this result exactly the same reasoning as given below (10.26) applies. For large environments the average energy exchange state (10.29) is typical in the sense described in Sect. 6 and thus can be expected to effectively play the role of an attractive equilibrium state. In Sect. 10.4 it will be demonstrated that for a certain (generically occurring, cf. Sect. 12.2) type of spectrum of the environment the average state (10.29) takes on the form of a standard canonical Gibbs state. Within such a scenario the above finding (10.32) essentially reflects the results of a paper by Goldstein et al., entitled "canonical typicality" [3] and is closely related to [4].

## 10.4 Canonical Conditions

In the last section we have investigated a situation for which an energy transfer between the system and its environment was allowed. With these constraints alone it does not seem possible to end up in an equilibrium state that does not depend on the initial state. For a canonical situation, however, the system should be found in the canonical equilibrium state, independent of the initial conditions. This behavior can be found, if we take a further condition into account: a special form of state density of the environment.

Let us assume for the moment an exponential increase of the number of eigenstates within the compartments $B$ of the environment E with increasing energy $E_B^E$, i.e.,

$$N_B = N_0 \, e^{\alpha E_B^E} , \tag{10.33}$$

where $N_0$ and $\alpha$ are some constants. The justification for this exponential increase of the environmental-level density for standard thermodynamic systems will be discussed later in Sects. 12.2 and 12.4.

Using this exponential increase of state density in the environment we find for the energy distribution of the system S as defined in (10.28)

$$W_A = N_A \sum_E \frac{W_E}{N_E} N_0 \, e^{\alpha E_B^E} M_{E,AB}$$

$$= N_A \, e^{-\alpha E_A} \sum_E \frac{N_0 \, e^{\alpha E} W_E}{N_E} , \tag{10.34}$$

where we have, exploiting the delta function $M_{E,AB}$ as defined in (9.9), replaced $E_B^E$ by $E - E_A$. Obviously, the sum does not depend on $A$ at all. Since $W_A$ has been constructed as some probability distribution it is still normalized by definition. Thus the sum has to reduce to a normalizing factor. This could also be shown in a rather lengthy calculation, which we skip here. Finally we get for the energy probability distribution of the system

$$W_A = \frac{N_A \, e^{-\alpha E_A}}{\sum_{A'} N_{A'} \, e^{-\alpha E_{A'}}} . \tag{10.35}$$

This result does no longer depend on the initial state. Furthermore, the energy probability distributions of almost all states from the respective accessible region is then the canonical distribution. Plugging (10.35) into (10.29) eventually yields the typical state

$$\hat{\omega} = \frac{1}{\sum_A N_A \, e^{-\alpha E_A}} \sum_A e^{-\alpha E_A} \hat{\Pi}_A \,. \tag{10.36}$$

Obviously, this is the well-known canonical equilibrium state with the inverse temperature $\beta = \alpha$. And of course its von Neumann entropy is simply the Shannon entropy (cf. (3.49)) under canonical conditions. The above typical canonical state is in accord and reflects results by Goldstein et al. [3].

For some more concrete illustrations of the implications, which the rather abstractly derived principles in this chapter bear on the dynamics of adequate systems, see Sect. 16.2.

## 10.5 Beyond Weak Coupling

If the interaction between any two systems, e.g., a considered system and its environment, becomes strong compared to the energy scales of the decoupled systems, a definition of the AR based on energy eigenspaces of the decoupled systems is no longer reasonable since those energy eigenspaces are no longer (approximately) invariant. So what can be expected then? Will there be, nevertheless, typicality of the reduced state of the considered system? Before we investigate this it should be noted here that this specific question is independent of the question whether or not the possibly typical state can indeed be concretely calculated for a given model. Or to rephrase: even if there is typicality, that does not imply that one can indeed calculate, e.g., equilibrium observables in strongly interacting systems.

Let us, however, focus on the question of typicality itself. For simplicity we assume the state of the total system to be fully restricted to some projective subspace represented by $\hat{\Pi}_\alpha$. We take this subspace to be invariant, thus $\langle \psi | \hat{\Pi}_\alpha | \psi \rangle = 1$ defines the AR. This invariance may result from the fact that the AR corresponds to a certain energy regime of the full system, i.e., including the interaction. (Note that then $\hat{\Pi}_\alpha$ may not be concretely accessible for a given system, which will result in the impossibility to calculate the average state concretely.) Or it may result from a fundamental conserved quantity in the system. However, following the definitions (10.1) and (10.2) the average reduced state may be written as

$$\hat{\omega} = [\![\mathrm{Tr}_E\{|\psi\rangle\langle\psi|\}]\!] = \mathrm{Tr}_E\{[\![|\psi\rangle\langle\psi|]\!]_{AR}\} = \frac{1}{N_\alpha} \mathrm{Tr}_E\{\hat{\Pi}_\alpha\} \,. \tag{10.37}$$

This, of course, is again the reduced state that one would also get from a hypothetical, maximum mixed state of the total system which is in accord with the Boltzmann a priori principle of equal weights. This means the average reduced state is the same state that one would get from a consideration based on a Boltzmann equilibrium ensemble for the total system. (By the way, this also applies to the microcanonical and energy exchange conditions.) But is this $\hat{\omega}$ typical? Again we have to analyze the mean squared distance as given by (10.13) here appearing as

$$[\![D^2(\hat{\rho}, \hat{\omega})]\!] = \sum_{ac}^{N_S} \left( \Delta_H^2(\langle \hat{X}^{a,c} \rangle) + \Delta_H^2(\langle \hat{Y}^{a,c} \rangle) \right). \qquad (10.38)$$

(Since there is no energy course graining of the considered system in this case the sums over $A, C$ vanish.) The indices $a, c$ run over all (relevant) states of the considered system, the total number of which we denote by $N_S$. From (8.21) we find an upper bound on the above addends

$$\Delta_H^2(\langle \hat{X}^{a,c} \rangle) \le \Delta_S^2(\hat{X}^{a,c}), \quad \Delta_H^2(\langle \hat{Y}^{a,c} \rangle) \le \Delta_S^2(\hat{Y}^{a,c}), \qquad (10.39)$$

where $\Delta_S^2(\hat{X}^{a,c})$ and $\Delta_S^2(\hat{Y}^{a,c})$ are the spectral variances according to (8.20). Note that the upper bound is independent of the indices $a$ and $c$. Thus, with the definitions (10.4) and (10.5) and again with (8.20) we find yet another upper bound on $\Delta_S^2(\hat{X}^{a,c})$ and $\Delta_S^2(\hat{Y}^{a,c})$ by

$$\Delta_S^2(\hat{X}^{a,c}) \le \frac{1}{N_S}, \quad \Delta_S^2(\hat{Y}^{a,c}) \le \frac{1}{N_S}. \qquad (10.40)$$

Now, plugging (10.40), (10.39), and (10.13) together eventually yields

$$[\![D^2(\hat{\rho}, \hat{\omega})]\!] \le 2\frac{N_S}{N_\alpha}. \qquad (10.41)$$

Hence, whenever the dimension of the subspace onto which the full system is confined is much larger than the dimension of the Hilbert space of the considered system, there is typicality of the reduced state with respect to this AR. This essentially reflects results by Popescu et al. [5]. Note, however, that here, (10.41), unlike the weak coupling results for microcanonical and energy exchange conditions (10.26), (10.32), the average squared distance $D^2$ does not only scale with the size of the environment. Thus if the considered system itself becomes fairly large obviously no typicality can be guaranteed.

# References

1. E. Lubkin, J. Math. Phys. **19**, 1028 (1978)
2. J. Gemmer, G. Mahler, Eur. Phys. J. D **17**, 385 (2001)
3. S. Goldstein, J.L. Lebowitz, R. Tumulka, N. Zanghi, Phys. Rev. Lett. **96**, 050403 (2006)
4. J. Gemmer, G. Mahler, Eur. Phys. J. B **31**, 249 (2003)
5. S. Popescu, A.J. Short, A. Winter, Nature Phys. **2**, 754 (2006)

# Chapter 11
# Entanglement, Correlations, and Local Entropy

*Any serious consideration of a physical theory must take into account the distinction between the objective reality, which is independent of any theory, and the physical concepts with which the theory operates. These concepts are intended to correspond with the objective reality, and by means of these concepts we picture this reality to ourselves.*
— A. Einstein, B. Podolsky and N. Rosen [1]

**Abstract** This chapter is essentially a comment on the role of entanglement and correlations in the previously described approach to relaxation in quantum systems (Chaps. 6–10). Within open system theory the concept of system and environment remaining uncorrelated under standard conditions seems to be a paradigm. Within the approach at hand increasing correlations may be viewed as the "source" of relaxation. We analyze this apparent contradiction in some detail. Furthermore, the issue of local entropy and purity is investigated.

## 11.1 Entanglement, Correlations, and Relaxation

During the last decades, entanglement has been in the focus of quantum research. Since it is one main ingredient of almost all recently investigated new quantum mechanical phenomena as, e.g., teleportation, cryptography, the quantum computer, there is a great variety of literature about entanglement (see, e.g., [2–4]). Here, it may be viewed as an essential ingredient to differentiate between a quantum approach to thermodynamics and any classical one. If one has a precisely given, pure microstate (no ensemble) the only way to have non-zero entropy, at least locally for a subsystem, is by means of entanglement. An analogue construction is impossible in the classical case. This is due to the fact that in this case all "thermodynamical fluctuations" (cf. Sect. 16.3) are eventually quantum uncertainties. (For an initial ensemble, tracing out some environment may lead to increasing local entropy in the classical case as well, though [5].) Think, for example, of a bipartite system in an EPR state [1] featuring zero total von Neumann entropy. However, considering the local state of one of the subsystems, one finds a totally mixed state, i.e., a state according to the maximum local entropy. In contrast, a product state with vanishing correlations features zero local entropy as well.

Gemmer, J. et al.: *Entanglement, Correlations, and Local Entropy*. Lect. Notes Phys. **784**, 119–127 (2009)
DOI 10.1007/978-3-540-70510-9_11

In the previous sections we have shown, that, for pertinent bipartite total system scenarios, the considered system tends toward maximum entropy. Since in these examples the state of the total system is always pure, the local increase of entropy can only be due to increasing entanglement. This picture, however, is in sharp contrast to the widespread idea in the field of open quantum systems, that weakly interacting systems should remain uncorrelated from their environments (the latter concept is sometimes referred to as "Born approximation").

There are various arguments that seem to support the idea of system and environment remaining (almost) uncorrelated during the relaxation process as discussed, e.g., in the context of open quantum systems, cf. Sect. 4.8.

1. Non-interacting systems cannot become correlated. Thus, at first sight, the assumption that weak interactions should only lead to negligible correlations seems reasonable [6, 7].
2. Controlled projective approximations for the local dynamics of the system yield to leading order in the interaction and for uncorrelated initial states autonomous equations of motion (cf. Sect. 4.8). These are in excellent accord with many, e.g., quantum optical experiments. Less controlled approaches that simply assume the strict factorizability of system and environment yield the very same local equations of motion [8, 9]. This seems to imply that, for weak interactions, the factorizability assumption should apply. Note, however, that such a conclusion cannot be drawn with any confidence.
3. Even if one leaves the comparison with those less controlled approaches aside, the projective approximations result in dynamical maps on the system of the form $\hat{\rho}(t + \tau) = \hat{\mathcal{V}}(\tau)\hat{\rho}(t)$ [cf. (4.22)] with $\hat{\mathcal{V}}(\tau_1 + \tau_2) = \hat{\mathcal{V}}(\tau_1)\hat{\mathcal{V}}(\tau_2)$ [10]. This means that even after $t + \tau_1$ the dynamics continues to be of the very same type that one found, requiring factorizability at $t$, for the time span from $t$ to $t + \tau_1$. Again, it is tempting to conclude that there would be factorizability at $t + \tau_1$ which, taking a closer look, may not necessarily be true.

In the following we thus shortly comment on the issue of relaxation and correlations in general. We prefer to investigate a question pointing in the opposite direction: Is it possible that a full bipartite system undergoes a unitary transformation, such that the purity of the considered system decreases, i.e., the entropy increases within its relaxation process, without substantial production of system–environment correlations being generated?

To investigate this question we specify the correlations $\hat{\rho}^c$ as an addend of the full system density matrix $\hat{\rho}$ as

$$\hat{\rho}^c := \hat{\rho} - \hat{\rho}^{(S)} \otimes \hat{\rho}^{(E)}, \tag{11.1}$$

with the local density operators

$$\hat{\rho}^{(S/E)} = \text{Tr}_{E/S}\{\hat{\rho}\} \tag{11.2}$$

of system and environment. Obviously $\hat{\rho}^{(S)} \otimes \hat{\rho}^{(E)}$ specifies the uncorrelated product part of the density matrix. To measure the size of the correlated and the uncorrelated parts we use the absolute value of an operator

$$p_i := \sqrt{P^i} = \sqrt{\text{Tr}\left\{(\hat{\rho}^i)^2\right\}}, \qquad (11.3)$$

with $i = c, S, E,$ or none. Evidently, $P = \text{Tr}\left\{\hat{\rho}^2\right\}$ is the purity of the full system and the local values $P^{S/E}$ the purities of the corresponding subsystems (cf. (2.22)). To decide whether or not correlations are negligible altogether, we consider the correlations-vs.-product-contributions coefficient

$$\eta := \frac{p_c}{p_S\, p_E}. \qquad (11.4)$$

If $\eta \ll 1$, correlations may safely be neglected.

Computing the size of the correlations yields

$$p_c^2 = p^2 - 2\,\text{Tr}\left\{\hat{\rho}\,\hat{\rho}^S\,\hat{\rho}^E\right\} + p_S^2\, p_E^2. \qquad (11.5)$$

Since the trace of a product of two Hermitian matrices fulfills the conditions on an inner product, one finds via Schwartz's inequality

$$\left|\text{Tr}\left\{\hat{\rho}\,\hat{\rho}_S\,\hat{\rho}_E\right\}\right| \le p\, p_S\, p_E. \qquad (11.6)$$

Inserting this into (11.5) yields

$$p_c^2 \ge (p - p_S\, p_E)^2, \qquad (11.7)$$

or for the coefficient (11.4)

$$\eta \ge \frac{p}{p_S\, p_E} - 1. \qquad (11.8)$$

Note that the total purity $P$ and thus $p$ is invariant under any unitary transformation. Often the environment is assumed to be exactly stationary, which might not precisely hold true, nevertheless $p_E(0) \approx p_E(t)$ should be a reasonable approximation for large, thermal reservoirs. Thus, the only quantity that may substantially change upon relaxation on the right-hand side of (11.8) is the purity of the system $P^S$, i.e., also $p_S$. And if $p_S$ decreases since entropy typically increases upon relaxation, the right-hand side of (11.8) obviously increases as well.

Hence, in the case of a stationary bath and an initial product state one finds

$$\eta \ge \frac{p_S(0)}{p_S(t)} - 1. \qquad (11.9)$$

This lower bound for $\eta$ may easily take on rather high values, e.g., for an $N$-level system coupled to a bath in the high-temperature limit ($k_B T$ much larger than the level spacing) one gets for an initially pure system state, $p_S(0) = 1$,

$$\eta \geq \sqrt{N} - 1. \tag{11.10}$$

In the case of a pure total state, which has been primarily addressed in Chap. 10, one has $p(t) = 1$ and $p_S(t) = p_E(t)$. Thus inserting into (11.4) we find

$$\eta \geq \frac{1}{p_S^2(t)} - 1. \tag{11.11}$$

This implies, again for the case of an $N$-level system relaxing to a high-temperature equilibrium state,

$$\eta \geq N - 1. \tag{11.12}$$

Thus, even for moderately sized systems, the correlations-vs.-product-contributions coefficient $\eta$ cannot be expected to remain small compared to one upon relaxation, neither in the scenario addressed by (11.10) nor in the case addressed by (11.12). This result is absolutely independent of the interaction strength. It only connects a decrease of purity (increase of entropy) to an increase of system–reservoir correlations, regardless of the timescale on which this relaxation process happens. Thus, we conclude that, quite contrary to the idea of system and bath remaining uncorrelated, correlations are generically generated upon relaxation.

To conclude this consideration we reconsider the "factorization arguments" given above point by point:

1. The weaker the interaction, the slower the overall buildup of correlations. But so is the environment-induced relaxation. So by the time the system is relaxed there will be substantial correlations, irrespective of the interaction strength.
2. There is nothing wrong with well-controlled approaches that produce correct results for local dynamics by projecting on factorizing states. Simply the conclusion that this would imply continuous factorizability is wrong. The coincidence of those local dynamics with dynamics based on a bold factorization assumption simply implies that in many cases (in which those approaches are successful) the buildup of correlations has no significant influence on the local dynamics. For an example where it does have influence and the above approaches thus fail, even for weak interactions, see Chap. 19.
3. If the initial state does not feature factorizability, but one nevertheless chooses to project onto factorizing states, the local equations of motion as obtained by the projective technique acquire a time-dependent inhomogeneity. The picture remains perfectly consistent if, for very many initial states, this inhomogeneity only has negligible influence on the local dynamics. More evidence in that direction comes from considerations presented in Chaps. 18 and 19.

Eventually we state that we do not intend to generally criticize the idea of completely positive maps on the basis of the above consideration that questions the absence of correlations, like, e.g., done to some extent in [6]. While the investigations presented there are surely correct for the respective scenario, the local dynamics of very many relevant scenarios in this context will most likely nevertheless result in completely positive maps.

## 11.2 Entropy and Purity

The properties entropy and purity have already played an important role in the last section. There we have learned that a thermodynamical relaxation process with increasing entropy, decreasing purity respectively, is connected with an increase of the correlations between system and heat bath as well. Here, we concentrate on the microcanonical contact scenario once more and analyze purity and entropy of possible states to show that $\hat{\omega}$ as given in (10.19) is indeed the unique maximum entropy and minimum purity state. In this context we focus here on off-diagonal elements (coherences) rather than on diagonal elements (probabilities). In principle the whole consideration could also be recapitulated for energy exchange conditions, however, focusing on conceptual issues, we are not going to display that here.

The average local state has been computed in Sect. 10.2.1 finding the state (10.19). The purity of this state is given by

$$P = \text{Tr}\left\{\hat{\omega}^2\right\} = \sum_A \frac{W_A^2}{N_A} . \tag{11.13}$$

Furthermore, we are aware of the fact that within the accessible region (AR) this diagonal state (10.19) is the typical state, i.e., mostly all reduced states of the system are close to this state. Below we show that the purity (11.13) is indeed the minimum purity or maximum entropy which is consistent with the microcanonical contact conditions.

To check that this is, indeed, the state with the smallest purity consistent with the given energy probability distribution $\{W_A\}$, we introduce a deviation D of the diagonal elements and a deviation E of the off-diagonal elements such that the resulting state is still consistent with $\{W_A\}$ and compute its purity. E is thus introduced as a matrix that does not have any diagonal elements. For the deviation

$$D = \sum_{A,a} D_{A,a} |A, a\rangle\langle A, a|, \tag{11.14}$$

the partial trace over one degenerate subspace $A$ has to vanish

$$\text{Tr}_a\{D\} = \sum_a D_{A,a} = 0, \tag{11.15}$$

because under microcanonical conditions the total probability distribution $\{W_A\}$ introduced by the initial state is fixed. The deviation D only redistributes the probability within a subspace $A$. E and D of course have to be Hermitian. Now with

$$\hat{\rho} = \hat{\omega} + D + E, \tag{11.16}$$

we find

$$\begin{aligned} P(\hat{\rho}) &= \text{Tr} \left\{ (\hat{\omega} + D + E)^2 \right\} \\ &= \text{Tr} \left\{ \hat{\omega}^2 + D^2 + E^2 \right\} + 2\text{Tr} \left\{ \hat{\omega}D + \hat{\omega}E + DE \right\}. \end{aligned} \tag{11.17}$$

Due to the properties of E and the diagonality of $\hat{\omega}$ and D the last two terms vanish. Using the definitions (10.19) and (11.14) we compute the term

$$\text{Tr} \left\{ \hat{\omega}D \right\} = \sum_{A,a} \frac{W_A}{N_A} D_{A,a} = \sum_{A} \frac{W_A}{N_A} \sum_{a} D_{A,a} = 0. \tag{11.18}$$

Thus, we find

$$P(\hat{\rho}) = \text{Tr} \left\{ \hat{\omega}^2 \right\} + \text{Tr} \left\{ D^2 \right\} + \text{Tr} \left\{ E^2 \right\}. \tag{11.19}$$

Since

$$\text{Tr} \left\{ D^2 \right\} \geq 0, \quad \text{Tr} \left\{ E^2 \right\} \geq 0, \tag{11.20}$$

the smallest purity is reached for

$$E = 0 \quad \text{and} \quad D = 0. \tag{11.21}$$

Thus, the smallest possible purity state is unique and consists only of $\hat{\omega}$.

The minimum purity corresponding to the maximum entropy as discussed in Sect. 2.2.4 is found to be

$$S_{\text{max}} = -k_B \sum_{A} W_A \ln \frac{W_A}{N_A}. \tag{11.22}$$

This reduces for sharp energy probability distribution $\{W_A\} = \delta_{AA'}$ to the standard form of

$$\hat{S}_{\text{max}} = k_B \ln N_{A'}. \tag{11.23}$$

For a numerical demonstration of several aspects of these considerations we refer to Sect. 16.1.

For "historical" reasons let us state here some further comments on the purity of the states within the AR. The idea behind the approach to thermodynamics from quantum mechanics at hand was initially centered around the average of the purity within the AR and the discussion of the landscape of the purity over this part of the Hilbert space (see, e.g., the first edition of this book or [11–17]). Within such an approach the Hilbert space average (HA) of the purity over the accessible region is computed. The purity itself is a quadratic function of the density operator and, thus, contains terms up to fourth order in the coordinates of the total system's wave vector. In that sense it is equivalent to the variances computed in the last chapter. Being equipped with the state of the lowest possible purity and finding the average of the purity already close to this absolute minimum, one concludes that the most probable state in AR is very close to the state of minimum purity.

This rather complex line of argument has been replaced here by a direct investigation of the distance between the average state in the AR and all actual reduced states. Thus, in this present straightforward approach the purity itself does not play such an outstanding role any longer. However, we would like to show the direct equivalence of the old and the new approach by a rather short discussion of the purity and its connection to Hilbert space variances (HVs).

The HA of the purity of the system can be computed according to

$$
\begin{aligned}
[\![P]\!] &= [\![\mathrm{Tr}\left\{\rho^2\right\}]\!] \\
&= \left[\!\!\left[ \sum_{AC}\sum_{ac} (\hat{\rho}^{Aa,Cc})^2 \right]\!\!\right] \\
&= \sum_{AC}\sum_{ac} [\![(\hat{\rho}^{Aa,Cc})^2]\!] \\
&= \sum_{AC}\sum_{ac}\left( \Delta_{\mathrm{H}}^2(\langle\hat{\rho}^{Aa,Cc}\rangle) + [\![\hat{\rho}^{Aa,Cc}]\!]^2 \right).
\end{aligned}
\tag{11.24}
$$

Both quantities, the Hilbert space average and the Hilbert space variance, have already been computed in Chap. 10. Using these results, we find for the average purity

$$
[\![P]\!] = \sum_{AC}\sum_{B} \frac{W_{AB}W_{CB}}{N_B} + \sum_{A}\sum_{BD} \frac{W_{AB}W_{AD}}{N_A} \\
- \sum_{AB} W_{AB}^2 \frac{N_A + N_B}{N_A N_B(N_A N_B + 1)},
\tag{11.25}
$$

where we have skipped the rather lengthy calculations here. However, this average of the purity is equivalent to a former result obtained, e.g., in the first edition of this book.

For initial product states, i.e., $W_{AB} = W_A W_B$, realizing that $\sum_A W_A = \sum_B W_B = 1$, since $W_A$ and $W_B$ are probabilities, we find

$$\llbracket P \rrbracket = \sum_B \frac{W_B^2}{N_B} + \sum_A \frac{W_A^2}{N_A} - \sum_{AB} W_A^2 W_B^2 \frac{N_A + N_B}{N_A N_B (N_A N_B + 1)} . \qquad (11.26)$$

In case of large $N_B \gg 1$ we may approximate

$$N_A N_B + 1 \approx N_A N_B , \qquad (11.27)$$

which yields

$$\llbracket P \rrbracket \approx \sum_B \frac{W_B^2}{N_B} + \sum_A \frac{W_A^2}{N_A} - \sum_{AB} W_A^2 W_B^2 \frac{N_A + N_B}{N_A^2 N_B^2} . \qquad (11.28)$$

Since the addends of the last term are all positive and we are interested in an upper bound for the purity we may safely neglect the last term. This implies

$$\llbracket P \rrbracket \leq \sum_A \frac{W_A^2}{N_A} + \sum_B \frac{W_B^2}{N_B} . \qquad (11.29)$$

Since the first part is just the minimal purity of our system and the second part is extremely small for large environments we find the average purity near the minimal one. That means that nearly any state of the AR already features a local reduced state of the system being near to the thermal equilibrium state and has thus a purity near the minimal one, as demanded.

# References

1. A. Einstein, B. Podolsky, N. Rosen, Phys. Rev. **47**, 777 (1935)
2. M. Nielsen, I. Chuang, *Quantum Computation and Quantum Information* (Cambridge University Press, Cambridge, 2000)
3. G. Mahler, V.A. Weberruß, *Quantum Networks*, 2nd edn. (Springer, Berlin, Heidelberg, 1998)
4. I. Bengtsson, K. Zyczkowski, *Geometry of Quantum States: An Introduction to Quantum Entanglement* (Cambridge University Press, 2008)
5. A. Kolovsky, Phys. Rev E **50**, 3565 (1994)
6. P. Pechukas, Phys. Rev. Lett. **73**, 1060 (1994)
7. U. Weiss, *Quantum Dissipative Systems, Series in Modern Condensed Matter Physics*, vol. 10, 2nd edn. (World Scientific, Singapore, New Jersey, London, Hong Kong, 1999)
8. M.O. Scully, M.S. Zubairy, *Quantum Optics* (Cambridge University Press, Cambridge, 1997)
9. K. Blum, *Density Matrix Theory and Applications*, 2nd edn. (Plenum Press, New York, London, 1996)
10. R. Alicki, K. Lendi, *Quantum Dynamical Semigroups and Applications* (Springer, Berlin, 2001)
11. J. Gemmer, *A Quantum Approach to Thermodynamics*. Dissertation, Universität Stuttgart (2003)
12. J. Gemmer, G. Mahler, Eur. Phys. J. B **31**, 249 (2003)
13. J. Gemmer, G. Mahler, Eur. Phys. J. D **17**, 385 (2001)
14. J. Gemmer, G. Mahler, Europhys. Lett. **59**, 159 (2002)

15. J. Gemmer, A. Otte, G. Mahler, Phys. Rev. Lett. **86**, 1927 (2001)
16. G. Mahler, J. Gemmer, A. Otte, in *Between Chance and Choice*, ed. by H. Atmanspacher, R. Bishop (2002), Imprint Academic, p. 279ff
17. G. Mahler, M. Michel, J. Gemmer, Physica E **29**, 53 (2005)

# Chapter 12
# Generic Spectra of Large Systems

*Experiments cannot be extrapolated, only theories.*
— D. J. Raine and E. G. Thomas [1]

**Abstract** Taking a closer look it is not obvious why entropy, as given by its standard microcanonical definition, should be an extensive quantity. It is demonstrated that this may nevertheless be expected if the full system is made up from a multitude of mutually non- or only weakly interacting identical subsystems. In the same limit an exponentially growing density of states within a finite energy range is shown to result. As an example the entropy of an ideal gas is computed on the basis of the presented concepts.

## 12.1 The Extensivity of Entropy

If a set of axioms is formulated as a basis of thermodynamics, one is usually told that entropy has to be an extensive quantity. This basically means that if two identical systems with entropy $S$ are brought in contact such as to form a system of twice the size of the original system, the entropy $S^{\text{tot}}$ of the joint system should double,

$$S^{\text{tot}} = 2S .\tag{12.1}$$

Formulated more rigorously this means that entropy should be a homogeneous function of the first order, or that it should be possible to write it as a function of the other extensive variables, say, energy $U$, volume $V$, and particle number $N$ as

$$S = N \, s(U/N, V/N) ,\tag{12.2}$$

where $s(U/N, V/N)$ is the entropy per particle. This is obviously an important property, since it guarantees, e.g., that temperature defined in the usual way (see (3.17))

$$T = \frac{\partial U}{\partial S}\tag{12.3}$$

remains the same under this procedure, i.e., temperature is an intensive quantity.

Gemmer, J. et al.: *Generic Spectra of Large Systems*. Lect. Notes Phys. **784**, 129–138 (2009)
DOI 10.1007/978-3-540-70510-9_12

However, this basic requirement faces severe problems for the standard definition of entropy. The classical definition of entropy for the microcanonical ensemble (see Sect. 3.3.2) reads

$$S = k_B \ln m \approx k_B \ln G(U) \,, \tag{12.4}$$

where $m$ denotes the number of microstates consistent with the energy $U$, i.e., the volume of the corresponding energy shell in phase space divided by the volume of some elementary cell, and $G(U)$ the state density. In our approach the same formula holds (for a sharp energy probability distribution) for the equilibrium entropy (see (11.23)), except $G(U)$ being the quantum mechanical state density at the energy $U$.

Regardless of whether we are following classical or quantum mechanical ideas, if one assumes that the thermal contact of two identical systems, while containing only negligible energy by itself, allows for energy exchange between the systems, the entropy $S^{tot}$ of the doubled system at the double energy could be calculated from the state density by the convolution

$$S^{tot} = k_B \ln \int_0^{2U} G(E) G(2U - E) \, dE \,. \tag{12.5}$$

It is obvious that this, in general, cannot be twice the entropy of one of the separate systems, for

$$k_B \ln \int_0^{2U} G(E) G(2U - E) \, dE \neq 2 k_B \ln G(U) \,. \tag{12.6}$$

This can only be true, if the function $G(E) G(2U - E)$ is extremely peaked at $E = U$. In general, however, there is no reason to assume this, even if $G(E)$ was a rapidly growing function. If $G(E)$ grows exponentially, the integrand of the convolution is flat, rather than peaked. The identity of (12.6) is often claimed in standard textbooks by referring to the ideal gas, for which it happens to be approximately true, or by complicated considerations based on the canonical ensemble [2]. All this, however, is not a straightforward, general extensivity proof for the microcanonical case. So, according to those definitions, one cannot claim without further study that entropy is an extensive quantity. (This problem is not to be confused with Gibbs' paradox that can be solved by using Boltzmann statistics of identical particles; here dividing the left-hand side of (12.6) by some function of $N$ will not fix the problem [3].)

Finally one is often referred to Shannon entropy

$$S^{(\mu)} = -k_B \sum_i W_i^{(\mu)} \ln W_i^{(\mu)} \,, \tag{12.7}$$

which appears to be extensive, since (12.1) holds, if $W_l^{(12)} = W_i^{(1)} W_j^{(2)}$. However, this means that the probabilities of finding the systems in their individual states should be uncorrelated. This is clearly not the case in the microcanonical ensemble.

If one system is found at the energy $E$, the other one necessarily has to be at the energy $U - E$.

It thus remains to be shown, if, and under what condition, $S$ can indeed be a homogeneous function of $U$.

## 12.2 Spectra of Modular Systems

Practically all of the matter we encounter in nature has some sort of modular structure. Gases are made of weakly interacting identical particles. Crystals are periodic structures of, possibly strongly interacting, identical units, even disordered matter, like glass or polymers, and can be split up into fairly small parts without changing the properties of the parts essentially.

Let us, as an example, consider the sequential buildup of some piece of solid material. First, we have one atom with some energy spectrum. If we bring two atoms together, the spectrum of the resulting molecule will be substantially different from the energy spectrum of the two separate atoms. The energy resulting from the binding can be as large as typical-level splitting within the spectrum of the separate atoms. However, the spectrum of the molecule will already be broader than the spectra of the separate atoms. If we now combine two 2-atom molecules to one 4-atom molecule, the spectrum of the 4-atom molecule will again be considerably different from the spectrum of the two separate 2-atom molecules. If we continue this process, at some point, say, if the separate parts contain a hundred atoms or so each, the separate parts will already have broad energy spectra, typically containing bands that stretch over a considerable energy region with a smooth state density. If we now combine these parts again, the energy contained in the binding will be negligible compared to the structures of the energy spectrum of the two separate parts. Most of the atoms in one part do not even feel the force of the atoms in the other part anymore, simply because they are too far away. Thus, the energy distortion of the separate spectra caused by the binding will be negligible. This is the limit beyond which the weak coupling limit applies. This limit is always assumed to hold in thermodynamics. For the contact between a system and its environment it is thus assumed that the spectra of the separate systems are almost undistorted by the contact. So, this principle should apply to the different identical parts of one system above some size. Here we assume that there are a lot of parts above this limit to make up a macroscopic system, as is the case in our example, where there are a lot of parts containing some hundred atoms, to be combined to form, e.g., a piece of metal, containing on the order of $10^{23}$ atoms.

The bottom line is that the spectrum or state density of any macroscopic system can be viewed as the spectrum of a system consisting of very many almost interaction-free parts, even if the basic particles are strongly interacting. In the case of a gas no additional consideration is necessary, for its spectrum can naturally be understood as the combined spectrum of all the individual gas particles.

We finally analyze the properties of spectra that result from very many identical non-interacting systems. Just as the state density of two non-interacting systems should be the convolution of the two individual state densities, the state density of the modular system, $G(U)$, should be the convolution of all individual state densities, $g(E)$. Defining

$$C_N\{g(E)\}(U) := \big(g(E) * g(E) * \cdots * g(E)\big)(U) \tag{12.8}$$

as the convolution of $N$ identical functions $g(E)$, where the convolution labeled by "$*$" is mathematically defined by the integration

$$C_N\{g(E)\}(U) := \int \cdots \int g(E_1)g(E_2 - E_1) \cdots g(E_i - E_{i-1})$$
$$\cdots g(U - E_{N-1}) \prod_{j=1}^{N-1} dE_j , \tag{12.9}$$

we can thus write

$$G(U) = C_N\{g(E)\}(U) . \tag{12.10}$$

To evaluate this convolution, we start by considering another convolution. We define

$$r(E) := \frac{e^{-\alpha E} g(E)}{\int e^{-\alpha E} g(E) dE}, \tag{12.11}$$

and the quantities

$$R := \int e^{-\alpha E} g(E) \, dE , \tag{12.12}$$

$$\bar{r} := \int E \, r(E) \, dE , \tag{12.13}$$

$$\sigma^2 := \int E^2 \, r(E) \, dE - \bar{r}^2 . \tag{12.14}$$

If the increase of $g(E)$ with energy is not faster than exponential, which we have to assume here, then all these quantities are finite and, since $r(E)$ is normalized, $\bar{r}$ is the mean value of $r(E)$ and $\sigma^2$ is the variance of $r(E)$. Now, consider the convolution of all $r(E)$ written as

$$C_N\{r(E)\}(U) = \frac{e^{-\alpha U} G(U)}{R^N} . \tag{12.15}$$

To evaluate $C_N\{r(E)\}(U)$ we exploit the properties of a convolution. Since the integral over a convolution equals the product of the integrals of the convoluted functions, we have

$$\int \mathcal{C}_N\{r(E)\}(U)\,\mathrm{d}U = 1 \; . \tag{12.16}$$

Since the mean value of a convolution of normalized functions is the sum of the mean values of the convoluted functions, we find

$$M := \int U\mathcal{C}_N\{r(E)\}(U)\,\mathrm{d}U = N\bar{r} \; . \tag{12.17}$$

As the square of the variance of a convolution of normalized functions is the sum of the squares of the convoluted functions, we finally get

$$\Sigma^2 := \int U^2 \mathcal{C}_N\{r(E)\}(U)\,\mathrm{d}U - N^2\,\bar{r}^2 = N\,\sigma^2 \; . \tag{12.18}$$

The Fourier transform of two convoluted functions equals the product of the Fourier transforms of the convoluted functions. If for simplicity we define the Fourier transform of a function $r(E)$ as $\mathcal{F}\{r(E)\}$, we thus find

$$\mathcal{F}\{\mathcal{C}_N\{r(E)\}\} = \left(\mathcal{F}\{r(E)\}\right)^N \; . \tag{12.19}$$

If $r(E)$ is integrable, $\mathcal{F}\{r(E)\}$ is integrable as well and it is very likely that the function $\mathcal{F}\{r(E)\}$ has a single global maximum somewhere. This maximum should become much more predominant, if the function is multiplied very many times with itself, regardless of how strongly peaked the maximum originally was. This means that the function $\mathcal{F}\{\mathcal{C}_N\{r(E)\}\}$ should get extremely peaked at some point, if $N$ becomes large enough. It may be shown (see Appendix D) that this peak, containing almost all of the area under the curve, is approximately Gaussian. This statement is essentially also implied by the central limit theorem. One can now split $\mathcal{F}\{\mathcal{C}_N\{r(E)\}\}$ up into two parts, the Gaussian and the rest. Since a Fourier transform is additive leaving the area under the square of the curve invariant, and transforming a Gaussian into a Gaussian, $\mathcal{C}_N\{r(E)\}$ should again mainly consist of a Gaussian and a small part that cannot be determined, but gets smaller and smaller as $N$ gets larger. In the region in which the Gaussian is peaked, $\mathcal{F}\{\mathcal{C}_N\{r(E)\}\}$ should be almost entirely dominated by the Gaussian part. At the edges, where the Gaussian vanishes, the small remainder may dominate. If we assume that the integral, the mean value, and the variance of $\mathcal{F}\{\mathcal{C}_N\{r(E)\}\}$ are entirely dominated by its Gaussian part, we can, using (12.16), (12.17), and (12.18), give a good approximation for $\mathcal{C}_N\{r(E)\}$ that should be valid at the peak, i.e., around $U = N\bar{r}$

$$\mathcal{C}_N\{r(E)\}(U) \approx \frac{1}{\sqrt{2\pi N\sigma^2}} \exp\left(-\frac{(U - N\bar{r})^2}{2N\sigma^2}\right) \; . \tag{12.20}$$

Solving (12.15) for $G(U)$ and inserting (12.20), evaluated at the peak, we thus find

$$G(N\bar{r}) \approx \frac{R^N e^{\alpha N \bar{r}}}{\sqrt{2\pi N \sigma^2}} \,, \tag{12.21}$$

where $\bar{r}$, $R$, and $\sigma$ are all functions of $\alpha$. Thus, we have expressed $G$ as a function of $\alpha$. Since we want $G$ as a function of the internal energy $U$, we define

$$U := N\bar{r}(\alpha) \quad \text{or} \quad \frac{U}{N} = \bar{r}(\alpha) \,. \tag{12.22}$$

Solving formally for $\alpha$ we get

$$\alpha = \alpha(U/N) = \bar{r}^{-1}(U/N) \,. \tag{12.23}$$

Now $R, \sigma$, and $\alpha$ are all functions of the argument $(U/N)$ and we can rewrite (12.21) as

$$G(U) \approx \frac{\left(R(U/N)\right)^N e^{\alpha(U/N)U}}{\sqrt{2\pi N}\sigma(U/N)} \,, \tag{12.24}$$

or, by taking the logarithm

$$\ln G(U) \approx N\left(\ln R(U/N) + \frac{U}{N}\alpha(U/N)\right) - \frac{1}{2}\ln(2\pi N) - \ln \sigma(U/N) \,. \tag{12.25}$$

If we keep $U/N$ fixed, but let $N \gg 1$, which amounts to a simple upscaling of the system, we can neglect everything except for the first part on the right-hand side of (12.25) to get

$$\ln G(U) \approx N\left(\ln R(U/N) + \frac{U}{N}\alpha(U/N)\right) \,. \tag{12.26}$$

This is obviously a homogeneous function of the first order and thus an extensive quantity. Therefore, (12.2) is finally confirmed.

The joint spectrum of a few non- or weakly interacting systems does not give rise to an extensive entropy, contrary to the standard definition of entropy; but the spectrum of very many such subsystems always does, regardless of the form of the spectrum of the individual subsystem of which the joint system is made.

## 12.3 Entropy of an Ideal Gas

To check (12.26) we consider a classical ideal gas, just taking the spectrum of a free particle in one dimension as the function to be convoluted. The total energy of a classical gas depends on $3N$ degrees of freedom, corresponding to the components of the momenta of all $N$ particles. From the dispersion relation of a classical free particle confined to one dimension

$$E = \frac{1}{2m} p^2 , \qquad (12.27)$$

where $m$ is the mass of a single particle, we find

$$\frac{dp}{dE} = \frac{m}{p} = \sqrt{\frac{m}{2E}} . \qquad (12.28)$$

Since there are two momenta corresponding to one energy and taking $h$ as the volume of an elementary cell, we get for a particle restricted to the length $L$ the state density

$$g(E) = \frac{L}{h} \sqrt{\frac{2m}{E}} . \qquad (12.29)$$

With this state density we find, using some standard table of integrals, for the quantities defined in Sect. 12.2

$$R = \frac{L}{h} \sqrt{\frac{2\pi m}{\alpha}} , \quad \bar{r} = \frac{1}{2\alpha} . \qquad (12.30)$$

Setting $\bar{r} = \frac{U}{N'}$ and writing $\alpha$ and $R$ as functions of this argument we get

$$\alpha = \frac{1}{2} \left( \frac{U}{N'} \right)^{-1} , \quad R = \frac{L}{h} \left( 4m\pi \left( \frac{U}{N'} \right) \right)^{\frac{1}{2}} . \qquad (12.31)$$

Inserting these results into (12.26) yields

$$\ln G(U) = N' \left( \ln \frac{L}{h} + \frac{1}{2} \ln \left( 4m\pi \left( \frac{U}{N'} \right) \right) + \frac{1}{2} \right) . \qquad (12.32)$$

Relating the number of degrees of freedom $N'$ to the number of particles $N$ by $N' = 3N$ we eventually find

$$\ln G(U) = N \left( 3 \ln \frac{L}{h} + \frac{3}{2} \ln \left( \frac{4}{3} m\pi \left( \frac{U}{N} \right) \right) + \frac{3}{2} \right) . \qquad (12.33)$$

This is exactly the standard textbook result (without the corrected Boltzmann statistics, see, e.g., [3]), which is usually calculated by evaluating the surface area of hyperspheres and using the Stirling formula.

## 12.4 Environmental Spectra and Boltzmann Distribution

In Sect. 10.4 it has been claimed that the environmental state density may more or less routinely be assumed to be of an exponential form. In the following we are going to justify this statement under the assumption that the spectrum of the environment has the standard structure of large modular systems as established in Sect. 12.2. Then it should be possible to write the logarithm of its state density according to (12.2) as

$$\ln G^{E}(E^{E}) = N s^{E}(E^{E}/N) \, , \tag{12.34}$$

where $N$ is now the number of some basic units of the environment. If one looks at the graph of such a homogeneous function for different $N$, it is clearly seen that increasing $N$ just amounts to an upscaling of the whole picture. This means that the graph becomes smoother and smoother within finite energy intervals (see Fig. 12.1).

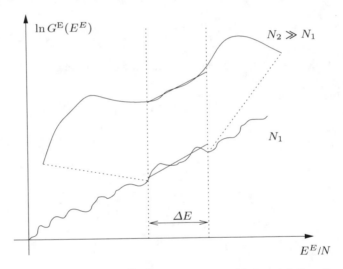

**Fig. 12.1** Upscaling of the graph $\ln G^{E}(E^{E})$ with increasing $N$; the original section within $\Delta E$ gets stretched. With respect to the same $\Delta E$ the new graph gets closer to a linear approximation (*straight line*)

This can be stated in a more mathematical form by checking the expansion of $\ln G^{E}(E^{E})$ around some point of fixed energy per unit, $E_0/N = \epsilon$

$$\ln G^{E}(E^{E}) \approx N s^{E}\big|_{\epsilon} + \frac{ds^{E}}{dE^{E}}\bigg|_{\epsilon} (E^{E} - E_0) + \frac{1}{2N} \frac{d^2 s^{E}}{d(E^{E})^2}\bigg|_{\epsilon} (E^{E} - E_0)^2 + O(\epsilon^3) \, . \tag{12.35}$$

Evidently, already the second-order term scales with $N^{-1}$, terms of order $n$ scale with $N^{1-n}$. Therefore, for large $N$, a truncation of the expansion after the linear term will be a valid approximation over a wide energy range, with the range of validity becoming larger with increasing $N$. Without $\ln G^E(E^E)$ being a homogeneous function of the first order such a truncation would remain without justification, although it is often routinely used [2].

Now, regarding (10.28), which has been the basis for the considerations in Sect. 10.4, the $N_B$ appearing there should be identified with the state density $G^E(E^E)$ as discussed here. If now the integrated weight of $W_E/N_E$ is primarily concentrated within the, as argued above, rather large region around $E_0$ in which a linearization of $G^E(E^E)$ applies (of course $\epsilon$ has to be chosen such as to possibly satisfy this), then $G^E(E^E)$ may be safely approximated by an expression of the form

$$G^E(E^E) \propto e^{\alpha E^E}, \quad \alpha \equiv \left.\frac{ds^E}{dE^E}\right|_{\epsilon} . \tag{12.36}$$

Such a form appears in (10.33), this type of environmental spectra gives rise to a Boltzmann distribution as explained in Sect. 10.4.

## 12.5 Beyond the Boltzmann Distribution?

In the last years the standard limits of thermodynamics have been challenged by exploiting the laws of quantum mechanics [4–6]. It should be pointed out here that within the framework of the ideas presented here, the Boltzmann distribution does not follow naturally from some basic principles like it does from the maximum entropy principle in the context of Jaynes' principle. Rather, it is due to the special structure of the spectra of the systems that represent the environment. If a system is in contact with a system, which is not built according to the scheme described in Sect. 12.2, it can have a stable equilibrium energy probability distribution that significantly differs from the Boltzmann distribution. In fact, any distribution described by (10.28) must be considered stable, as long as the state density of the container system is large enough. Thus, if one could build a system with a high state density, but not of modular origin, one could get a non-standard equilibrium distribution. However, realizing such a system is probably very hard, it would either have to be impossible to split up into identical parts or, alternatively, the parts would have to interact strongly over large distances. Furthermore, one would have to decouple this system entirely from any further system, including the electromagnetic field. Although all this seems rather unrealistic, such effects might be seen in some future experiments.

## References

1. E.G. Raine, D. J. Thomas, *An Introduction to the Science of Cosmology* (IOP Publishing, London, 2001)
2. W. Brenig, *Statistische Theorie der Wärme*, vol. 1, 4th edn. (Springer, Berlin, 1996)

3. M. Toda, R. Kubo, N. Saito, *Statistical Physics I: Equilibrium Statistical Mechanics*, 2nd edn. No. 30 in Solid-State Sciences (Springer, Berlin, Heidelberg, New York, 1978)
4. A.E. Allahverdyan, T.M. Nieuwenhuizen, Phys. Rev. Lett. **85**, 232 (2000)
5. A.E. Allahverdyan, T.M. Nieuwenhuizen, Phys. Rev. Lett. **85**, 1799 (2000)
6. M.O. Scully, Phys. Rev. Lett. **88**, 050602 (2002)

# Chapter 13
# Temperature

> *All concepts ... have a definite and limited applicability*
> *... Such a case is that of temperature, defined as the mean*
> *kinetical energy of the random linear motion of the component*
> *particles of a many-particle system in thermal equilibrium.*
> *This notion is difficult to apply if there are too few particles in*
> *the system, or if the temperature is so low that thermal*
> *equilibrium takes a long time to establish itself, or if the*
> *temperature is so high that the nature of particles changes*
> *with small changes of the temperature.*
>
> — Th. Brody [1]

**Abstract** According to Chap. 5 a certain behavior of a quantity which may meaningfully be termed "temperature" is required, at least in equilibrium. Here a definition of temperature based on the quantum microstate is given. On the basis of this definition the above-mentioned behavior is demonstrated for various standard processes and setups.

If it is hard to define entropy as a function of the microstate on the basis of classical mechanics, it is even harder to do so for the temperature. One could claim that temperature should only be defined for equilibrium and thus there is no need to define it as a function of the microstate. Based on this reasoning temperature would then simply be defined as

$$\frac{1}{k_B T} = \frac{\partial S}{\partial E} = \frac{\partial}{\partial E} \ln G(E) = \frac{1}{G(E)} \frac{\partial G(E)}{\partial E} , \qquad (13.1)$$

with $G(E)$ being the state density cf. (3.48). In this way one would neglect all dynamical aspects (see [2]), since this definition is based on the Hamiltonian of the system rather than on its state. Strictly speaking, this definition would exclude all situations in which temperature appears as a function of time or space, because those are non-equilibrium situations. To circumvent this restriction it would, at least, be convenient to be able to express temperature as a function of the microstate. There have been several attempts in this direction.

As already explained in Chap. 5, a quantity like temperature is essentially determined by two properties. It should take on the same value for two systems in energy

Gemmer, J. et al.: *Temperature*. Lect. Notes Phys. **784**, 139–148 (2009)
DOI 10.1007/978-3-540-70510-9_13      © Springer-Verlag Berlin Heidelberg 2009

exchanging contact, and if the energy of a system is changed without changing its volume, it should be a measure for the energy change per entropy change.

Most definitions rely on the second property. Maxwell connected the mean kinetic energy of a classical particle with temperature. In the canonical ensemble (Boltzmann distribution) it is guaranteed that the energy change per entropy change equals temperature (cf. Sect. 13.3.1). And the ensemble mean of the kinetic energy of a particle equals $k_B T$ in this case. Thus, if ergodicity is assumed, i.e., if the time average equals the ensemble average, temperature may indeed be defined as the time- averaged kinetic energy. Similar approaches have been proposed on the basis of the microcanonical ensemble [3, 4]. However, temperature is eventually not really given by an observable, but by a time average over an observable, leaving open the question of the averaging time and thus the question on what minimum timescale temperature may be defined. Furthermore, the definition is entirely based on ergodicity. Nevertheless, it allows, at least to some extent, for an investigation of processes, in which temperature varies in time and/or space, since that definition is not necessarily restricted to full equilibrium.

To avoid those problems of standard temperature definitions, we want to present yet another, entirely quantum mechanical definition here.

## 13.1   Definition of Spectral Temperature

We define the inverse of spectral temperature as

$$
\frac{1}{k_B T} := - \left( 1 - \frac{W_0 + W_M}{2} \right)^{-1}
$$
$$
\sum_{i=1}^{M} \left( \frac{W_i + W_{i-1}}{2} \right) \frac{\ln W_i - \ln W_{i-1} - (\ln N_i - \ln N_{i-1})}{E_i - E_{i-1}} , \tag{13.2}
$$

where $W_i$ is the probability of finding the quantum system within an energy compartment with mean energy $E_i$, $M$ is the number of the highest energy compartment, while the lowest one corresponds to $E = 0$. The $N_i$ are the corresponding amounts of eigenstates within the compartments. The formula may also be applied to spectra with no upper bound since in this case $W_i$ has to vanish with increasing energy. This formula is motivated by the following consideration: Assume a system features a standard Boltzmann distribution. Then one gets for the occupation probabilities

$$
W_i \propto N_i \exp(-\beta E_i) . \tag{13.3}
$$

Plugging this into (13.2) yields $1/k_B T = \beta$, i.e., the above-defined spectral temperature is in accord with the standard notion of a temperature in the canonical ensemble. However, the spectral temperature is always defined, even for states far from any kind of equilibrium ensemble. Since it is a function of the energy occupation probabilities only, it cannot change in time for an isolated system. In the following we are going to examine the properties and implication of this spectral temperature in some detail.

If the spectrum of a system is very dense and if it is possible to describe the energy probability distribution, $\{W_i\}$, as well as the "state densities," $\{N_i\}$, by smooth continuous functions $(W(E), N(E))$ with well-defined derivatives, (13.2) may be approximated by

$$\frac{1}{k_B T} \approx - \int_0^{E_{max}} W(E) \left( \frac{d}{dE} \ln W(E) - \frac{d}{dE} \ln N(E) \right) dE \ . \tag{13.4}$$

This can further be simplified by integrating the first term to yield

$$\frac{1}{k_B T} \approx W(0) - W(E_{max}) + \int_0^{E_{max}} W(E) \frac{d}{dE} \ln N(E)\, dE$$

$$\approx W(0) - W(E_{max}) + \int_0^{E_{max}} \frac{W(E)}{N(E)} \frac{dN(E)}{dE}\, dE \ . \tag{13.5}$$

Since for larger systems typically neither the lowest nor the highest energy level is occupied with considerable probability (if the spectra are finite at all), it is the last term on the right-hand side of (13.5) that basically matters. This term can be interpreted as the average over the standard definition of the inverse temperature in the context of the microcanonical ensemble (13.1). Thus for all systems that feature significant $W(E)$ only in an energy interval within which $\frac{d}{dE} \ln N(E)$ is more or less constant, one again finds that the spectral temperature is in accord with the above notion of a temperature, cf. (13.1). Note that it has been shown in Sect. 12.4 that the respective energy interval can be expected to be rather large for standard modular systems.

## 13.2  The Equality of Spectral Temperatures in Equilibrium

The equality of temperatures in equilibrium is usually shown based on entropy being extensive, i.e., additive for two systems in contact, on entropy approaching a maximum in equilibrium, and on the standard definition of temperature as given by (13.1). If we were exclusively dealing with large modular systems as described in Chap. 12, we could also introduce the equality this way, exploiting the corresponding properties derived so far. In the following, however, it will be demonstrated that the concept of equal equilibrium temperatures holds for even more general situations if based on spectral temperatures. To do so we analyze spectral temperatures of subsystems in energy exchanging contact. Due to the reasoning given in Sect. 10.3 we always base our calculations on the typical energy probability distributions (10.28) or corresponding formulations for the environment.

If two systems are in heat contact at the total energy $E = E^S + E^E$, we expect the energy probability distribution of some considered system to be given by the typical distribution as given by (10.28). Due to probability distributions now being possibly smooth functions we, as already mentioned, change the notation used in (10.28) in the following way: $W_A \rightarrow W^S(E^S)$ where now $E^S$ of course has to correspond to the energy at the former "grain" $A$. For the environmental quantities and the state densities, we change the notation in an analogous way. Doing so we may rewrite the typical distribution for the system from (10.28) as

$$W^S(E^S) = N^S(E^S) \sum_E \frac{N^E(E - E^S)W(E)}{N(E)} . \tag{13.6}$$

Since the considerations leading to (10.28) are completely symmetric with respect to an environment–system exchange, the typical energy probability distribution for the environment may be simply found from exchanging the indices

$$W^E(E^E) = N^E(E^E) \sum_E \frac{N^S(E - E^E)W(E)}{N(E)} . \tag{13.7}$$

We check now if and under what circumstances those typical energy probability distributions yield the same temperature according to the definition (13.2) or (13.5).

First we consider the case of a small discrete system S, coupled to a large, continuous environmental system E. The latter is assumed to have a spectrum which may be routinely expected for large, modular systems as described in Sect. 12.2. We, furthermore, assume $W(E)/N(E)$ to be rather sharply concentrated around some value $E_0$ as done in Sect. 12.4. As argued in Sects. 12.4 and 10.4, the discrete system S then typically features a Boltzmann distribution with the parameter $\alpha$ as given in (12.36). As argued below (13.3) this parameter then equals the spectral temperature. If E is much bigger than S it is intuitively conceivable that almost all energy should be contained in E. This is confirmed by taking a closer look at (13.7): If $W(E)/N(E)$ is peaked at $E_0$ then $W^E(E^E)$ is also peaked closely to $E_0$. But in this case from the reasoning below (13.5) follows that the spectral temperature of E will also equal $\alpha$ as given in (12.36). Hence in this scenario we find the same local temperatures for S and E for almost all pure states from the pertinent AR of the full system.

Now we examine the case of two large systems with continuous spectra in contact. In this case, as will be seen, we do not even need the assumption of the spectra being generic spectra of modular systems. Formulating (13.6) for a continuous spectrum yields

$$W^S(E^S) = N^S(E^S) \int_0^\infty \frac{N^E(E - E^S)W(E)}{N(E)} \, dE . \tag{13.8}$$

Applying (13.5) to this distribution yields

$$\frac{1}{k_B T^S} = W^S(0) - W^S(E_{max})$$

$$+ \int_0^{E_{max}} \int_0^\infty \frac{N^E(E - E^S) W(E)}{N(E)} \frac{dN^S(E^S)}{dE^S} dE \, dE^S . \tag{13.9}$$

The smallest energy value for which $N^E(E^E)$ takes on non-zero values at all is $E^E = 0$. Thus we can, after reversing the order of integrations, replace $E_{max}$ as a boundary by $E$. Furthermore, we assume both the probability densities to find the system in the ground state $W^S(0)$ and at the highest possible energy (if there is one), $W^S(E_{max})$, to vanish. We can then rewrite (13.9) as

$$\frac{1}{k_B T^S} = \int_0^\infty \frac{W(E)}{N(E)} \int_0^E \frac{dN^S(E^S)}{dE^S} N^E(E - E^S) \, dE^S \, dE , \tag{13.10}$$

and apply product integration to the inner integral to find

$$\frac{1}{k_B T^S} = \int_0^\infty \frac{W(E)}{N(E)} \left( N^S(E) N^E(0) - N^S(0) N^E(E) \right.$$

$$\left. - \int_0^E N^S(E^S) \frac{dN^E(E - E^S)}{dE^S} \, dE^S \right) dE . \tag{13.11}$$

Since state densities are supposed to vanish at zero energy, we get $N^S(0) = N^E(0) = 0$. Substituting $E - E^S = E^E$ and reversing the boundaries of the integration yields

$$\frac{1}{k_B T^S} = \int_0^\infty \frac{W(E)}{N(E)} \int_0^E \frac{dN^E(E^E)}{dE^E} N^S(E - E^E) \, dE^E \, dE . \tag{13.12}$$

One would have obtained exactly this result, if one had applied (13.5) to the container system. This may be seen from a comparison with (13.10), obviously only the subsystem indices are reversed.

If two large systems with continuous spectra are in heat contact, almost all microstates from the AR yield the same local spectral temperatures for the subsystems, regardless of whether the spectra feature the generic form for modular systems or how broad the energy probability distribution of the full system is.

## 13.3 Spectral Temperature as the Derivative of Energy with Respect to Entropy

As already explained, we do expect the temperature not only to take on the same values for systems in contact but also to be a measure for the energy change per entropy change, if all other extensive variables are kept fixed, since this is basically what the Gibbsian fundamental form states. Evidently, there are situations in which the temperature as defined by (13.2) will not show this behavior. If, e.g., one considered an isolated system controlled by a time-dependent Hamiltonian, one would find that energy may very well change while entropy is strictly conserved. Nevertheless, one could compute a finite temperature for this system, which would obviously not be in agreement with the temperature appearing in the first law. However, this is probably not the situation one has in mind, when trying to apply the Gibbsian fundamental form. Here we want to distinguish two processes, for which the first law should be applicable. First, we investigate the process of transferring energy into an arbitrarily small system by bringing it into contact with, according to our definition, a hotter environment and, second, the case of slowly depositing energy into a large system by any kind of procedure (running current through it, stirring it, etc.).

### 13.3.1 Contact with a Hotter System

In this case we consider a discrete system in equilibrium, the entropy of which is given by

$$S = -k_B \sum_i W_i \ln W_i \quad \text{with} \quad W_i = \frac{W(E_i)}{N(E_i)} , \qquad (13.13)$$

where $W_i$ is now the probability of finding the system in one of the $N(E_i)$ energy eigenstates of the energy compartment $E_i$, not the probability of finding the system somewhere at the energy $E_i$. The internal energy of the system is now given by

$$U = \sum_i W_i E_i . \qquad (13.14)$$

The energy probability distribution of the system in contact with a larger system reads, according to (12.36)

$$W_i = \frac{\exp\left(-\frac{E_i}{k_B T}\right)}{\sum_j \exp\left(-\frac{E_j}{k_B T}\right)} , \qquad (13.15)$$

where $T$ is the temperature for the surrounding system as well as for the system considered. If the surrounding area gets hotter, $T$ increases and $S$ and $U$ change. Thus we compute

$$\frac{\partial U}{\partial S} = \frac{\frac{\partial U}{\partial T}}{\frac{\partial S}{\partial T}} . \qquad (13.16)$$

For the derivative in the numerator we get

$$\frac{\partial U}{\partial T} = \sum_i \frac{\partial W_i}{\partial T} E_i \ .$$ (13.17)

Computing the derivate in the denominator yields

$$\frac{\partial S}{\partial T} = -k_B \sum_i \left( \frac{\partial W_i}{\partial T} \ln W_i + \frac{\partial W_i}{\partial T} \right) \ .$$ (13.18)

Because the order for the summation and the derivative can be exchanged on the right-hand side of (13.18) and as $\sum_i W_i = 1$, the last term vanishes. Together with (13.15) we thus get

$$\frac{\partial S}{\partial T} = -k_B \sum_i \frac{\partial W_i}{\partial T} \left( -\frac{E_i}{k_B T} - \ln \sum_j \exp\left( -\frac{E_j}{k_B T} \right) \right) \ .$$ (13.19)

Since the second term in the large brackets does not carry the index $i$, the same argument as before applies and the term vanishes. We thus find

$$\frac{\partial S}{\partial T} = \frac{1}{T} \sum_i \frac{\partial W_i}{\partial T} E_i \ .$$ (13.20)

Inserting (13.17) and (13.20) into (13.16) eventually yields

$$\frac{\partial U}{\partial S} = T \ ,$$ (13.21)

which means that for this kind of process our temperature exhibits the desired behavior.

### 13.3.2 Energy Deposition

Now we consider a large system in isolating contact with an environment, into which energy is deposited by any kind of process. The internal energy of such a system reads

$$U = \int W(E) E \ dE \ ,$$ (13.22)

where $W(E)$ is again the probability of finding the system at some energy, not in a single energy eigenstate. The von Neumann entropy of such a system in micro-canonical equilibrium is, as may be calculated (10.19),

$$S = -k_B \int W(E) \, \ln \frac{W(E)}{N(E)} \, dE \, . \tag{13.23}$$

According to Sect. 12.2 we can assume that the width of the energy distribution of the system is small enough so that the state density $N(E)$ is well described by some exponential within the region, where $W(E)$ takes on substantial values. As has already been explained, this region can be fairly broad, if the system is large. In this case we can replace

$$N(E) \approx N(U) \, e^{\beta(E-U)} \, . \tag{13.24}$$

Doing so we find

$$S \approx -k_B \int W(E) \Big( \ln W(E) - \ln N(U) + \beta(E - U) \Big) \, dE, \tag{13.25}$$

and after integrating the last two terms

$$S \approx k_B \ln N(U) - k_B \int W(E) \ln W(E) \, dE \, . \tag{13.26}$$

As an instructive example we consider the case of energy probability $W(E)$ being uniformly distributed over an interval of length $\epsilon$. In this case we find from (13.26)

$$S \approx k_B \ln N(U) + k_B \ln \epsilon \, . \tag{13.27}$$

The change of entropy $\delta S$ that arises in such a situation from a change of the mean energy, $\delta U$, and a change of the width of the distribution by a factor $C$ is

$$\delta S \approx k_B \frac{\partial}{\partial U} \ln N(U) \delta U + k_B C \, . \tag{13.28}$$

To get an idea for the orders of magnitude involved we set

$$\frac{\partial}{\partial U} \ln N(U) \delta U =: \frac{1}{k_B T_{emp}} \, , \tag{13.29}$$

where $T_{emp}$ is the empirical temperature as defined in (13.1), yielding

$$\delta S \approx \frac{\delta U}{T_{emp}} + k_B C \, . \tag{13.30}$$

This may become more elucidating by plugging in numbers and dimensions

$$\delta S \approx \frac{\delta U [J]}{T_{emp}[K]} + 1.38 \times 10^{-23} [J/K] C \, . \tag{13.31}$$

From this equation it is obvious that for state changes involving macroscopic energy changes $\delta U$ at reasonable temperatures the second term, corresponding to the change of the width of the energy probability distribution, becomes negligible, unless the width is increased by a factor of $C > 10^{15}$ or so. Such a change of the width, however, seems implausible from what we know about, say, mechanical energy depositing processes, even if they do not proceed as described by adiabatic following (see Chap. 14). A very similar picture will result for non-uniform energy probability distributions. Thus it is safe to drop the second term on the right-hand side of (13.26), so that

$$S \approx k_B \ln N(U) . \tag{13.32}$$

Thus we are eventually able to calculate the entropy change per energy change for generic processes:

$$\frac{\partial S}{\partial U} = k_B \frac{\partial}{\partial U} \ln N(U) . \tag{13.33}$$

This result has now to be compared with the spectral temperature for this situation. With the definition of the inverse spectral temperature (13.5) we obtain

$$\frac{1}{T} = k_B \int W(E) \frac{\mathrm{d}}{\mathrm{d}E} \ln N(E) \, \mathrm{d}E, \tag{13.34}$$

or, consistently assuming the same situation as above (exponential growth of state density) and approximating the logarithm of the state density around the internal energy $U$,

$$\frac{1}{T} = k_B \int W(E) \frac{\mathrm{d}}{\mathrm{d}E} \bigg( \ln N(E)\big|_U$$
$$+ \frac{\partial}{\partial E} \ln N(E)\bigg|_U (E - U) + O(E^2) \bigg) \mathrm{d}E . \tag{13.35}$$

The first term is constant and therefore the derivative vanishes, leading us to

$$\frac{1}{T} = k_B \frac{\partial}{\partial E} \ln N(E)\bigg|_U \int W(E) \frac{\mathrm{d}}{\mathrm{d}E} (E - U) \, \mathrm{d}E . \tag{13.36}$$

After integration we find

$$\frac{1}{T} = k_B \frac{\partial}{\partial E} \ln N(E)\bigg|_U = k_B \frac{\partial}{\partial U} \ln N(U) , \tag{13.37}$$

which is evidently the same as the entropy change per energy change as given by (13.33). Thus, we finally conclude that the temperature according to our definition features the properties needed to guarantee agreement with the Gibbsian fundamental form.

# References

1. T. Brody, *The Philosophy behind Physics* (Springer, Berlin, Heidelberg, 1993)
2. W. Weidlich, *Thermodynamik und statistische Mechanik* (Akademische Verlagsgesellschaft, Wiesbaden, 1976)
3. G. Rickayzen, J.G. Powles, J. Chem. Phys. **114**, 4333 (2001)
4. H.H. Rugh, Phys. Rev. Lett. **78**, 772 (1997)

# Chapter 14
# Pressure and Adiabatic Processes

> *... the laws of macroscopic bodies are quite different from those of mechanics or electromagnetic theory. They do not afford a complete microscopic description of a system. They provide certain macroscopic observable quantities, such as pressure or temperature. These represent averages over microscopic properties.*
>
> — F. Mandl [1]

**Abstract** According to Chap. 5 a quantity, which may meaningfully be termed "pressure," is required to exist, at least in equilibrium. This variable relies on a process, in which some extensive quantity changes while entropy remains invariant (adiabatic processes). Here a definition of pressure based on the quantum microstate is given. On the basis of this definition the above-mentioned behavior is demonstrated for various standard processes. It is shown that the adiabatic process may even be stabilized by the influence of the environment.

It appears as if one could introduce pressure within classical statistical mechanics as an observable, i.e., as a function of the microstate. The momentary change of the momenta of all particles that occurs due to the interaction with some wall has to equal the force exerted onto that wall and could thus be interpreted as pressure. And indeed, there are simple models of ideal gases which can account for some of their properties in this way [2, 3]. In general, however, this is not the way pressure is calculated within statistical mechanics. No ensemble average over such a "pressure observable" is taken. Instead one calculates the internal energy $U$ as a function of entropy $S$ and volume $V$. The derivative of the internal energy with respect to volume, while keeping entropy constant, is then identified with negative pressure (cf. (3.18))

$$\left(\frac{\partial U}{\partial V}\right)_{S=\text{const.}} := -P \; . \tag{14.1}$$

This amounts to identifying the pertinent force with the change of energy per change of length, which appears quite convincing, but the claim is that the change appears in such a way that entropy does not change. The internal energy of the system could, in principle, change in many ways but it is assumed that a process is selected that

Gemmer, J. et al.: *Pressure and Adiabatic Processes.* Lect. Notes Phys. **784**, 149–155 (2009)
DOI 10.1007/978-3-540-70510-9_14　　　　　　　　　 © Springer-Verlag Berlin Heidelberg 2009

keeps entropy constant. Without this assumption the above definition (14.1) would be meaningless.

In this way pressure is defined by an infinitesimal step of an adiabatic process. It has to be examined if, and under what conditions, adiabatic processes occur at all. In the case of temperature it is rather obvious that processes exist during which entropy changes while the volume is kept constant; in this case, however, it is far from obvious that processes exist during which the volume changes while entropy remains constant.

## 14.1 On the Concept of Adiabatic Processes

At first sight, isentropic processes may appear almost trivial: If the influence of the environment on the system under consideration, S, would be described by means of a time-dependent change of some parameter $a(t)$ entering the Hamiltonian of the system S, i.e., if the environment could be reduced to a changing "effective potential," a classical control by $\hat{H}^S(a(t))$ would result. Irrespective of $a(t)$, the von Neumann entropy of S would necessarily remain constant.

However, in the context of the present theory, such a reduction is considered "unphysical." The environment, regardless of whether or not it gives rise to a changing Hamiltonian for the considered system, will always become correlated with the considered system, thus causing the local entropy of the latter to increase (see Sect. 10.2.2). To understand the combined effect of an "adiabatic process inducing" environment onto the system, we divide the continuous evolution into steps alternating between two different mechanisms: during one step type the effect of the environment is modeled only by the changing parameter in the local Hamiltonian, $a(t)$, and during the other only by the inevitable relaxation into the microcanonical equilibrium as described in Chap. 6 and Sect. 10.2.2. Letting the step duration go to zero should result in the true combined effect. Since the relaxation to microcanonical equilibrium makes the off-diagonal elements (in energy representation) of the local density operator, $\hat{\rho}^S$, vanish, the remaining entropy is controlled by the energy occupation probabilities. Thus, if those change during the "parameter changing steps," entropy changes inevitably as well under the full evolution. Therefore, adiabatic processes are not trivial at all in a true physical process. The invariance of entropy, however, can be guaranteed if the occupation numbers do not change during the parameter changing steps. (They will obviously not be changed during the "relaxation steps," for we assume microcanonical conditions.) In quantum mechanics such a behavior can be found within the scheme of adiabatic following.

Under the conditions of adiabatic following not only the entropy, but all occupation numbers of states remain constant. Similar to the classical picture, for adiabatic following to work, the speed of change must be low enough. This is shortly explained in the following.

The adiabatic approximation (see [4–6] and for the classical version remember Sect. 4.5) is a method of solving the time-dependent Schrödinger equation with a time-dependent Hamiltonian. If a Hamiltonian contains a parameter $a(t)$ like length or volume that varies in time, it will have the following form:

$$\hat{H}(a(t)) = \sum_i E_i(a(t)) |i, a(t)\rangle\langle i, a(t)| = \sum_i E_i(t) |i, t\rangle\langle i, t| . \tag{14.2}$$

At each time $t$ a momentary Hamiltonian with a momentary set of eigenvectors and eigenvalues is defined. If the wave function is expanded in terms of this momentary basis with an adequate phase factor, i.e., with the definition

$$\psi_i := \langle i, t|\psi\rangle \exp\left(\frac{1}{i\hbar}\int_0^t E_i dt'\right), \tag{14.3}$$

the time-dependent Schrödinger equation can be transformed to the form

$$\frac{\partial \psi_j}{\partial t} = -\sum_i \psi_i \langle j, t|\left(\frac{\partial}{\partial t}|i, t\rangle\right) \exp\left(\frac{1}{i\hbar}\int_0^t (E_i - E_j) dt'\right) . \tag{14.4}$$

The bracket term on the right-hand side of (14.4) scales with the velocity of the parameter change, $da(t)/dt$; this term gets multiplied by a rotating phase factor that rotates faster the larger the energy distance $E_i - E_j$. This means that if the initial state is a momentary eigenstate of the Hamiltonian $|\psi(0)\rangle = |i, 0\rangle$, the transition rate to other eigenstates will be extremely small if the velocity of the parameter change is low, and it will fall off like $(E_i - E_j)^{-1}$ for transitions to eigenstates that are energetically further away. Thus in this case of slow parameter change we have as an approximate solution

$$|\psi(t)\rangle \approx |i, t\rangle . \tag{14.5}$$

Obviously, for such an evolution entropy is conserved. This is what is called the adiabatic approximation or the adiabatic following. However, it is not easy to decide whether or not the adiabatic approximation applies to a given scenario. For example, in the context of a compression of a macroscopic amount of gas the relevant level spacing is almost arbitrarily small. It is thus not obvious if a given, finite compression rapidity is indeed low enough to justify the adiabatic approximation. Furthermore, we so far entirely neglected the most likely inevitable decohering effect of an environment. To clarify the influence of such a decohering effect on the range of validity of the adiabatic approximation we turn to the following consideration.

To account for environmental decohering effects, the process of a changing local system parameter, like, e.g., volume, should really be described by a Hamiltonian of the following form:

$$\hat{H}(t) = \hat{H}_S(a(t)) + \hat{H}_E + \hat{I}(t) . \tag{14.6}$$

To implement an adiabatic process, one still wants to have a thermally insulating contact with the environment. The full energy of the system S, however, cannot be a strictly conserved quantity anymore, since without a changing energy one cannot get a finite pressure. However, any change of energy is induced by the parameter $a(t)$, thus, if $a(t)$ stopped changing at some time, energy should no longer change either. Demanding this behavior we get as a condition for the interaction $\hat{I}(t)$

$$[\hat{H}_S(t), \hat{I}(t)] = 0 . \tag{14.7}$$

As described in Sect. 10.2.2, the effect of a suitable coupled environment system is to reduce purity within the gas system down to the limit set by the conserved quantities derived from (14.7). This amounts to making the off-diagonal elements of $\hat{\rho}^S$, represented in the basis of the momentary eigenvectors of $\hat{H}_S(t)$, vanish. In order to get a qualitative understanding of the type of evolution that a Hamiltonian like the one defined in (14.6) will typically give rise to, we refer to the same scheme as introduced at the beginning of this section, i.e., we decompose the continuous evolution into two different types of (infinitesimal) time steps. In one type of step we imagine the interaction to be turned off and the system to develop according to its local Hamiltonian $\hat{H}_S(t)$, this evolution being described by the corresponding von Neumann equation. During the other type of step, we imagine the interaction to be turned on, but constant in time as well as the local Hamiltonian. During this period the evolution is described by the Schrödinger equation for the full system and will result in quenching the momentary off-diagonal elements. These two types of steps are now supposed to interchange. In the limit of the steps becoming infinitesimally short, the true, continuous evolution results.

For the first type the von Neumann equation reads

$$i\hbar \frac{\partial \hat{\rho}^S}{\partial t} = [\hat{H}_S(t), \hat{\rho}^S] . \tag{14.8}$$

The probability $W_i$ of the system to be found in a momentary eigenstate $|i, t\rangle$ of $\hat{H}_S(t)$ is

$$W_i = \langle i, t | \hat{\rho}^S(t) | i, t \rangle . \tag{14.9}$$

If those probabilities do not change, the adiabatic approximation holds exactly true. Therefore, we calculate the derivatives with respect to time finding

$$\frac{\partial}{\partial t} W_i = \left( \frac{\partial}{\partial t} \langle i, t | \right) \hat{\rho}^S(t) | i, t \rangle + \langle i, t | \frac{\partial \hat{\rho}^S(t)}{\partial t} | i, t \rangle + \langle i, t | \hat{\rho}^S(t) \left( \frac{\partial}{\partial t} | i, t \rangle \right) . \tag{14.10}$$

Splitting up $\hat{\rho}^S$ into a diagonal part and an off-diagonal part E

$$\hat{\rho}^S =: \sum_i W_i |i, t\rangle \langle i, t| + \mathsf{E}, \tag{14.11}$$

and inserting (14.8) and (14.11) into (14.10) yields

$$
\frac{\partial}{\partial t} W_i = W_i \left( \left( \frac{\partial}{\partial t} \langle i, t| \right) |i, t\rangle + \langle i, t| \left( \frac{\partial}{\partial t} |i, t\rangle \right) \right)
$$
$$
+ \left( \frac{\partial}{\partial t} \langle i, t| \right) \mathsf{E} |i, t\rangle + \langle i, t| \mathsf{E} \left( \frac{\partial}{\partial t} |i, t\rangle \right)
$$
$$
+ \langle i, t| [\hat{H}_\mathrm{S}, \mathsf{E}] |i, t\rangle .
\tag{14.12}
$$

The first part on the right-hand side of (14.12) vanishes since

$$
\left( \left( \frac{\partial}{\partial t} \langle i, t| \right) |i, t\rangle + \langle i, t| \left( \frac{\partial}{\partial t} |i, t\rangle \right) \right) = \frac{\partial}{\partial t} \langle i, t|i, t\rangle = 0 .
\tag{14.13}
$$

Thus (14.12) reduces to

$$
\frac{\partial}{\partial t} W_i = \left( \frac{\partial}{\partial t} \langle i, t| \right) \mathsf{E} |i, t\rangle + \langle i, t| \mathsf{E} \left( \frac{\partial}{\partial t} |i, t\rangle \right) + \langle i, t| [\hat{H}_\mathrm{S}, \mathsf{E}] |i, t\rangle .
\tag{14.14}
$$

Obviously, this derivative vanishes, if E vanishes. This means that if, during the intermediate step in which the interaction is active, the off-diagonal elements were completely suppressed, the rate of change of the probability would vanish at the beginning of each step of the von Neumann equation type. It would take on non-zero values during this step, especially if the step was long and $\hat{\rho}^\mathrm{S}(t)$ changed quickly. If we made the steps shorter, the interaction with the environment might not erase the off-diagonal elements completely. Thus, this situation is controlled by a sort of antagonism. A rapidly changing $\hat{\rho}^\mathrm{S}(t)$ tends to make the adiabatic approximation fail, while the contact with the environment that quickly reduces the off-diagonal elements stabilizes such a behavior. To some extend this stabilization is comparable to the so-called "quantum Zeno effect."

   This principle can also be found from a different consideration. Instead of solving the full Schrödinger equation one can introduce a term into the von Neumann equation of the local system, which models the effect of the environment the way it was found in Sect. 10.2.2. Such an equation reads

$$
i\hbar \frac{\partial \hat{\rho}^\mathrm{S}}{\partial t} = [\hat{H}_\mathrm{S}(t), \hat{\rho}^\mathrm{S}] - i\hbar \sum_{i,i'} \langle i, t| \hat{\rho}^\mathrm{S} |i', t\rangle \, C_{ii'} \, |i, t\rangle \langle i', t|
\tag{14.15}
$$

with

$$
C_{ii'} = 0 \quad \text{for} \quad i = i', \quad C_{ii'} \geq 0 \quad \text{for} \quad i \neq i' .
\tag{14.16}
$$

This equation obviously leaves the diagonal elements invariant and reduces the off-diagonal elements. The bigger the $C_{ii'}$'s, the quicker the reduction. To analyze this equation we define

$$\hat{\rho}^S(t) = \sum_i W_i |i,t\rangle\langle i,t| + r(t)E(t) , \quad \text{Tr}\{E^2\} = 1 , \quad r \geq 0 , \quad (14.17)$$

where the $W_i$'s are time independent and E may now, other than before, also contain diagonal elements. Thus, the second term on the right-hand side of (14.17) now contains all deviations from the adiabatic behavior. With this definition $r$ is a measure for those deviations. Taking the derivative of $\hat{\rho}^S(t)$ according to (14.17) and observing (14.13) one finds

$$\frac{\partial}{\partial t}\hat{\rho}^S(t) = \sum_i W_i \left( \left(\frac{\partial}{\partial t}|i,t\rangle\right)\langle i,t| + |i,t\rangle\left(\frac{\partial}{\partial t}\langle i,t|\right) \right) + \frac{\partial r}{\partial t}E + r\frac{\partial E}{\partial t}$$

$$= \frac{\partial r}{\partial t}E + r\frac{\partial E}{\partial t} . \qquad (14.18)$$

Inserted into (14.15) we get

$$i\hbar\left(\frac{\partial r}{\partial t}E + r\frac{\partial E}{\partial t}\right) = [\hat{H}_S(t), \hat{\rho}^S] - i\hbar \sum_{i,i'} \langle i,t|\hat{\rho}^S|i',t\rangle C_{ii'}|i,t\rangle\langle i',t| . \quad (14.19)$$

Multiplying (14.19) from the left by E, taking the trace and realizing that

$$\text{Tr}\left\{E\frac{\partial}{\partial t}E\right\} = \frac{1}{2}\frac{\partial}{\partial t}\text{Tr}\{E^2\} = 0 , \quad \text{Tr}\{E[\hat{H}_S, rE]\} = 0, \qquad (14.20)$$

one finds, solving finally for $\partial r/\partial t$

$$\frac{\partial r}{\partial t} = -r\sum_{i,i'}|\langle i,t|E|i',t\rangle|^2 C_{ii'}$$

$$- \sum_i W_i \left(\langle i,t|E\left(\frac{\partial}{\partial t}|i,t\rangle\right) + \left(\frac{\partial}{\partial t}\langle i,t|\right)E|i,t\rangle\right) . \qquad (14.21)$$

If the right-hand side of (14.21) consisted exclusively of the first sum, $r$ could only decrease in time; the decrease would be faster the bigger the $C_{ii'}$. Only the second sum can cause $r$ to deviate from its stable value $r = 0$ and this sum would be large if $\hat{H}_S(t)$ changed quickly. Thus, a fast local decoherence should stabilize the adiabatic approximation even for rapidly changing Hamiltonians.

To conclude we can say that within the context of our approach adiabatic, i.e., entropy conserving, processes are very likely to happen, if the decoherence induced by the environment proceeds fast compared with the change of the local Hamiltonian. The evolution of the state will then have the following form:

$$\hat{\rho}^S(t) \approx \sum_i W_i |i(a(t))\rangle\langle i(a(t))| . \qquad (14.22)$$

If we identify the parameter $a$ by the volume $V$, we eventually find for the pressure

$$\frac{\partial U}{\partial V} = \frac{\partial}{\partial V} \text{Tr} \left\{ \hat{\rho}^S \hat{H}_S \right\} = \sum_i W_i \frac{\partial E_i}{\partial V} = -P . \tag{14.23}$$

In this way, pressure, or any other conjugate variable (except for temperature), may be defined, whenever a local Hamiltonian $\hat{H}_S$ can be specified such that the weak coupling limit applies (see Sect. 9.3) and the change of the system proceeds in such a way that, with a changing local Hamiltonian, the whole system remains within the weak coupling limit. If this is guaranteed, pressure can be defined by (14.23), regardless of whether the system is changing or not and regardless of whether the system is thermally insulated or not. The infinitesimal process step is just a virtual one, it is a mathematical construction to define pressure in such a way that it fits the whole scheme of thermodynamics, rather than something that is necessarily physically happening.

# References

1. F. Mandl, *Statistical Physics* (Wiley, London, New York, 1971)
2. H. Paus, *Physik in Experimenten und Beispielen*, 2nd edn. (Hanser, München, Wien, 2002)
3. H. Pick, J.M. Spaeth, H. Paus, *Kursvorlesung Experimentalphysik I* (Universität Stuttgart, 1969)
4. P. Ehrenfest, Ann. Phys. **51**, 327 (1916)
5. L.I. Schiff, *Quantum Mechanics*, 3rd edn. (McGraw-Hill, Duesseldorf, 1968)
6. J. Du, L. Hu, Y. Wang, J. Wu, M. Zhao, D. Suter, Phys. Rev. Lett. **101**, 060403 (2008)

# Chapter 15
# Quantum Mechanical and Classical State Densities

*Any real measurement involves some kind of coarse-grained average which will eventually obscure the quantum effects, and it is this average that obeys classical mechanics.*
— L. E. Ballentine [1]

**Abstract** According to standard statistical approach to thermodynamics (cf. Sect. 3.3) all equilibrium properties of a system may be inferred from, e.g., the entropy given as a function of all extensive variables. The "entropy function," however, depends on the density of states. Thermodynamic properties of systems which are fundamentally quantum are often well described by computations based on classical models. Thus the question arises whether quantum mechanical and classical state densities are generally similar for given systems (that have a classical counterpart). In this section we analyze this question in some detail.

Regardless of its foundation or justification Boltzmann's "recipe" to calculate thermodynamic behavior from a classical Hamilton function of a system works extremely well. This recipe essentially consists of his entropy definition and the first and second laws (cf. Sect. 3). Using this recipe, not only the thermodynamic behavior of gases but also thermodynamic properties of much more complicated systems, like liquid crystals, polymers, which are definitely quantum, may (surprisingly) be computed to good precision.

If now, like in a particular approach at hand, a fully quantum mechanical entropy definition (von Neumann entropy of the reduced, considered system) is suggested, the question arises whether this other definition produces equally good (if not better) results. Thus one should check whether or not the classical and the quantum mechanical definitions of entropy yield approximately the same numbers for given systems, i.e., if

$$S^{\text{class}} \approx S^{\text{qm}} ,\tag{15.1}$$

as already claimed in Sect. 5.2. With the microcanonical equilibrium entropy definitions

Gemmer, J. et al.: *Quantum Mechanical and Classical State Densities.* Lect. Notes Phys. **784**, 157–171 (2009)
DOI 10.1007/978-3-540-70510-9_15
© Springer-Verlag Berlin Heidelberg 2009

$$S^{class} = k_B \ln G^{class}(U, V) , \qquad S^{qm} = k_B \ln G^{qm}(U, V) , \qquad (15.2)$$

it remains to investigate if

$$G^{class}(U, V) \approx G^{qm}(U, V) . \qquad (15.3)$$

Here $G^{class}(U, V)$ is according to Boltzmann the number of classical microstates that is consistent with the macrostate specified by $U, V$; stated more mathematically, the volume of the region in phase space that contains all microstates of the system that feature the energy $U$ and is restricted to the (configuration space) volume $V$. This region is also referred to as the energy shell.

$G^{qm}(U, V)$ is the quantum mechanical density of energy eigenstates at the energy $U$, given that the whole system is contained within the volume $V$. With this definition $S^{qm}$ is the equilibrium entropy we found for the case of microcanonical conditions and sharp energies, as may be inferred from (10.19). If the validity of (15.3) cannot be established, a theory relying on $S^{qm}$ would remain highly problematic from a practical point of view, regardless of its theoretical plausibility.

From an operational point of view, the validity of (15.3) is far from obvious, because both quantities are evidently computed in entirely different ways. And of course, in general, $G^{qm}(U, V)$ is discrete, while $G^{class}(U, V)$ is a smooth continuous function. There are indeed cases where the recipe based on $G^{qm}(U, V)$ works better than the one based on $G^{class}(U, V)$. If one, e.g., changes from an ideal to a molecular gas, the deficiencies of the classical entropy definition, $S^{class}$, become visible at low temperatures; the heat capacity deviates significantly from the predicted behavior. This effect is referred to as the "freezing out of internal degrees of freedom." It is due to the fact that the quantum mechanical level spacing of the spectrum arising from the internal (vibrational, rotational) degrees of freedom is much larger than that arising from the translational degrees of freedom. This behavior is described correctly by calculations based on $S^{qm}$. Nevertheless, if one claims $S^{qm}$ to be the "correct" definition, the striking success of the classical entropy definition needs explanation. This can only be done by showing the validity of (15.3) for a reasonably large class of cases.

For some simple systems, for which both types of spectra can be calculated exactly, there is a striking similarity between $G^{class}(U, V)$ and $G^{qm}(U, V)$; for a free particle they are the same. If the free particle is restricted to some volume, $G^{qm}(U, V)$ becomes discrete, but as long as $V$ is large, the level spacing is small and if the energy interval is chosen to contain many levels – it may still be extremely small compared to macroscopic energies – $G^{class}(U, V)$ and $G^{qm}(U, V)$ are still almost identical. This, eventually, is the reason why both methods lead to almost identical thermodynamic state functions for an ideal gas. A very similar situation is found for the harmonic oscillator. The quantum energy spectrum $G^{qm}(U)$ of the harmonic oscillator consists of an infinite number of equidistant energy levels. The volume of the classical energy shell, $G^{class}(U)$, of a harmonic oscillator is constant with respect to energy. Thus, if the level spacing is small, as is the case for small frequencies, the number of levels that is contained within a given interval is almost independent of the energy $U$ around which the interval is centered.

In the following we want to analyze whether $G^{\text{class}}(U, V)$ and $G^{\text{qm}}(U, V)$ can be viewed as approximations for each other, at least for a large set of cases.

## 15.1 Bohr–Sommerfeld Quantization

One hint toward a possible solution in that direction comes from the Bohr–Sommerfeld quantization [2, 3]. This theory from the early days of quantum mechanics states that energy eigenstates correspond to closed trajectories in classical phase space that enclose areas of the size $jh$

$$\oint p \, dq = jh , \tag{15.4}$$

$j$ being an integer. This integration over a region in phase space could be transformed into an integral over the classical state density with respect to energy

$$\int_0^{E_j} G^{\text{class}}(E) \, dE = jh , \tag{15.5}$$

with $E_j$ denoting the respective energy level. If this theory is right, the desired connection is established and the quantum mechanical spectrum can be calculated from $G^{\text{class}}(E)$ by (15.5), integrating only up to an energy level $E_j$.

A simple example for which the Bohr–Sommerfeld quantization produces good results is the harmonic oscillator in one dimension. Possible trajectories are ellipses in phase space (see Fig. 15.1). The classical phase space volume, the area of the ellipse enclosed by the trajectory, is according to (15.4)

$$\oint p \, dq = \pi \sqrt{2mU} \sqrt{\frac{2U}{m\omega^2}} = \frac{U}{\nu} , \tag{15.6}$$

where $\nu = \omega/2\pi$ is the frequency of the oscillation. From standard quantum mechanics we know that $E_j = (j + \frac{1}{2})h\nu$. Applying (15.5) yields $U_j = jh\nu$ and is thus almost precisely correct.

Unfortunately, the Sommerfeld theory is not always applicable and the above formula holds true for some special cases only.

## 15.2 Partition Function Approach

Some more evidence for the similarity of $G^{\text{class}}$ and $G^{\text{qm}}$ can be obtained from a consideration which is usually done in the context of the partition function [4]. The partition function which, within standard classical mechanics, completely determines the thermodynamic properties of a system reads for the quantum mechanical case

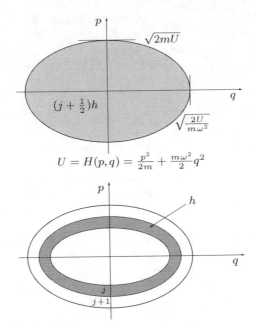

**Fig. 15.1** Bohr–Sommerfeld quantization: phase space of a one-dimensional harmonic oscillator. The elliptic trajectory $j$ includes a volume of $(j + \frac{1}{2})h$. Between two trajectories a volume of $h$ is enclosed

$$Z^{\mathrm{qm}} = \mathrm{Tr}\left\{ \exp\left( -\frac{\hat{H}}{k_B T} \right) \right\} = \int G^{\mathrm{qm}}(E) \exp\left( -\frac{E}{k_B T} \right) \mathrm{d}E, \qquad (15.7)$$

and for the classical case

$$Z^{\mathrm{class}} = \iint \exp\left( -\frac{H(\mathbf{q}, \mathbf{p})}{k_B T} \right) \prod_\mu \mathrm{d}\mathbf{q}\,\mathrm{d}\mathbf{p}$$

$$= \int G^{\mathrm{class}}(E) \exp\left( -\frac{E}{k_B T} \right) \mathrm{d}E . \qquad (15.8)$$

(If one sets $1/k_B T = \alpha$, the partition function becomes equal to the function $R(\alpha)$, which is crucial for the spectrum of large modular systems as described in Sect. 12.2.) In the literature [4] one finds

$$Z^{\mathrm{qm}}\left( \beta := \frac{1}{k_B T} \right) \approx$$

$$Z^{\mathrm{class}}(\beta) - \frac{\hbar \beta^3}{24m} \iint \exp(\beta H(\mathbf{q}, \mathbf{p})) \sum_i \left( \frac{\partial H}{\partial q_i} \right)^2 \prod_\mu \mathrm{d}\mathbf{q}\,\mathrm{d}\mathbf{p} , \qquad (15.9)$$

where the correction term is basically the leading order term of an expansion in terms of powers of $\hbar$, but higher order terms will also involve higher orders of $\beta$, the gradient of the Hamiltonian, and inverse mass $1/m$.

If $Z^{qm}(\beta)$ and $Z^{class}(\beta)$ were exactly the same, $G^{class}$ and $G^{qm}$ would have to be equal as well, since by taking derivatives with respect to $\beta$ of the partition function, all moments of $e^{-\beta E}G(E)$ can be produced and if all moments of the two functions are the same, the two functions have to be the same. This, however, cannot be the case since one knows that $G^{qm}$ is discrete while $G^{class}$ is not. Thus, strictly speaking, the correction terms can never really vanish. Nevertheless (15.9) already provides a strong indication that for a large class of systems (the class of systems for which the correction term is small) at least the "rough structure" of $Z^{qm}(\beta)$ and $Z^{class}(\beta)$ could be the same.

Unfortunately, the smallness of the correction term does not provide a necessary criterion for the equality of $G^{class}$ and $G^{qm}$. If one thinks, e.g., of a wide potential well with a sawtooth-shaped bottom, it is obvious that if one makes each tooth smaller and smaller (but keeps the width the same by introducing more and more teeth), the spectrum should approach the spectrum of a flat bottom potential well for which the similarity of $G^{class}$ and $G^{qm}$ can be shown explicitly. The correction term for the sawtooth bottom potential well, however, does not decrease with the teeth getting smaller, it might indeed be arbitrarily big if the edges of the teeth are arbitrarily steep. Thus, there are systems for which the expansion in (15.9) does not even converge, even though $G^{class}$ and $G^{qm}$ of those systems may be very similar.

## 15.3 Minimum Uncertainty Wave Package Approach

In order to avoid the insufficiencies of the above analysis, we present here yet another treatment, which might help to clarify the relation between $G^{class}$ and $G^{qm}$.

The basic idea is the following. Rather than analyzing the spectrum of the Hamiltonian directly, one can analyze the spectrum of a totally mixed state ($\hat{1}$-state) subject to this Hamiltonian. Since a system in the totally mixed state occupies every state with the same probability, it can be found in a certain energy interval with a probability proportional to the number of energy eigenstates within this interval. If the $\hat{1}$-state is given as an incoherent mixture of many contributions, its spectrum will result as the sum of the individual spectra of the contributions. Here, the $\hat{1}$-state will be given as a mixture of minimum momentum–position uncertainty wave packages, thus each of them corresponds to a point in classical phase space. If it is then possible to show that only those wave packages contribute to $G^{qm}(U)$, which correspond to points in phase space that feature the classical energy $U$, i.e., if the energy spread of those packages is small, a connection between $G^{class}$ and $G^{qm}$ can be established.

Before we set up this complicated approximation scheme in full detail for arbitrary systems, we consider, again, as an instructive example, the one-dimensional harmonic oscillator. In classical phase space the energy shells are ellipses as shown in Fig. 15.1. To each point (volume element) within this energy shell, a quantum mechanical state of minimum position–momentum uncertainty may be assigned. For the harmonic oscillator these states are known as "coherent" or "Glauber states" (see Fig. 15.2). These states are known to have an energy probability distribution centered around the energies of their classical counterparts. Furthermore, the widths of these distributions decrease, relative to their mean energies, with increasing mean energies. Thus, each quantum state corresponding to a point within the energy shell may add the same "weight" within the same energy interval to the quantum mechanical energy spectrum, see Fig. 15.3. In this case one will, eventually, find as many states in a certain energy interval in the quantum spectrum as there are points in the corresponding classical energy shell. This obviously establishes the similarity between $G^{\text{class}}$ and $G^{\text{qm}}$ that we are looking for. If and to what extent such a scheme yields reasonable results in general, will be investigated in the following.

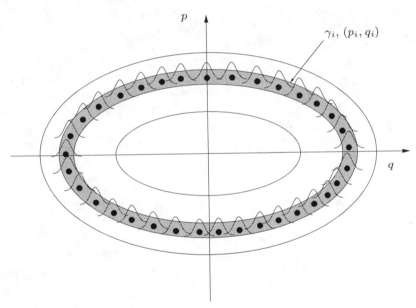

**Fig. 15.2** Minimum uncertainty wave packages: phase space of the harmonic oscillator. Within the loop of volume $h$ a minimum momentum–position uncertainty wave package $\gamma_i$ is defined at every point $(p_i, q_i)$

Now we try to apply these ideas to a more general Hamilton model. We start off by rewriting the quantum mechanical state density $G^{\text{qm}}$. Therefore, we consider the respective Hamiltonian in energy basis which reads

$$\hat{H} = \sum_E E \hat{P}(E) , \qquad (15.10)$$

where $\hat{P}(E)$ is the projector, projecting out the energy eigenspace with energy $E$. The quantum mechanical state density at energy $E$ can then be written as the trace

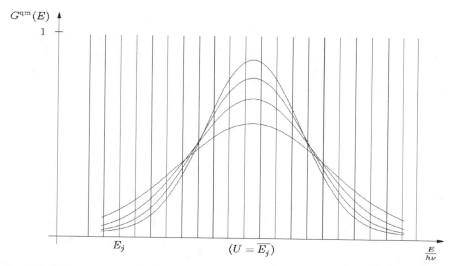

**Fig. 15.3** Minimum uncertainty wave packages in energy space. Only wave packages within the phase space volume of $h$ contribute to the state density $G^{\mathrm{qm}}(E)$ at energy $E_j$. All these packages add energy space to the respective level. Note that in the case of the harmonic oscillator all packages have additionally the same shape

over the projector $\hat{P}(E)$

$$\mathrm{Tr}\left\{\hat{P}(E)\right\} = G^{\mathrm{qm}}(E) \,. \tag{15.11}$$

Using a complete but not necessarily orthogonal basis $|\boldsymbol{\gamma}\rangle$, with

$$\sum_{\gamma}|\boldsymbol{\gamma}\rangle\langle\boldsymbol{\gamma}| = \hat{1}\,, \qquad \langle\boldsymbol{\gamma}|\boldsymbol{\gamma}'\rangle \neq 0\,, \tag{15.12}$$

we find for the quantum mechanical state density, carrying out the trace operation

$$G^{\mathrm{qm}}(E) = \mathrm{Tr}\left\{\hat{P}(E)\right\} = \sum_{\gamma}\langle\boldsymbol{\gamma}|\hat{P}(E)|\boldsymbol{\gamma}\rangle \,. \tag{15.13}$$

Furthermore, defining

$$g(\boldsymbol{\gamma}, E) := \langle\boldsymbol{\gamma}|\hat{P}(E)|\boldsymbol{\gamma}\rangle, \tag{15.14}$$

we get

$$G^{\mathrm{qm}}(E) = \sum_{\gamma} g(\boldsymbol{\gamma}, E) \,. \tag{15.15}$$

According to its definition in (15.14) $g(\boldsymbol{\gamma}, E)$ is the energy spectrum of a single $|\boldsymbol{\gamma}\rangle$ that contributes to the spectrum of the Hamiltonian or, as mentioned before, to the energy spectrum of the $\hat{1}$-operator subject to this Hamiltonian (for a visualization see Fig. 15.4). The full spectrum $G^{qm}(E)$ of the Hamiltonian is the sum of all individual spectra of those contributions $g(\boldsymbol{\gamma}, E)$, as stated in (15.15).

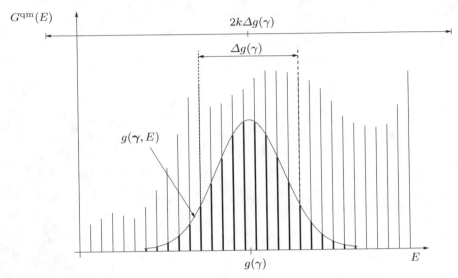

**Fig. 15.4** Minimum uncertainty wave packages: contribution of a single $g(\boldsymbol{\gamma}, E)$ to the spectrum of the Hamiltonian, for an arbitrary quantum mechanical state density

Now we choose the special complete basis as introduced in (15.12), which spans the whole Hilbert space. First we restrict ourselves to a two-dimensional phase space $(q, p)$, an extension to $6N$ phase space coordinates can easily be done later. In position representation this basis is defined as

$$\langle x|\boldsymbol{\gamma}\rangle := \langle x|q, p\rangle := \frac{1}{\sqrt[4]{2\pi}}\sqrt{\frac{\Delta q\, \Delta p}{\hbar \Delta x}} \exp\left(-\frac{(x-q)^2}{4(\Delta x)^2} - i\frac{p}{\hbar}x\right) . \qquad (15.16)$$

Obviously, this basis consists of Gaussian (minimum position–momentum uncertainty) wave packages with variance $\Delta x$, each of which corresponds to a point $\boldsymbol{\gamma} = (q, p)$ in phase space. The wave packages are defined on a lattice in phase space with distance $\Delta q$ and $\Delta p$, respectively (see Fig. 15.5), thus the coordinates $q$ and $p$ are integer multiples of $\Delta q$, $\Delta p$ only. In "standard" quantum mechanics one often tries to create a complete basis consisting of true quantum mechanical states by choosing the lattice grid such that $\Delta q\, \Delta p = h$. Thus one gets exactly one normalized state per "Planck cell" (see [5]). Note that the Gaussian wave packages defined in (15.16) are not normalized, if one does not choose this special subset. We are interested in the case $\Delta q \to 0$ and $\Delta p \to 0$, an "infinitesimal" Gaussian wave package basis. In this limit the norm of the basis states $\langle \boldsymbol{\gamma}|\boldsymbol{\gamma}\rangle$ will also vanish, but

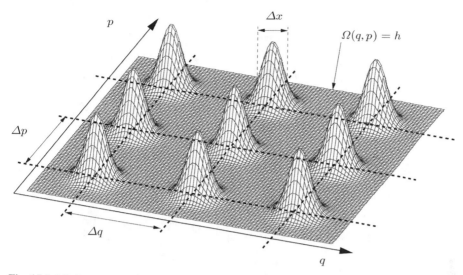

**Fig. 15.5** Minimum uncertainty wave packages defined on a lattice in phase space with distance $\Delta q$ and $\Delta p$, respectively

the basis remains complete. To prove this, we have to show that the following holds true:

$$\langle x|\sum_{\boldsymbol{\gamma}}|\boldsymbol{\gamma}\rangle\langle\boldsymbol{\gamma}||x'\rangle \overset{!}{=} \langle x|\hat{1}|x'\rangle =$$

$$\sum_{\boldsymbol{\gamma}}\langle x|\boldsymbol{\gamma}\rangle\langle\boldsymbol{\gamma}|x'\rangle \overset{!}{=} \delta(x-x'). \tag{15.17}$$

Using the position representation of the basis states (15.16) the left-hand side turns into

$$\sum_{\boldsymbol{\gamma}}\langle x|\boldsymbol{\gamma}\rangle\langle\boldsymbol{\gamma}|x'\rangle$$

$$= \frac{1}{\sqrt{2\pi}\,\hbar\Delta x}\sum_{q,p}\exp\left(-\frac{(x-q)^2+(x'-q)^2}{4(\Delta x)^2}-\frac{\mathrm{i}}{\hbar}\,p\,(x-x')\right)\Delta q\,\Delta p\,. \tag{15.18}$$

Within this equation, we perform the limit $\Delta q \to 0$, $\Delta p \to 0$ and switch from the sum to an integration

$$= \frac{1}{\sqrt{2\pi}\,\hbar\Delta x}\iint \mathrm{d}q\mathrm{d}p\,\exp\left(-\frac{(x-q)^2+(x'-q)^2}{4(\Delta x)^2}-\frac{\mathrm{i}}{\hbar}\,p\,(x-x')\right), \tag{15.19}$$

allowing for infinitesimal contributions. The integration over $p$ yields a $\delta$-function in $(x-x')$ that can be pulled out of the $q$ integration. Since the $\delta$-function is zero everywhere except for $x = x'$, we can set $x = x'$ in the integrand and carry out the integration over the remaining Gaussian, finding

$$= \frac{1}{\sqrt{2\pi}\,\Delta x}\,\delta(x - x')\int dq\,\exp\left(-\frac{(x - q)^2}{2(\Delta x)^2}\right) = \delta(x - x')\,, \qquad (15.20)$$

which obviously shows that the chosen infinitesimal basis is complete. (This scheme is also known from the context of quantum optics, where it is usually called a "P-representation," see [6].)

The normalization of the basis states can be investigated by using again definition (15.16)

$$\langle \gamma | \gamma \rangle = \langle \gamma | \int |x\rangle\langle x|\mathrm{d}x\,|\gamma\rangle$$

$$= \frac{1}{\sqrt{2\pi}}\frac{\Delta q\,\Delta p}{\hbar\Delta x}\int \mathrm{d}x\,\exp\left(-\frac{(x - q)^2}{2(\Delta x)^2}\right) = \frac{\Delta q\,\Delta p}{2\pi\,\hbar} = \frac{\Delta q\,\Delta p}{h}\,. \qquad (15.21)$$

From this equation it can be seen that the $\gamma$-wave packages become normalized if one chooses $\Delta q\,\Delta p = h$. In the case of $\Delta q \to 0$ and $\Delta p \to 0$ the normalization vanishes. If one sums the "weights" of the infinitesimal contributions in one "Planck cell," i.e., if one sums over a volume $\Omega(q, p) = h$ in phase space and does the limit later, one gets

$$\lim_{\Delta q \to 0}\lim_{\Delta p \to 0}\sum_{\substack{\gamma \\ \Omega(q,p)=h}}\langle \gamma | \gamma \rangle = \lim_{\Delta q \to 0}\lim_{\Delta p \to 0}\sum_{q}\sum_{\substack{p \\ \Omega(q,p)=h}}\frac{\Delta q\,\Delta p}{h}$$

$$= \frac{1}{h}\iint\limits_{\Omega(q,p)=h} \mathrm{d}q\,\mathrm{d}p = 1\,. \qquad (15.22)$$

This means that all contributions coming from $\gamma$-wave packages corresponding to a phase space volume $h$ ("Planck cell") will add up to yield the weight one together. (For a more detailed discussion of the basis characteristics of Gaussian wave packages see [7, 8].)

Now we jump back to the considerations about the spectrum of the Hamiltonian and the $\hat{1}$-operator, respectively, to analyze the "weight" of a single $|\gamma\rangle$ wave package to the full spectrum. As mentioned before the contribution of a single wave package corresponds to $g(\gamma, E)$. Since such a wave package might contribute to many energy levels we have to sum over all energies

$$\sum_{E}g(\gamma, E) = \sum_{E}\langle \gamma |\hat{P}(E)|\gamma \rangle = \langle \gamma | \gamma \rangle \quad \text{since} \quad \sum_{E}\hat{P}(E) = \hat{1}\,, \qquad (15.23)$$

and therefore, together with (15.21)

$$\sum_E g(\boldsymbol{\gamma}, E) = \frac{\Delta q \, \Delta p}{h} \, . \tag{15.24}$$

This means that while the spectrum is built up from contributions coming from $\boldsymbol{\gamma}$-wave packages, the contributions from wave packages corresponding to a phase space volume $h$ will add exactly the weight 1 to the spectrum (see (15.22)). If this weight was now concentrated in one single energy level, one could say that the Sommerfeld quantization was exactly right and that a volume $h$ would correspond to exactly one energy eigenstate. This, however, is typically not the case. Nevertheless, (15.24) means that if the contributions of wave packages corresponding to classical states up to a certain energy fall entirely into some energy range of the quantum mechanical spectrum, and no other wave packages lead to contributions within this range, the total number of states within this range has to equal the classical phase space volume up to the corresponding energy divided by $h$. This situation might very well be satisfied to good accuracy, as we will see. This is the reason why in some simple cases like the particle in a box or the harmonic oscillator this simple relation holds exactly true.

It is straightforward to generalize the above-introduced scheme to $N$ particles, or rather $3N$ degrees of freedom. $|\boldsymbol{\gamma}\rangle$ then has to be chosen as a product of the above-described wave packages so that the contributions corresponding to a phase space region of volume $h^{3N}$ will add the weight one to the spectrum.

To decide now whether or not the approximation of the number of quantum mechanical states within the respective energy interval by the volume of the classical phase space divided by $h$ is a reasonable approximation, we analyze (15.15) more thoroughly. Since the exact form of $g(\boldsymbol{\gamma}, E)$ can only be calculated with huge effort, we cannot evaluate this equation exactly. However, some properties of $g(\boldsymbol{\gamma}, E)$ can be estimated with pretty good precision. The mean value, e.g., reads

$$\overline{g(\boldsymbol{\gamma})} := \frac{\sum_E E \, g(\boldsymbol{\gamma}, E)}{\sum_E g(\boldsymbol{\gamma}, E)} = \frac{\langle \boldsymbol{\gamma}|\hat{H}|\boldsymbol{\gamma}\rangle}{\langle \boldsymbol{\gamma}|\boldsymbol{\gamma}\rangle} \approx \frac{p^2}{2m} + \frac{1}{2m}\left(\frac{\hbar}{\Delta x}\right)^2 + V(q) \, , \tag{15.25}$$

where $V$ is the potential energy part of the Hamiltonian and the right-hand side holds if $\Delta x$ is small compared to the curvature of the potential. Also the energy spread, the variance of $g(\boldsymbol{\gamma}, E)$, can roughly be estimated as

$$\Delta g(\boldsymbol{\gamma}) := \frac{\sqrt{\langle \boldsymbol{\gamma}|\hat{H}^2|\boldsymbol{\gamma}\rangle - \langle \boldsymbol{\gamma}|\hat{H}|\boldsymbol{\gamma}\rangle^2}}{\langle \boldsymbol{\gamma}|\boldsymbol{\gamma}\rangle} \approx \frac{p\hbar}{m\Delta x} + \left.\frac{\partial H}{\partial x}\right|_{q,p} \Delta x \, . \tag{15.26}$$

(The second equality, of course, only holds for the structures of $V(q)$ being small compared to $\Delta x$.) We thus find that the mean of $g(\boldsymbol{\gamma}, E)$ will be quite close to the energy of the corresponding classical state $U = H(q, p)$ and that its width will be reasonably small, at least for systems with large mass and small potential gradients. Or, stated in other words, a minimum momentum–position uncertainty wave package has a spectrum concentrated around the energy of its corresponding classical state.

Let us assume that the spectra of those packages have a form, in which almost all of their weight is concentrated within a range of some variances $\Delta g(\gamma)$ from the mean, like for Gaussians or Poissonians (see Fig. 15.4). Summing over the relevant energy interval only, we can approximate (15.24) by

$$\sum_{E=U-k\Delta g(\gamma)}^{E=U+k\Delta g(\gamma)} g(\gamma, E) \approx \frac{\Delta q\,\Delta p}{h}\,, \tag{15.27}$$

where $k$ is some small integer and $U$ the classical energy $H(q, p)$. Around the classical energy $U$ we introduce a typical spread $\Delta g(U)$ of our wave packages in the energy region (e.g., an average over all spreads in the energy region). We can then approximately calculate the number of energy eigenstates within a small energy region around $U$. We sum over an energy interval $E = U - l\Delta g(U)\ldots U + l\Delta g(U)$, where $l$ is an integer, so that we sum over a region of some typical spreads $\Delta g(U)$. According to the connection of $g(\gamma, E)$ and $G^{qm}(U)$ in (15.15) it is possible to write

$$\sum_{E=U-l\Delta g(U)}^{E=U+l\Delta g(U)} G^{qm}(E) = \sum_{E=U-l\Delta g(U)}^{E=U+l\Delta g(U)} \sum_{\gamma} g(\gamma, E)\,. \tag{15.28}$$

Now we want to rearrange this double sum, to use the approximation (15.27). Because of the dependencies on the different parameters, this is not so easy. In Fig. 15.6 we try to illustrate the situation. The idea is that, if $l$ is large enough, almost all $g(\gamma, E)$ from this energy region fall entirely into that part of the spectrum that is summed over and to which (15.27) applies (area I in Fig. 15.6). No

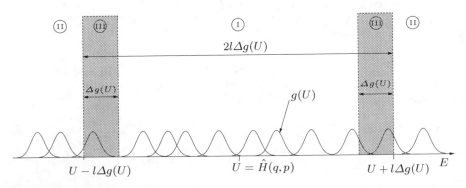

**Fig. 15.6** Summation of wave packages. Large region I: all wave packages fall entirely into that part of the spectrum that is summed over. Region II: all packages are totally outside of the considered energy region. They do not contribute to the sum. Region III: problematic packages are at the edges

packages that are totally outside of the considered energy region contribute to the sum, therefore we can forget about them (area II in Fig. 15.6). Indeed there could be some packages at the edges of the considered region, for which we do not know their contribution to the sum, because they are inside as well as outside of the region (area III in Fig. 15.6). Within this interval of $2k\,\Delta g(\gamma)$ at the edges of the considered energy region, (15.27) can obviously not be applied, therefore we also neglect it. As a reasonable approximation for (15.28) one may accept to use

$$\sum_{E=U-l\Delta g(U)}^{E=U+l\Delta g(U)} \sum_{\gamma} g(\gamma, E) \approx \sum_{\substack{\gamma \\ U-l\Delta g(U)\le H(\gamma)\le U+l\Delta g(U)}} \frac{\Delta q\,\Delta p}{h} \,. \tag{15.29}$$

This is a sum, for which only those $\gamma$'s contribute with $\Delta q\,\Delta p/h$ that are totally inside the considered energy interval around $U$. We need to estimate the error induced by the edges of the region. If the gradient of the classical Hamilton function in (15.26) does not vary much within the energy region (no extremely different spreads of the packages within and at the edges of the considered region), the relative error, $\tilde{\epsilon}$ can be estimated as the ratio of the energy interval at the edges, for which the $g(\gamma, E)$ do not fall entirely into the summed over part of the spectrum to the total summed energy interval

$$\tilde{\epsilon} \approx \frac{2\Delta g(U)}{2l\,\Delta g(U)} = \frac{1}{l} \,. \tag{15.30}$$

Keeping this in mind and with $\Delta q \to 0$ and $\Delta p \to 0$ in (15.29) we get

$$\lim_{\Delta q\to 0}\lim_{\Delta p\to 0} \sum_{\substack{q,p \\ U-l\Delta g(U)\le H(q,p)\le U+l\Delta g(U)}} \frac{\Delta q\,\Delta p}{h}$$

$$= \iint_{U-l\Delta g(U)\le H(q,p)\le U+l\Delta g(U)} \frac{\mathrm{d}q\,\mathrm{d}p}{h} \,, \tag{15.31}$$

where the integral in phase space is restricted to an energy shell with the spread $2l\,\Delta g(U)$. In classical mechanics such an integral over an energy shell in phase space can be transformed into an integral of the classical state density $G^{\mathrm{class}}(E)$ over the energy itself. We conclude that

$$\sum_{E=U-l\Delta g(U)}^{E=U+l\Delta g(U)} G^{\mathrm{qm}}(E) \approx \frac{1}{h} \int_{U-l\Delta g(U)}^{U+l\Delta g(U)} \mathrm{d}E\, G^{\mathrm{class}}(E) \,, \tag{15.32}$$

with a relative error given by (15.30).

We have thus found that the number of states within a certain energy region is equal to the phase space volume corresponding to that energy region, divided by $h$. The energy region that has to be summed over is proportional to $l\Delta g(U)$. The relative error occurring is inversely proportional to $l$. Thus, this method is definitely not adequate to identify individual energy eigenstates with satisfactory precision. However, the number of states within an interval of, say 100 times the energy spread of a minimum uncertainty wave package, can be predicted with pretty good accuracy.

## 15.4  Implications of the Minimum Uncertainty Wave Package Approach

The method introduced in the last section essentially yields an equivalence of classical and quantum mechanical state densities on a certain gross scale: The amounts of states within an energy interval $l\Delta g$ are the same for classical and quantum mechanical descriptions, up to a relative error of size $1/l$ (under the presupposition that a volume element in phase space of scale $h$ contains "one" state). Thus the degree of their similarity is set by the size (or rather smallness) of the energy interval $l\Delta g$. This in turn is small for systems with small potential gradients and big particle masses. However, if the entropy (or rather the partition function) is calculated on the basis of the canonical ensemble, the details of the state density become more and more irrelevant in the limit of high temperatures. It is straightforward to show that, based on such a reasoning, a criterion may be expressed in the following way:

$$l\Delta g \ll k_{\mathrm{B}}T. \tag{15.33}$$

If this condition is fulfilled very similar outcomes of thermodynamic potentials will result from classical and quantum mechanical descriptions. For larger temperatures this condition may apply even for somewhat smaller masses and larger potential gradients. Note that this condition, unlike the one that may be deduced from (15.9), does not necessarily scale directly with the magnitude of the potential gradient.

## 15.5  Correspondence Principle

It is quite comforting to see that the quantum approach to thermodynamics is compatible with classical procedures – in those cases where classical models are available at all. This fact can hardly be overestimated, as there are many branches of physics, in which thermodynamics is routinely applied while a quantum foundation would neither be feasible nor of practical interest. Pertinent examples would include, e.g., macroscopic physical and biophysical systems. The situation is different, e.g., for the concept of chaos: Here the exponential increase of distance between initially nearby states has no direct counterpart in quantum mechanics (where the distance is invariant under unitary transformation). This "failure of the correspondence principle," has led to considerable discussions in the last decades (Joos, e.g., writes "the Schrödinger equation is not the appropriate tool for analyzing the route from quantum to classical chaos" [9]). It is, however, of no relevance for the similarity of thermodynamical potentials as computed from classical or quantum descriptions: If (15.33) holds a classical description applies to the quantum system, irrespective of the absence of exponential instability, chaos, ergodicity, mixing, etc., in quantum dynamics.

# References

1. L.B. Ballentine, *Quantum Mechanics*, 2nd edn. (World Scientific, Singapore, 1998)
2. W. Brenig, *Statistische Theorie der Wärme*, vol. 1, 4th edn. (Springer, Berlin, 1996)
3. M. Toda, R. Kubo, N. Saito, *Statistical Physics I: Equilibrium Statistical Mechanics*, 2nd edn. No. 30 in Solid-State Sciences (Springer, Berlin, Heidelberg, New York, 1978)
4. H. Römer, T. Filk, *Statistische Mechanik* (VCH, Weinheim, New York, Basel, Cambridge, Tokyo, 1994)
5. J.V. Neumann, *Mathematischen Grundlagen der Quantenmechanik*. Die Grundlehren der Mathematischen Wissenschaften (Springer, Berlin, Heidelberg, New York, 1968)
6. M.O. Scully, M.S. Zubairy, *Quantum Optics* (Cambridge University Press, Cambridge, 1997)
7. J.R. Klauder, E.C. Sudarshan, *Fundamentals of Quantum Optics* (Benjamin, New York, 1968)
8. J. Schnack, Eurphys. Lett. **45**, 647 (1999)
9. E. Joos, in *Decoherence and the Appearance of a Classical World in Quantum Theory*, ed. by D. Giulini, E. Joos, C. Kiefer, J. Kupsch, I.O. Stamatescu, H. Zeh, 2nd edn. (Springer, Berlin, Heidelberg, 2003), chap. 3, p. 41

# Chapter 16
# Equilibration in Model Systems

*Statistical mechanics is not the mechanics of large,*
*complicated systems; rather it is the mechanics of limited not*
*completely isolated systems.*

— J. M. Blatt [1]

**Abstract** To illustrate the relaxation principles that have been analyzed in an abstract way in Chaps. 6–10, we now turn to some numerical data, based on a certain type of models. These models are still rather abstract and may thus be viewed as extremely reduced and simplified models for a whole class of concrete physical situations. The models are always taken to consist of (cf. Fig. 9.1) the considered system S, the environment E, and the interaction. Straightforward numerical solutions of the corresponding time-dependent Schrödinger equations already demonstrate local equilibration.

The models addressed in the following are discrete, closed, quantum mechanical Hamilton models, for which a discrete, finite-dimensional Schrödinger equation has to be solved. Local spectra are translated into discrete diagonal matrix Hamiltonians that describe the decoupled bipartite system. As described in Sects. 9.3 and 8.1 the interaction strength sets the width of the local energy subspaces. The occupation probabilities of those subspaces or pertinent sums of them have to be conserved (cf. (9.15), (9.4)) in order to have a well-defined accessible region (AR). Now, the most simple way to get such conserved occupation probabilities is to concentrate all states within such an energy subspace in one degenerate level and have the adjacent subspaces (also formed by degenerate levels) well separated by finite energies. This way the (local or global) subspace occupation probabilities are to a good approximation conserved as long as the elements of the interaction matrix are small compared to the finite energy distances between adjacent levels. Thus, just for simplicity, we construct our models according to this scheme. This way systems which exhibit fully developed typicality are easily numerically accessible on a standard computer.

The form of the interaction depends in principle on the concrete physical subsystems and their interactions. However, since our theory indicates that for the (equilibrium) quantities considered here (entropy, occupation probabilities, etc.) the concrete form of the interaction should not matter, the interaction is taken as some random matrix. Many of the situations analyzed in this chapter are very similar to

Gemmer, J. et al.: *Equilibration in Model Systems*. Lect. Notes Phys. **784**, 173–188 (2009)
DOI 10.1007/978-3-540-70510-9_16      © Springer-Verlag Berlin Heidelberg 2009

those known from the context of quantum master equations. However, note that in order to apply the theories at hand to those systems, neither a Markovian nor a Born assumption has to hold. And indeed the environmental autocorrelation functions of the above-described models do not decay at all (due to the degeneracy). Furthermore we do not necessarily impose factorizing initial conditions.

## 16.1 Microcanonical Entropy

All data in our first example refer to the situation depicted in Fig. 16.1 (cf. [2]). The system under consideration consists of a two-level system ($N_0^S = N_1^S = 1$), while the environment consists of just one energy level with degeneracy $N^E = 50$. This is necessarily a microcanonical situation regardless of the interaction $\hat{I}$. The environment cannot absorb any energy; therefore, energy cannot be exchanged between the subsystems.

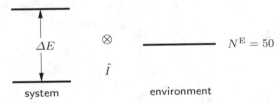

**Fig. 16.1** Microcanonical scenario: a two-level system is weakly coupled to a system with one energy level of degeneracy $N^E = 50$. This is a model for a system in contact with a much larger environment such that no energy can be exchanged

In this situation the probabilities of finding the system in the ground (excited) state are conserved quantities and in this example chosen as $W_0^S = 0.85$ ($W_1^S = 0.15$). The probability to find the environment in its only level is of course $W^E = 1$. Since the assumptions made in Sect. 10.2.2 apply, we may, considering (10.25), expect typicality of the average, i.e., the maximum entropy state if

$$\sum_B \frac{(W_B^E)^2}{N_B^E} = \frac{1}{N^E} = 0.02 \tag{16.1}$$

is small compared to 1, which is obviously the case. Thus, drawing states in accord with the given probabilities at random, we expect most of them to exhibit the maximum local entropy which is in accord with those probabilities.

To examine this expectation, a set of random states, uniformly distributed over the AR, has been generated. Their local entropies have been calculated and sorted into a histogram. Since those states are distributed uniformly over the AR, the number of states in any "entropy bin" reflects the relative size of the respective Hilbert space compartment.

The histogram is shown in Fig. 16.2. The maximum local entropy in this case is $S_{\text{max}}^{\text{S}} = 0.423\,k_{\text{B}}$. Obviously, almost all states have local entropies close to $S_{\text{max}}^{\text{S}}$. Thus compartments corresponding to entropies of, say, $S^{\text{S}} > 0.4\,k_{\text{B}}$ indeed fill almost the entire AR, just as theory predicts. Local pure states ($S^{\text{S}} = 0$) are practically of measure zero.

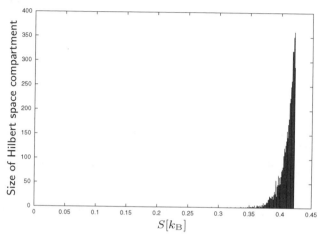

**Fig. 16.2** Relative size of Hilbert space compartments: this histogram shows the relative frequency of states with a given local entropy $S$, among all states from the accessible region. In this case the maximum possible entropy is $S_{\text{max}}^{\text{S}} - 0.423\,k_{\text{B}}$. Obviously, almost all states feature entropies close to the maximum

In order to examine the dynamics, a coupling $\hat{I}$ is introduced. To keep the concrete example as general as possible, $\hat{I}$ has been chosen as a random matrix in the basis of the energy eigenstates of the uncoupled system, with Gaussian-distributed real and imaginary parts of the matrix elements of zero mean and a standard deviation of

$$\Delta I = 0.01\,\Delta E\,. \tag{16.2}$$

This coupling is weak, compared to the Hamiltonian of the uncoupled system. Therefore, the respective interaction cannot contain much energy. The spectrum of the system (see Fig. 16.1) does not change significantly due to the coupling, and after all the environment is not able to absorb energy.

Now the Schrödinger equation for this system, including a realization of the interaction, has been solved for initial states consistent with the specifications given above (16.1). Then the local entropy at each time has been calculated, thus resulting in a picture of the entropy evolution. The result is shown in Fig. 16.3. Obviously the entropy approaches $S_{\text{max}}^{\text{S}}$ within a reasonable time, regardless of the concrete initial state. Thus, the tendency toward equilibrium is obvious. The concrete form of the interaction $\hat{I}$ only influences the details of this evolution, the equilibrium value is always the same. If the interaction is chosen to be weaker, the timescale on which

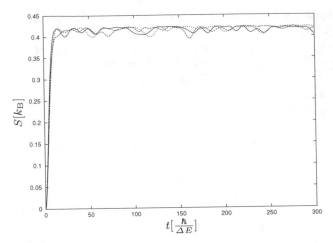

**Fig. 16.3** Evolution of the local entropy for different initial states. A universal state of maximum entropy (equilibrium) is reached, independent of the initial state

equilibrium is reached gets longer, but, eventually, the same maximum entropy will be reached in any case.

## 16.2 Canonical Occupation Probabilities and Entropy

Just like in the last section we present some numerical data (cf. [2]) to support the principles derived in Sects. 10.3 and 10.4. The first model, which has been analyzed numerically to illustrate the above-mentioned principles, is depicted in Fig. 16.4. The considered system S, again, consists only of a two-level system with $E_0 = 0$, $E_1 = 1$. The environment E in this case is a three-level system with an exponential state density: $N_B^E = 50 \cdot 2^B$ (i.e., $\alpha = \ln 2$) with $B = 0, 1, 2$. This degeneracy

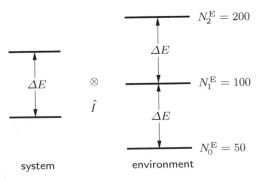

**Fig. 16.4** Canonical scenario: a two-level system is weakly coupled to a three-level environment, such that energy can be exchanged. The exponential degeneracy scheme of the environment guarantees a full independence of the equilibrium state from the initial state

scheme (environmental spectrum) has been chosen for two reasons: (i) If we restrict ourselves to initial states featuring arbitrary states for the system, but environment states that only occupy the intermediate level, theory predicts an equilibrium state of the system, which should be independent of its initial state [see (10.35)]. (ii) A straightforward calculation (which we omit here) shows that the average distance to the average state as calculated from (10.31) is always small compared to 1. Thus typicality and equilibration may be expected. (This also holds for the following examples.)

In this case we find from (10.35) for the typical occupation probabilities:

$$W_0^S = \frac{e^{-\alpha E_0}}{e^{-\alpha E_0} + e^{-\alpha E_1}} = \frac{1}{1 + e^{-\alpha}} = \frac{2}{3},$$

$$W_1^S = 1 - W_0^S = \frac{1}{3}. \tag{16.3}$$

To keep the situation as general as possible, $\hat{I}$ was, like in Sect. 16.1, chosen to be a matrix with random Gaussian-distributed entries in the basis of the eigenstates of the uncoupled system, but now with energy transfer allowed between the subsystems.

For this system the Schrödinger equation has been solved and the evolution of the probability of finding the system in its ground state, $W_0^S$, is displayed in Fig. 16.5. The different curves correspond to different interaction strengths, given by the standard deviation $\Delta I$ of the distribution of the matrix elements of $\hat{I}$ by

$$\Delta I_{\text{solid, dashed}} = 0.0075 \Delta E, \quad \Delta I_{\text{dotted}} = 0.002 \Delta E. \tag{16.4}$$

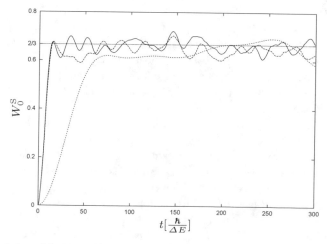

**Fig. 16.5** Evolution of the ground-level occupation probability for three different random interactions. The *dotted line* corresponds to a weaker interaction. Even in this case the same equilibrium value, $W_0^S = \frac{2}{3}$, is approached, only on a longer timescale

Obviously, the equilibrium value of $W_0^S = 2/3$ is reached independently of the concrete interaction $\hat{I}$. Within the weak coupling limit the interaction strength only influences the timescale on which equilibrium is reached.

Figure 16.6 displays the evolution of the same probability, $W_0^S$, but now for different initial states, featuring different probabilities for the ground state, as can be seen in the figure at $t = 0$. The equilibrium value is reached for any such evolution, regardless of the special initial state, thus we confirm the effective attractor behavior which generically occurs in thermodynamics.

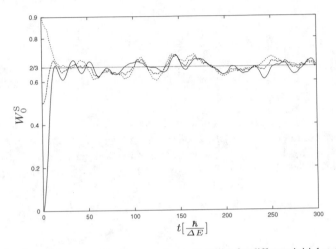

**Fig. 16.6** Evolution of the ground-level occupation probability for different initial states. The theoretically predicted equilibrium value is reached, independent of the initial states, as expected for canonical conditions

Figure 16.7 displays the evolution of the local entropy of the system for the same three initial states as used for Fig. 16.6. The maximum entropy, consistent with the equilibrium value of the energy probabilities, is $S_{max}^S = 0.637 \, k_B$. This is also the value one finds, if one maximizes entropy for fixed mean energy (Jaynes' principle). Obviously, this value is reached for any initial state during the concrete dynamics of this model. This supports the validity of (10.29), which states that the density matrix of the equilibrium state is diagonal in the basis of the local energy eigenstates.

To analyze the formation of a full Boltzmann distribution, we finally investigate the system depicted in Fig. 16.8. Here the system is an equidistant five-level system and the environment a five-level system with degeneracies $N_B^E = 6 \cdot 2^B$ ($B = 0, \ldots, 4$), which should lead to a Boltzmann distribution with $\beta = \alpha = \ln 2$ [cf. (10.35)]. For numerical analysis we restrict ourselves to initial states, where for both subsystems only the intermediate energy level is occupied (symbolized by the black dots in Fig. 16.8). Due to energy conservation other states of the environment would not play any role in this case even if they were present, just like in the previous

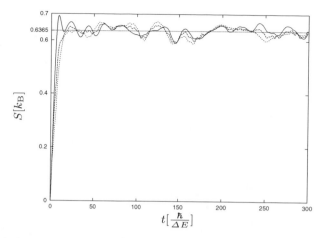

**Fig. 16.7** Evolution of the local entropy for different initial states. $S = 0.637\,k_\mathrm{B}$ is the maximum entropy that is consistent with the equilibrium energy probabilities. This maximum entropy state is reached in all cases

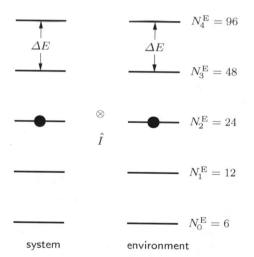

**Fig. 16.8** Canonical multilevel scenario: a five-level system is weakly coupled to a five-level environment with an exponential degeneracy scheme, such that energy may be exchanged. *Black dots* symbolize the initial state. This setup should lead to a Boltzmann distribution

model. Figure 16.9 shows the probabilities $W_A^S$ of the different energy levels to be occupied. While the system starts in the intermediate (third) energy level, soon the predicted Boltzmann distribution develops: Each probability becomes twice as high as the one for the level above.

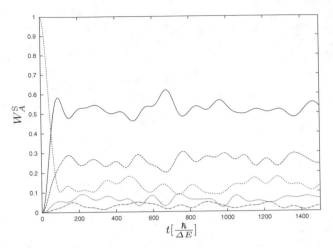

**Fig. 16.9** Evolution of the energy occupation probabilities. After some relaxation time a Boltz-mann distribution is reached. Each probability is twice as high as the one for the next higher energy level, as theory predicts

## 16.3 Probability Fluctuations

Unfortunately the term "fluctuations" has various meanings in the field of physics [3–5]. In the context of thermostatistics one speaks of thermal fluctuations, meaning that the probability distribution with respect to some extensive variable, e.g., energy, might not be exactly sharp for fixed intensive variable, e.g., temperature. Instead one gets an energy probability distribution peaked at some value, having a certain width. This width is taken to characterize "fluctuations," but the distribution itself is constant in time, i.e., does not fluctuate.

In the context of quantum mechanics fluctuations also refer to the width of (quantum mechanical) probability distributions ("uncertainties"). The so-called "vacuum fluctuations" refer to the fact that the probability to measure some finite electro-magnetic field intensity, say, does not vanish even in vacuum (i.e., a pure state). Nevertheless again the probability distribution itself is constant in time.

The fluctuations we want to discuss in this section are of a different kind. In our approach all occupation probabilities are explicit functions of time, which reflects the eternal motion of the pure state of the total system within its accessible region. Since the HV of the probabilities is not zero, probabilities will vary in time as the state vector wanders around in Hilbert space. These fluctuations in time will be studied numerically and compared to the pertinent HV as given in (8.19) concretely.

To those ends a system almost like the one depicted in Fig. 16.4 is analyzed, but now with a degeneracy scheme given by

$$N_B^{\mathrm{E}} = \frac{N_0^{\mathrm{E}}}{2} 2^B . \tag{16.5}$$

The ratios between the degrees of degeneracy of the different reservoir levels are thus the same as for the system sketched in Fig. 16.4, but the overall size of the environment is tunable by $N_0^E$. For various $N_0^E$, the Schrödinger equation has been solved numerically, and the following measure of the fluctuations of the occupation probability of the ground level of the system has been computed as

$$\Delta_t^2 W_0^S := \frac{1}{t_f - t_i} \left( \int_{t_i}^{t_f} \left(W_0^S(t)\right)^2 dt - \left( \int_{t_i}^{t_f} W_0^S(t) dt \right)^2 \right), \qquad (16.6)$$

for initial states with

$$W_0^S(t_i) = 0.2, \quad W_1^S(t_i) = 0.8. \qquad (16.7)$$

Figure 16.10 shows the dependence of the size of these fluctuations on the size of the environment $N_0^E$. The small crosses are the computed data points, the dashed line is a least square fit with a function proportional to $1/N_0^E$. Obviously, this function fits very well, confirming that fluctuations vanish like the square root of the effective system size $N_{eff}^E$. This fitting procedure yields

$$\Delta_t W_0^S = \sqrt{\frac{0.053}{N_0^E}} = \frac{1}{\sqrt{N_{eff}^E}}. \qquad (16.8)$$

Plugging the above numbers into (8.19) yields for the Hilbert space variance (HV) of the ground state occupation probability, or rather its square root

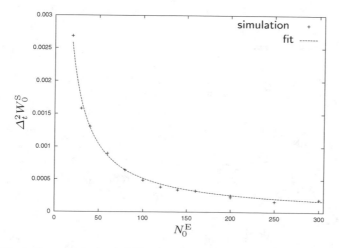

**Fig. 16.10** Fluctuations of the probability for the considered system to be in the ground state $\Delta_t W_0^S$, in dependence on the number of eigenstates of the environment system $N_0^E$ (for otherwise fixed degeneracy structure). Fluctuations decrease with increasing environment

$$\Delta_{\mathrm{H}}(W_0^{\mathrm{S}}) = \sqrt{\frac{0.053}{N_0^{\mathrm{E}}}}. \tag{16.9}$$

The excellent agreement between the concrete temporal fluctuations (16.8) and the HV (8.14) is remarkable. One would have expected such an agreement if the trajectories were ergodic with respect to the full AR which they are not. Nevertheless fluctuations scale here as if they were.

## 16.4 Spin Systems

So far we have illustrated that thermodynamic aspects can be observed in a great variety of bipartite few-level systems, just as theory predicts. The bipartite system consists of a small system, the observed system, and a larger system with some hundreds of energy levels, the environment. We have chosen the coupling between system and environment to be a random interaction, to avoid any bias.

Now we will show that based on theoretical concepts developed in this book, thermodynamic behavior can be found also in another class of systems – a class of modular systems (see Sect. 12.2) with only pair interaction. These special interactions of subsystems considered here are far from being unbiased like a total random interaction would be. We will deal mainly with linear chains, with an identical interaction between each subsystem, e.g., a Heisenberg interaction. Nevertheless, we will find that even in these special models, a thermodynamic behavior can be observed, without any further coupling to an environment. The investigations here are structured in an analysis of global and local properties of such systems. Furthermore, we will observe the local behavior of these chains additionally coupled to an environment.

### 16.4.1 Global Observables

Probably the first example in which dynamical equilibration of some observable in a small quantum system has been demonstrated by simple numerical integration is due to Jensen and Shankar [6]. In 1985 they considered a chain of $N = 7$ subsystems ($n = 2$ levels each), i.e., spins, coupled by a next neighbor interaction. Investigating the Hamiltonian given by

$$\hat{H} = \sum_{\mu=1}^{N} \left( \theta_1 \, \hat{\sigma}_1^{(\mu)} + \theta_2 \, \hat{\sigma}_3^{(\mu)} \right) + \lambda \sum_{\mu=1}^{N} \hat{\sigma}_3^{(\mu)} \, \hat{\sigma}_3^{(\mu+1)}, \tag{16.10}$$

with the Pauli operators $\hat{\sigma}_i^{(\mu)}$ of the spin $\mu$ (see Sect. 2.2.2) and subject to cyclic boundary conditions ($\theta_1$, $\theta_2$, and $\lambda$ are constants), they found hints of thermodynamic behavior in such a modular system. After a numerical integration of the Schrödinger equation for the full system, they found an equilibrium value for some expectation values of global observables of the system. One interesting observable in this context is, e.g., the magnetization in $x$-direction of the whole system

$$\hat{M}_1 = \sum_{\mu=1}^{N} \hat{\sigma}_1^{(\mu)}.$$  (16.11)

They chose to consider pure initial states that occupy a certain energy region (e.g., $E = 4$) The time dependence of the corresponding magnetization is shown in Fig. 16.11. Here one finds for a more or less time-independent magnetization (mean magnetization $\overline{M_1} = 2.46$; solid line in Fig. 16.11). This mean value is independent of the details of the initial state of the system, but note that, of course, the equilibrium value depends on the total initial energy of this state. Concretely, this magnetization turns out to be in accord with the magnetization of a corresponding microcanonical equilibrium state.

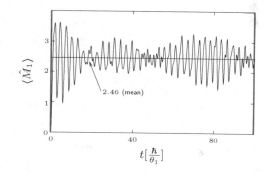

**Fig. 16.11** Equilibrium behavior of the global magnetization of a spin system. Total energy of the initial state $E = 4$. The *horizontal line* marks the average magnetization $\overline{M_1} = 2.46$ ($\theta_1 = 1, \theta_2 = 0.5$, and $\lambda = 0.5$, cf. [6])

According to Sect. 8.3 at least hints of equilibration may be expected (as observed) in this consideration. The magnetization is an observable with a bound spectrum. The dimension of the subspace that defines the AR in this case is given by the number of energy eigenstates within the region from which the initial state is chosen. Since the dimension of the full Hilbert space is in the current example "only" $n^N = 128$, the number of eigenstates corresponding to a fraction of the full spectrum is not extremely large. This is why the example still exhibits rather large equilibrium fluctuations (in the sense of Sect. 16.3). Nevertheless equilibration as discussed in Sect. 8 is illustrated.

## 16.4.2 Local Properties

Furthermore, one can observe the local properties within such a modular system, i.e., the behavior of a single spin in the chain (see Fig. 16.12). Hence, the rest of the system constructs the environment proper. The $N - 1$ uncoupled remaining subsystems together feature $N$ different energy levels, each of which with a binomial distributed degeneracy, as shown in the right-hand side of Fig. 16.12. For a weak coupling everywhere in the chain, the interaction provides for a small lifting of the

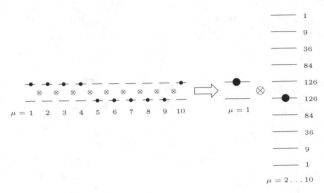

**Fig. 16.12** Spectrum of an $N = 10$ spin system. On the left side, the chain of coupled spins is shown (the *dots* mark the initial product state). The right-hand side shows the complete spectrum of $N = 9$ spins as an environment for the first spin of the chain. (Again, the dots mark the initial state)

degeneracy in the system, but leaves the spectrum qualitatively unchanged. Altogether the system is now in the form required for the theory. However, compared with the randomly chosen interactions, the coupling between the spin of interest and the rest of the system (environment) is a very special coupling, resulting from the structure of the whole system.

Concretely, the spins are coupled by a Heisenberg type of interaction with the total Hamiltonian

$$\hat{H} = \frac{1}{2} \sum_{\mu=1}^{N} \hat{\sigma}_3^{(\mu)} + \lambda \sum_{\mu=1}^{N-1} \sum_{i=1}^{3} \hat{\sigma}_i^{(\mu)} \hat{\sigma}_i^{(\mu+1)} , \tag{16.12}$$

where we choose $\lambda$ to be very small in comparison to the local energy gap $\Delta E = 1$ of each spin ($\lambda = 0.002$).

In the middle of the spectrum we find the largest number of states and, since we need a sufficiently large Hilbert space compartment to gain thermodynamical behavior, we initially start in this subspace (see Fig. 16.12). As an initial state we use a product state of the uncoupled system with total energy $E = 5$. Again, the Schrödinger equation is solved here for a system of $N = 10$ spins. We observe the occupation probability of the excited energy level of the first (see Fig. 16.13a) and the fifth (see Fig. 16.13b) spin of the chain. Both spins develop into the totally mixed state, which refers to an infinite temperature, since the environment consists of two bands with the same amount of levels. This result is in perfect agreement with the theoretical predictions.

It is also possible to find such a local equilibrium behavior for a lower total energy, but due to more fluctuations, because of the smaller Hilbert space compartments used. However, if we could include some hundreds of spins into our consideration, we expect to find a local thermodynamical behavior of a single subsystem with a finite temperature and due to very small fluctuations.

**Fig. 16.13a** Probability of finding the first spin in a chain of 10 spins excited. *Horizontal line* marks the mean value of the probability

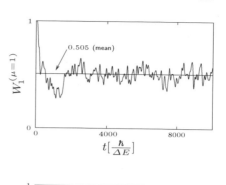

**Fig. 16.13b** Probability of finding the fifth spin in a chain of 10 spins excited. *Horizontal line* marks the mean value of the probability

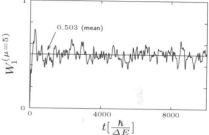

### 16.4.3 Chain Coupled Locally to a Bath

After we have studied several properties of isolated chains let us now investigate the local coupling of an environment to the edge of a modular system. Interesting questions in this context are whether the whole system is still to be found in the theoretically predicted equilibrium state and whether also the single subsystems are in the respective equilibrium state.

In Sect. 4.8 we have introduced a formalism how to model the influence of an environment, assuming an infinite environment, within the theory of open quantum systems. All the problems which appear by comparing a finite bath situation as presented here with an infinite one will be discussed later in Sect. 20. Let us here investigate the full solution of the Schrödinger equation in comparison to this open system approach and show the equivalence of these two approaches.

First, we consider the solution of the full Schrödinger equation of the system, here two spins, coupled to an environment. The two subsystems in the considered system together form a four-level system. For the environment we use a nine-level system with a degeneracy increasing exponentially with energy (Fig. 16.14). This environment is weakly coupled to one spin only. Both interactions, the internal system interaction and the external environment coupling, are chosen randomly. For such an environment theory would predict a canonical equilibrium state on the four-level system, with a temperature $T^{(E)}$ (see Sect. 10.4) for arbitrarily initial states. The temperature of this four-level system should be the spectral temperature imposed by the level structure of the environment.

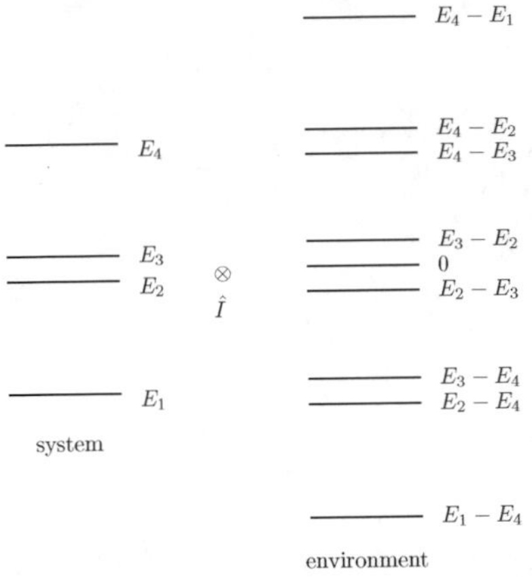

——— $E_4 - E_1$

——— $E_4$          ——— $E_4 - E_2$
                   ——— $E_4 - E_3$

——— $E_3$          ——— $E_3 - E_2$
——— $E_2$  $\otimes$   ——— 0
           $\hat{I}$   ——— $E_2 - E_3$

——— $E_1$          ——— $E_3 - E_4$
                   ——— $E_2 - E_4$

system

——— $E_1 - E_4$

environment

**Fig. 16.14** Modular system built up of two spins weakly coupled to an environment, such that energy can be exchanged. Starting with the environment being in the fifth level an arbitrary initial state of the system is possible

In Fig. 16.15 we show the probability of finding the full four-level system in one of its eigenstates according to the solution of the full Schrödinger equation in the finite bath case. The dynamics in Fig. 16.16 refers to the solution of a quantum master equation for an infinite bath. Obviously, both approaches reach the same equilibrium state, which is in perfect agreement with the predictions of the theory developed above.

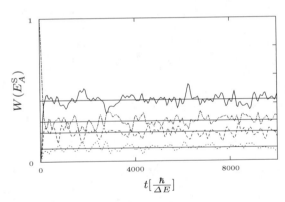

**Fig. 16.15** Probabilities of finding the two-spin system in one of its four eigenstates according to the solution of the full Schrödinger equation. *Solid lines* refer to the theoretically predicted values [7]

Furthermore, we are interested in the local temperatures of the single subsystems in comparison with the global temperature of the four-level system. Only if these two temperatures are equal, temperature would be an intensive and local quantity. As the pertinent temperature we use the spectral temperature of Chap. 13. Since the interaction between the two subsystems does not enter our temperature measure,

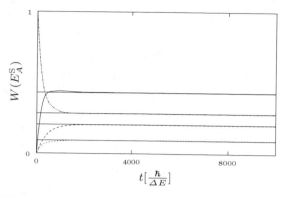

**Fig. 16.16** Probabilities of finding the two-spin system in one of its four eigenstates according to a quantum master equation. *Solid lines* refer to the theoretically predicted values [7]

we expect that we get a good agreement between global and local temperatures for internally weak couplings. Therefore, we investigate the local temperature and its dependence on the internal coupling strength. In Fig. 16.17a we show the result for a full integration of the Schrödinger equation for given internal coupling strength $\lambda$ and in Fig. 16.17b the same investigation due to a quantum master equation, together with the global fixed temperature (*solid* line). As can be seen in both cases, local and global temperatures are approximately the same for a small internal coupling strength $\lambda$, whereas for a larger coupling, global and local temperatures differ.

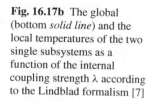

**Fig. 16.17a** The global (bottom *solid line*) and the local temperatures of the two single subsystems as a function of the internal coupling strength $\lambda$ according to the solution of the full Schrödinger equation [7]

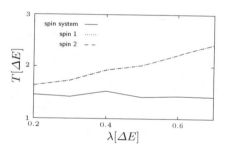

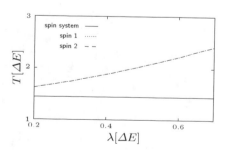

**Fig. 16.17b** The global (bottom *solid line*) and the local temperatures of the two single subsystems as a function of the internal coupling strength $\lambda$ according to the Lindblad formalism [7]

# References

1. J.M. Blatt, Progr. Theor. Phys. **22**(6), 745 (1959)
2. P. Borowski, J. Gemmer, G. Mahler, Eur. Phys. J. B **35**, 255 (2003)
3. A. Münster, *Statistische Thermodynamik* (Springer, Berlin, Heidelberg, 1956)
4. A. Münster, *Prinzipien der statistischen Mechanik, Handbuch der Physik*, vol. III/2 (Springer, Berlin, 1959)
5. A. Münster, *Statistical Thermodynamics*, vol. 2, 1st edn. (Springer, Berlin, New York, Wien, 1974)
6. R. Jensen, R. Shankar, Phys. Rev. Lett. **54**, 1879 (1985)
7. M. Henrich, Dynamische und thermische Eigenschaften abgeschlossener modularer quanten-mechanischer Systeme. Master's thesis, Universität Stuttgart (2004)

# Part III
# Non-equilibrium

# Chapter 17
# Brief Review of Relaxation and Transport Theories

*Professor Prigogine's search for "irreversibility at the microscopic level" can be seen as an attempt to extend thermodynamics all the way down to the level of quantum mechanics ... He was hoping for a fundamental, almost deterministic partial differential equation which could describe thermodynamics at the quantum level.*
— K. Gustafson [1]

**Abstract** In this chapter we will rather briefly mention well-established approaches to relaxation and transport behavior in quantum systems. This is not meant to be a comprehensive review of all approaches concerning those complex fields of theoretical physics, but an overview of concepts and central ideas. To learn more about the discussed topics, the interested reader is referred to the substantial secondary literature which is cited in this chapter.

During the last years thermal relaxation and transport theory have regained a lot of attention. The reasons for this renaissance are of practical as well as of fundamental nature. Novel practical aspects of the theory are often linked to the ongoing miniaturization of all sorts of devices. If one scales down an electronic circuit, for instance, at some point the question arises whether a "nanowire" may still be described by the corresponding macroscopic transport equation (Ohm's law). Molecular electronics, spintronics, and a lot of other concepts that may be subsumed under the term "nanotechnology" rely on the controlled transport of quantities like energy, charge, mass, magnetization, and the understanding of relaxation and loss processes.

The question for relaxation and transport behavior in nanodevices can only be answered by going back to the foundation and to the derivation of general equations of motion. Only from the foundations of a theory its limits of applicability may be inferred. Thus, the ongoing technological progress brings one back to rather fundamental questions. And those are far from being well understood: "Fourier's law, a challenge to theorists" reads the title of a recent paper by Bonetto et al. [2], and in a recent issue of *Nature Physics* the statement "No one has yet managed to derive Fourier's Law truly from fundamental principles" [3] can be found. Therefore, from both practical and theoretical points of view there is room and need to revisit the

Gemmer, J. et al.: *Brief Review of Relaxation and Transport Theories*. Lect. Notes Phys. **784**, 191–199 (2009)
DOI 10.1007/978-3-540-70510-9_17

foundations of relaxation and transport theory, i.e., its derivation from microscopic principles.

## 17.1 Relaxation

Almost everything stated so far referred to situations of fully developed equilibrium. We have described means to predict or guess the value that some variable, e.g., the probability of a system to be found at some energy level, will take on after some, possibly long, relaxation time. However, nothing has been said about how long such a relaxation time will be or what dynamics can be expected at all during relaxation. This is an important aspect of non-equilibrium thermodynamics. The most common scenario for the route to equilibrium is an exponential decay scheme. One typically thinks of merely statistical processes, simply controlled by transition rates rather than of some coherent evolutions. Nevertheless such a behavior should, starting from first principles, eventually follow from the Schrödinger equation. Of course there are many approaches to such a "derivation."

### *17.1.1 Fermi's Golden Rule*

In the literature "Fermi's golden rule" is mentioned frequently to account for exponential decay scenarios, often in the context of spontaneous emission [4]. In order for excited states to decay the considered system must be coupled to a large environment. Together with this environment, the full system is assumed to feature a smooth and rather high state density. Fermi's golden rule (cf. Sect. 2.5.2) yields a transition rate $\gamma$, i.e., the system is found to disappear from its excited state $|i\rangle$ with a probability $P_i$ proportional to the time passed and the probability to have been in the excited state (2.93). If such a behavior can be established at *any* infinitesimal time step during the evolution, an exponential decay results

$$P_i(t + dt) = -\gamma P_i(t) \, dt \quad \Rightarrow \quad P_i(t) = P_i(0) \, e^{-\gamma t} . \tag{17.1}$$

However, if Fermi's golden rule is established on the basis of a time-dependent perturbation theory, there are specific conditions, limiting the time of its applicability. On the one hand the time passed ($dt$) has to be long enough to give rise to a linear growth of $P_i(t + dt)$ with $dt$, on the other hand it has to be short enough to justify a truncation of the Dyson series at first order (2.86). Thus, the true coherent evolution during the full relaxation process can definitely not be reduced to an exponential decay behavior, controlled only by transition rates, without further considerations.

Implicitly it is often assumed that the application of Fermi's golden rule could somehow be iterated. The idea is to describe the respective evolution during the limited time of its applicability and then take the resulting final state for the new initial state. This, however, is not a feasible concept either, as will be explained in the following.

By means of a Dyson series, a short time evolution of the system can be described [see (2.82)]

$$|\psi(t+dt)\rangle = \hat{U}_I(t,dt)|\psi(t)\rangle .$$
(17.2)

Thus, the probability for the system to be found in the energy eigenstate $|i\rangle$ at $t+dt$ is given by

$$
\begin{aligned}
P_i &= |\langle i|\psi(t+dt)\rangle|^2 \\
&= \langle\psi(t)|\hat{U}_I^\dagger(t,dt)|i\rangle\langle i|\hat{U}_I(t,dt)|\psi(t)\rangle \\
&= \sum_{j,k}\langle\psi(t)|j\rangle\langle j|\hat{U}_I^\dagger(t,dt)|i\rangle\langle i|\hat{U}_I(t,dt)|k\rangle\langle k|\psi(t)\rangle .
\end{aligned}
$$
(17.3)

However, only in the case of $j = k$ this reduces to

$$|\langle i|\psi(t+dt)\rangle|^2 = \sum_j |\langle i|\hat{U}_I(t,dt)|j\rangle|^2 |\langle j|\psi(t)\rangle|^2,$$
(17.4)

which essentially is Fermi's golden rule describing the transition probability from state $|j\rangle$ to state $|i\rangle$, just as (2.87) does with

$$W_{ij}(dt) = |\langle i|\hat{U}_I(t,dt)|j\rangle|^2 .$$
(17.5)

Note that only for the special case of $j = k$ is it true that the new probability distribution $|\langle i|\psi(t+dt)\rangle|^2$ depends only on the old probability distribution $|\langle j|\psi(t)\rangle|^2$ and that the transition rate $|\langle i|\hat{U}_I(t,dt)|j\rangle|^2$ does not depend on $t$. In general neither is true, but these properties are needed for an autonomous iteration. Thus, if in the beginning the initial state of the system is a simple energy eigenstate, $j = k$ holds, (17.3) may be replaced by (17.4), and Fermi's golden rule applies. However, if the state of the system is a superposition of energy eigenstates, as will be the case after the first time step, the evolution has to be described by (17.3). In this case, the final probability distribution depends on all details of the initial state as well as on time, and Fermi's golden rule does not apply. Therefore, in general, an iteration of this rule cannot be justified in any simple, straightforward manner.

### 17.1.2 Weisskopf–Wigner Theory

An approach that is based on a (approximate) continuous solution of the Schrödinger equation for all times, rather than on an iteration scheme, is the Weisskopf–Wigner theory. A detailed discussion of this theory is definitely beyond the scope of this text, thus we want to refer the unfamiliar reader to [5], or other standard textbooks. At this point we just want to add a few remarks.

The situation analyzed in this theory is, to some extent, the same as the one in the section above. Whether or not an approximate solution can be found depends strongly on a non-degenerate ground state of the environment and on, in the considered limit, infinitely high state density, combined with an infinitesimally weak coupling. These conditions are met, for example, by an atom coupled to the electromagnetic field in open space, but makes the method difficult to apply to many other situations.

### 17.1.3 Open Quantum Systems

The pertinent ideas, although basically invented to describe the relaxation process rather than the fully developed equilibrium, have already been explained in Sect. 4.8. Here, we do not want to repeat these explanations, but mention again that those theories basically explain how a system will be driven toward equilibrium due to an environment already being in equilibrium. Furthermore, the environment (as already being in equilibrium) is assumed to be non-changing. Neither is necessarily the case for the situations analyzed in the following chapters.

## 17.2 Transport

In this section we concentrate on two central theories of transport behavior within non-equilibrium statistical mechanics. First, on particle and quasi-particle diffusion following from a Boltzmann equation approach (cf. Sect. 4.1) and subsequently on linear response theory and the Kubo formula.

In the following our focus will be primarily on "gradient-driven" transport, i.e., transport that results from a gradient of the spatial density of the transported quantity. This should be contrasted with "force-driven" transport where an external force induces the current. While gradient-driven transport may occur for any transported quantity, it is the only transport mechanism in the context of heat conduction. For this reason heat conduction will be our prime example. Note, however, that a direct proportionality of the conductances occurring in gradient-driven and force-driven transport is the main claim of the famous "Einstein relation" [6].

As will be explained in some detail below, both of the above approaches feature specific peculiarities. While the Kubo formula is derived on the basis of a force-driven transport scenario, which raises the question of its applicability to heat conduction, the dynamics of a quantum system has to be mapped onto a pertinent Boltzmann equation, prior to using the latter as an investigation tool. To those ends one usually has to rely on some sort of quasi-particle concept, such as e.g., phonons.

Let us start with a consideration of a Boltzmann equation as a tool to investigate diffusion and consider the implementation of quasi-particles thereafter.

## 17.2.1 Boltzmann Equation

In 1872 Boltzmann investigated the macroscopic dynamics of dilute gases. For the description of a gas he introduced the $\mu$-space, which is essentially a one-particle phase space. An $N$ particle gas would thus technically be represented by $N$ points in $\mu$-space rather than one point in standard Hamiltonian phase space. However, instead of using $N$ points in $\mu$-space for the description of the gas, Boltzmann introduced a distribution function $\Phi(\mathbf{r}, \mathbf{v}, t)$ in a somewhat "coarse-grained" $\mu$-space which is supposed to give the number of particles being in a cell around $\mathbf{dr}\mathbf{dv} := d^3r d^3v$. Instead of trying to describe the motion of every single particle which is impossible due to the huge numbers of particles in a gas, Boltzmann suggested his famous equation which describes the time evolution of $\Phi(\mathbf{r}, \mathbf{v}, t)$ in $\mu$-space and is, in the absence of any external force, given by

$$\frac{\partial}{\partial t}\Phi(\mathbf{r}, \mathbf{v}, t) + \mathbf{v} \cdot \nabla_{\mathbf{r}}\,\Phi(\mathbf{r}, \mathbf{v}, t) = \left.\frac{\partial \Phi(\mathbf{r}, \mathbf{v}, t)}{\partial t}\right|_{\mathrm{coll}}. \tag{17.6}$$

The left-hand side is supposed to account for the dynamics due to particles that do not collide, whereas the right-hand side describes the dynamics arising from collisions. Those dynamics are only taken into account in terms of transition rates $R$ and the (coarse-grained) particle densities $\Phi$, neglecting all correlations and structures on a finer scale. Thus, this type of dynamics implements the famous "assumption of molecular chaos" or the so-called "Stoßzahlansatz." According to the Stoßzahlansatz particles are not correlated before collisions, even though they get correlated by collisions, due to very many intermediate collisions before the same particles collide again. This treatment of collisions introduces irreversibility into the Boltzmann equation that is not present in the underlying Hamiltonian equations (for more details, cf. Sect. 4.1). If either $\Phi$ is close to equilibrium or for systems in which particles only collide with external scattering centers without particle–particle collisions, the Boltzmann equation may be linearized yielding

$$\frac{\partial}{\partial t}\Phi(\mathbf{r}, \mathbf{v}, t) + \mathbf{v} \cdot \nabla_{\mathbf{r}}\,\Phi(\mathbf{r}, \mathbf{v}, t) = \int d\mathbf{v}'\, R(\mathbf{v}, \mathbf{v}')\,\Phi(\mathbf{r}, \mathbf{v}', t). \tag{17.7}$$

A "velocity discretized" version of this linear Boltzmann equation reads

$$\frac{\partial}{\partial t}\Phi_i(\mathbf{r}, t) + \mathbf{v}_i \cdot \nabla_{\mathbf{r}}\,\Phi_i(\mathbf{r}, t) = \sum_j R_{ij}\,\Phi_j(\mathbf{r}, t), \tag{17.8}$$

with the rates $R_{ij}$.

In the limit of small density gradients (17.8) may feature diffusive solutions, i.e., solutions that fulfill the diffusion equation

$$\frac{\partial}{\partial t}\rho(\mathbf{r}) = D\,\Delta\rho(\mathbf{r}), \tag{17.9}$$

where the particle density $\rho(\mathbf{r})$ is given by

$$\rho(\mathbf{r}) = \sum_i \Phi_i(\mathbf{r}, t) \,. \tag{17.10}$$

Then the diffusion coefficient in (17.9) is found to be

$$D = -\sum_{i,j} v_i \, R_{ij}^{-1} \, v_j \, \Phi_j^0 \,, \tag{17.11}$$

with $\Phi_j^0$ being the equilibrium distribution to which $R_{ij}\Phi_j^0 = 0$ applies. The coefficients $R_{ij}^{-1}$ denote the inverse of the rate matrix (within the relevant space) from which the equilibrium eigenspace has been removed [7–10]. In the absence of any external scatterers $D$ may diverge, indicating that no diffusive behavior can be expected.

### 17.2.2 Peierls Approach and Green's Functions

Boltzmann derived his equation for dilute gases consisting of classical particles, i.e., point particles. To use this celebrated approach for the investigation of heat transport in solids, one somehow has to map the corresponding quantum dynamics onto a Boltzmann equation. To this end Peierls suggested to consider the phonons within a crystal as a gas, too [11]. Thus, in a classical picture transport is interpreted to result from quasi-particles like phonons diffusing through the solid. In this picture a finite heat conductivity results from scattering processes of the quasi-particles, in analogy to the particles' proper experiencing collisions. Eventually the above mapping has to provide pertinent quasi-particle velocities and their mutual scattering rates. Peierls simply suggested ad hoc, without any deeper justification, to compute the velocities from the dispersion relations of the phonons (group velocities) and the scattering rates from Fermi's golden rule. To do so he treated the lattice anharmonicities as perturbations. Since the lowest beyond harmonic order is the third order, the dominant scattering processes are such that two particles "collide" to form a third one, or vice versa. Besides energy conservation in such a process also quasi-momentum conservations has to hold:

$$\omega_1 + \omega_2 = \omega_3 \,, \tag{17.12}$$

$$\mathbf{k}_1 + \mathbf{k}_2 = \mathbf{k}_3 + n\mathbf{g} \,, \quad n \in \mathbb{Z} \,, \tag{17.13}$$

where $(\omega_1, \mathbf{k}_1)$ and $(\omega_2, \mathbf{k}_2)$ correspond to the two incoming phonons, $(\omega_3, \mathbf{k}_3)$ to the outgoing one, and $\mathbf{g}$ is the appropriate reciprocal lattice vector (cf. [12]).

It is not simple to generate a quantitative result in the sense of, say, (17.11) from this approach, nevertheless this theory succeeds in explaining qualitatively the basic heat transport properties of electrically isolating solids (cf. [12]). Crucial is the fact that all processes with $\mathbf{g} = 0$ do not give rise to a finite conductivity, since they conserve the total heat current. Of vital importance for the existence of a thermal resistance are thus the scattering processes with $\mathbf{g} \neq 0$ called Umklapp processes. According to the Boltzmann distribution of phonons, the number of high-energy phonons decreases by $\exp\{-\Theta/T\}$ for low temperatures. Thus for very low temperatures ($T < \Theta$, $\Theta$ Debye temperature) most of the two "incoming" phonons do not have enough joint momentum (and energy) to fulfill the above scattering conditions for Umklapp process, thus, the latter die out rapidly. Thus, only considering the frequency of Umklapp processes one would expect conductivity to decrease with increasing temperature. However, (squared) phonon velocities (which enter (17.11)) increase with increasing temperature, so does the heat capacity that occurs in converting energy density gradients (which appear in the Peierls–Boltzmann equation) into temperature gradients (which appear in the definition of heat conductivity, cf. (3.35), (17.14)). All these effects have, roughly speaking, to be multiplied. This qualitatively explains standard heat conductivities that feature a maximum at some temperature. However, in order to get a comprehensive picture of the conduction properties of solids at low temperatures, scattering processes at impurities and at the boundaries of the crystal have to be taken into account (see e.g., [12–16]), too. Indeed, there are even doubts about the importance of Umklapp processes at all for a finite conductivity in solids (see [17, 18]).

Regardless of the technical difficulties that occur when concretely applying the Peierls–Boltzmann theory, the ad hoc mapping of the extremely complicated quantum dynamics onto a comparatively simple Boltzmann equation has to be questioned. In this mapping essentially all quantum features are boldly discarded. However modern, more carefully developed theories based on so-called Green's functions essentially yield the same result, i.e., justify "Peierls' mapping" [19]. Those theories are, roughly speaking, rather subtle, time-dependent perturbation theories in which the particle–particle interactions (phonon–phonon scattering) are treated perturbatively up to second order. Thus, in the above example, the Peierls–Boltzmann approach may in principle yield reasonable results as long as the anharmonicities are small.

### 17.2.3  Linear Response Theory

In Sect. 3.2 it has been suggested that the energy current should be proportional to the temperature gradient

$$\mathbf{j} = -\kappa(T)\,\nabla T \,, \tag{17.14}$$

where $\kappa$ is the heat conductivity. This equation, called Fourier's law, defines the temperature-dependent heat conductivity $\kappa$ (see [6, 20]).

To obtain an explicit expression for the heat conductivity a similar approach as applied in standard direct current (dc) electric conductivity is often used. In the latter case this approach is based on the idea that the electric potential is an external perturbation to the system under consideration, i.e., this perturbative theory aims at a force-driven current. For a periodic time-dependent perturbation, this theory makes use of time-dependent perturbation theory of first order, in which one has to introduce a Dyson series expansion of the time evolution, truncated after the leading order term. Like in the case of Fermi's golden rule this perturbation leads to transitions in the system. For a perturbation constant in time and space one has to consider the zero frequency limit of this theory. This way, one is able to deduce the dc electric conductivity, given by the famous Kubo formula which is basically a current–current autocorrelation function (for a complete derivation see [6, 21–23]).

A similar approach for the thermal case has first been undertaken by Luttinger [6, 24]; he essentially boldly replaced the electrical current by heat current, finding

$$\kappa = -T \lim_{s \to 0} \int_0^\infty dt\, e^{-st} \int_0^\beta d\beta' \, \langle \mathbf{j}(-t - i\beta')\mathbf{j} \rangle \ . \tag{17.15}$$

(The (complex) time dependence of the current operator $\mathbf{j}(-t - i\beta')$ refers to the Heisenberg picture of the unperturbed system, $\beta$ denotes the inverse temperature, and the brackets indicate a canonical average.) However, a temperature gradient cannot be described as a part of the Hamiltonian of the system as it is the case for any electrical perturbation potential. Or, like already mentioned heat current is always gradient driven. Therefore, the above replacement requires concrete justification.

Nevertheless, the thermal Kubo formula is widely accepted. Kubo writes in his book *"Statistical Physics II"* (p. 183 in [6]):

> It is generally accepted, however, that such formulas exist and are of the same form as those for responses to mechanical disturbances.

Thus, the thermal Kubo formula has become a widely employed technique and it allows for a straightforward application to any quantum system, once its Hamiltonian is (approximately) diagonalized. For a comprehensive overview over the application to spin systems, e.g., see [25] and the references given therein. Furthermore, we would like to refer to some central works in connection to the thermal Kubo formula, see [26–31].

# References

1. K. Gustafson, Mind and Matter **1**, 9 (2003)
2. F. Bonetto, J. Lebowitz, L. Rey-Bellet, *Mathematical Physics 2000* (World Scientific Publishing Company, 2000), chap. "Fourier's Law: A Challenge to Theorists", pp. 128–150. Also published at arXiv:math-ph/0002052
3. M. Buchanan, Nat. Phys. **1**, 71 (2005)
4. W.R. Theis, *Grundzüge der Quantentheorie* (Teubner, Stuttgart, 1985)
5. M.O. Scully, M.S. Zubairy, *Quantum Optics* (Cambridge University Press, Cambridge, 1997)
6. R. Kubo, M. Toda, N. Hashitsume, *Statistical Physics II: Nonequilibrium Statistical Mechanics*, 2nd edn. No. 31 in Solid-State Sciences (Springer, Berlin, Heidelberg, New-York, 1991)

7. R. Balescu, *Equilibrium and Nonequilibrium Statistical Mechanics* (John Wiley & Sons, New York, 1975)
8. W. Brenig, *Statistical Theory of Heat: Nonequilibrium Phenomena* (Springer, New York, 1989)
9. W. Brenig, *Statistische Theorie der Wärme*, vol. 1, 4th edn. (Springer, Berlin, 1996)
10. M. Kadiroglu, J. Gemmer, Phys. Rev. B **76**, 024306 (2007)
11. R. Peierls, *Quantum Theory of Solids* (Clarendon Press, Oxford, 2001)
12. C. Kittel, *Einführung in die Festkörperphysik* (R. Oldenbourg Verlag, München, Wien, 1969)
13. H. Michel, M. Wagner, Phys. Status Solidi B **75**, 507 (1976)
14. H. Michel, M. Wagner, Ann. Phys. (Leipzig) **35**, 425 (1978)
15. H. Michel, M. Wagner, Phys. Status Solidi B **85**, 195 (1978)
16. J. Ziman, *Electrons and Phonons* (Oxford: Clarendon Press, Oxford, 2001)
17. M. Wagner, Phil. Mag. B **79**, 1839 (1999)
18. E. Fermi, J. Pasta, S. Ulam, *Note e Memorie (Collected Papers)* (The University of Chicago Press, Rom, 1965), vol. II (United States 1939-1954), chap. "Studies of non Linear Problems", pp. 977–988
19. L.P. Kadanoff, G. Baym, *Quantum Statistical Mechanics* (Benjamin, Verlag, New York, 1962)
20. S. de Groot, P. Mazur, *Non-equilibrium Thermodynamics* (North-Holland Publ. Comp., Amsterdam, 1962)
21. G.D. Mahan, *Many-Particle Physics*, 3rd edn. (Plenum Press, New York, London, 2000)
22. R. Kubo, J. Phys. Soc. Jpn. **12**, 570 (1957)
23. D. Forster, *Hydrodynamic Fluctuations, Broken Symmetry, and Correlation Functions* (W.A. Benjamin, Inc.; Advanced Book Program Reading, Massachusetts, London, 1975)
24. J.M. Luttinger, Phys. Rev. **135**(6A), A1505 (1964)
25. F. Heidrich-Meisner, *Transport Properties of Low-Dimensional Quantum Spin Systems*. Dissertation, Technische Universität Braunschweig (2005)
26. Z. Zotos, J. Phys. Soc. Jpn. **74**, 173 (2005)
27. F. Heidrich-Meisner, A. Honecker, D. Cabra, W. Brenig, Phys. Rev. B **68**, 134436 (2003)
28. A. Klümper, K. Sakai, J. Phys. A **35**, 2173 (2002)
29. X. Zotos, F. Naef, P. Prelovsek, Phys. Rev. B **55**, 11029 (1997)
30. P. Jung, R. Helmes, A. Rosch, Phys. Rev. Lett. **96**, 067202 (2006)
31. J. Gemmer, R. Steinigeweg, M. Michel, Phys. Rev. B **73**, 104302 (2006)

# Chapter 18
# Projection Operator Techniques and Hilbert Space Average Method[1]

*From the physical point of view the simplifying assumptions are unfortunately rather restrictive and it is not clear how one should remove them. There remains the basic problem in which sense the master equation approximates the Liouville equation.*

— G. E. Uhlenbeck [2]

**Abstract** A common feature of quantum systems for which "non-equilibrium statistical mechanics" is applied is that they are too complex to be boldly solved by the corresponding time-dependent Schrödinger equation. What else can be done instead? Among a range of approaches there are projection techniques and the Hilbert space average method (HAM). Both implicitly assume that the (possibly few) observables of interest obey an autonomous equation of motion of their own. Thus they essentially aim at finding this autonomous equation of motion for given initial states, etc. In this section we describe both methods and discuss in which sense they may be considered equivalent.

In the following we are going to apply a projection method to two scenarios:

1. Relaxation: all relevant observables correspond to a comparatively simple system which is connected to a more complex system that then acts as an equilibrating reservoir [3].
2. Transport: the relevant observable is the spatial density of some quantity that is conserved on the full (complex, interacting) system such that, e.g., a diffusive motion may possibly be expected [4, 5].

These applications to pertinent, simple models may be found in Chap. 19 concerning the system–reservoir situation and in Chap. 20, which deals with transport investigations. Here we develop the formal mathematical basis, focusing on topics like the convergence (see also [6, 7]) and the connection of different approaches.

---

[1]This chapter is based on [1]. Here, the authors would like to acknowledge the outstanding and extensive work done by H.-P. Breuer on quantum relaxation processes. Its impact on the work at hand can hardly be overestimated.

Gemmer, J. et al.: *Projection Operator Techniques and Hilbert Space Average Method*. Lect. Notes Phys. **784**, 201–213 (2009)
DOI 10.1007/978-3-540-70510-9_18

## 18.1 Interaction Picture of the von Neumann Equation

The dynamics of a closed quantum system is in general described by the Schrödinger equation

$$\frac{d}{dt}\hat{\rho} = -i[\hat{H}, \hat{\rho}], \tag{18.1}$$

written for the density operator here. This equation of motion is sometimes also called the Liouville–von Neumann equation, cf. (2.49) The central operator determining the dynamics of the system is the Hamiltonian $\hat{H}$. It is often convenient to separate the Hamiltonian in to two parts, an unperturbed part eventually called $\hat{H}_0$, which is under total control in the sense that we know its eigensystem, and a perturbation $\hat{V}$. One could think, e.g., of a small system coupled to a large environment, where the perturbation $\hat{V}$ could be the interaction between those two parts. However, note that the discussion in the following is of more general nature and not restricted to such system–reservoir situations.

To describe the dynamical properties of complex quantum systems it is often beneficial to transform into the interaction picture (see Sect. 2.5.1), by using the transformation

$$\hat{V}(t) = e^{i\hat{H}_0 t}\hat{V}e^{-i\hat{H}_0 t}. \tag{18.2}$$

This transforms the Liouville–von Neumann equation into

$$\frac{d}{dt}\hat{\rho} = -i[\hat{V}(t), \hat{\rho}] \equiv \hat{\mathcal{L}}(t)\hat{\rho}, \tag{18.3}$$

where we have introduced the Liouvillian $\hat{\mathcal{L}}(t)$, a superoperator acting on the density operator.

In general the dynamical equation (18.3) is a complex high-dimensional system of differential equations and therefore not easily solvable. Thus, let us discuss a few approaches to extract the dynamics of some relevant quantities.

## 18.2 Projection on Arbitrary Observables

A number of analytical and computational strategies are known which allow the description of Markovian and non-Markovian relaxation and transport phenomena in complex quantum systems. An important method is the projection operator technique, in which the transition from the complete microscopic description to an effective set of relevant variables is formalized through an appropriate projection superoperator $\hat{\mathcal{P}}$. Given the projection $\hat{\mathcal{P}}$ one derives closed systems of effective dynamic equations for the relevant variables through a systematic perturbation expansion. In the literature (see, e.g., [3]) there are two important variants of this scheme which lead to equations of motion with entirely different mathematical structure, namely the Nakajima–Zwanzig (NZ) [8, 9] already introduced in Sect. 4.8 and the time-convolutionless (TCL) projection operator techniques [10–13]. In the following let us concentrate on the latter.

In a formal mathematical sense a projection superoperator is a linear map $\hat{\mathcal{P}}$ that operates on the states of the system under consideration. The latter are represented by density matrices, i.e., by positive operators $\hat{\rho}$ on the underlying Hilbert space $\mathscr{H}$ with unit trace,

$$\hat{\rho} \geq 0, \qquad \mathrm{Tr}\{\hat{\rho}\} = 1. \tag{18.4}$$

We write the map $\hat{\mathcal{P}}$ in terms of a linear independent set of relevant observables $\hat{B}_n = \hat{B}_n^{\dagger}$ as follows:

$$\hat{\mathcal{P}}\hat{\rho} = \sum_n \mathrm{Tr}\left\{\hat{B}_n\,\hat{\rho}\right\}\hat{B}_n = \sum_n b_n\hat{B}_n. \tag{18.5}$$

Hence, any state $\hat{\rho}$ is mapped onto a linear combination of a certain set of relevant observables which span the range of $\hat{\mathcal{P}}$. Those observables are in principle some quantities of interest, but note that we are not always free to choose whatever we want here. Sometimes it may even be unclear what kind of operators we should take at all. However, let us assume we have decided to take a special set of observables, discussing the complications that could occur later.

Without restriction we may assume that the $\hat{B}_n$ are orthonormal in the sense that

$$\mathrm{Tr}\left\{\hat{B}_n\,\hat{B}_m\right\} = \delta_{nm}. \tag{18.6}$$

This condition guarantees that the projection superoperator (18.5) defines a linear map with the property of a projection

$$\hat{\mathcal{P}}^2 = \hat{\mathcal{P}}. \tag{18.7}$$

The complementary projection to the irrelevant quantities is then defined according to

$$\hat{\mathcal{Q}} = \hat{\mathcal{I}} - \hat{\mathcal{P}}, \tag{18.8}$$

with $\hat{\mathcal{I}}$ being the identity map. Hence, we have $\hat{\mathcal{Q}}^2 = \hat{\mathcal{Q}}$ and $\hat{\mathcal{Q}}\hat{\mathcal{P}} = \hat{\mathcal{P}}\hat{\mathcal{Q}} = 0$.

Standard operators in Hilbert space can be written as "vectors" within the Liouville space. This superspace contains the above-defined superoperators acting on operators of the Hilbert space. Defining the notation $|\hat{A})$ for an operator vector in the superspace we introduce the Hilbert–Schmidt scalar product defined by

$$(\hat{A}|\hat{B}) = \mathrm{Tr}\left\{\hat{A}^{\dagger}\hat{B}\right\}. \tag{18.9}$$

Thus, we can write the representation (18.5) in the suggestive form:

$$\hat{\mathcal{P}} = \sum_n |\hat{B}_n)(\hat{B}_n| .\tag{18.10}$$

Of course, one may also consider other scalar products which lead to different projection superoperators. For instance, one can introduce a scalar product that depends explicitly on certain system parameters like the temperature or directly on the system's thermal equilibrium state (see, e.g., [14–18]). A number of further modifications and extensions of the formalism have been proposed in the literature (see, e.g., [6, 19–22]).

## 18.3 Time-Convolutionless Master Equation

The emergence of the time integration over a memory kernel in the Nakajima–Zwanzig equation (cf. Sect. 4.8) is often regarded as the characteristic feature which expresses the non-Markovian nature of the dynamics. It is said that this describes a pronounced memory effect, i.e., the present dynamics depends on the complete history of the system. However, there exists an alternative method for the development of effective equations of motion for the relevant variables which is known as time-convolutionless (TCL) projection operator method [3, 10–13]. By contrast to the NZ approach, the TCL method leads to a time-local differential equation of the general form

$$\frac{\mathrm{d}}{\mathrm{d}t}\hat{\mathcal{P}}\rho(t) = \hat{\mathcal{K}}(t)\hat{\mathcal{P}}\rho(t) .\tag{18.11}$$

Here, $\hat{\mathcal{K}}(t)$ is a time-dependent superoperator, called the TCL generator. To explicitly derive such a time-local equation starting from the Liouville–von Neumann equation of the full system is beyond the scope of this text. The interested reader is referred to the excellent summary of all projection operator techniques in the book by Breuer and Petruccione [3].

It must be emphasized that the TCL equation (18.11) describes non-Markovian dynamics, although it is local in time and does not involve an integration over the system's past. In fact, it takes into account all memory effects through the explicit time dependence of the generator $\hat{\mathcal{K}}(t)$. To obtain the time-local form of (18.11) one eliminates the dependence of the future time evolution on the system's history through the introduction of the backward propagator into the Nakajima–Zwanzig equation (4.14). This enables one to express the density matrix at previous times $t_1 < t$ in terms of the density matrix at time $t$ and to derive an exact time-local equation of motion. We remark that the backward propagator and, hence, also the TCL generator may not exist, typically at isolated points of the time axis. This may happen for very strong system–environment couplings and/or long integration times; an example is discussed in [3].

One can develop a systematic perturbation expansion for the TCL generator, which takes the form

$$\hat{\mathcal{K}}(t) = \sum_i \hat{\mathcal{K}}^{(i)}(t), \tag{18.12}$$

where all odd terms of the expansion vanish, see [1, 3]. The various orders of this expansion can be expressed through the ordered cumulants [23–26] of the Liouville superoperator $\hat{\mathcal{L}}(t)$ with the Liouvillian given in (18.3). For instance, the contributions of second order to the TCL generator are given by

$$\hat{\mathcal{K}}^{(2)}(t) = \int_0^t dt_1 \hat{\mathcal{P}} \hat{\mathcal{L}}(t) \hat{\mathcal{L}}(t_1) \hat{\mathcal{P}}. \tag{18.13}$$

One can formulate a simple set of rules that enables one to write down immediately these and corresponding expressions for any higher order (see [3]). The TCL technique has been applied to many physical systems. Examples, which include the determination of higher orders of the expansion, are the spin star model [27], relaxation processes in structured reservoirs [19], decoherence phenomena in open systems [28], the damped Jaynes–Cummings model [3, 29], and the dynamics of the atom laser [30].

The equations of motion for the expectation values $b_n(t)$ introduced in (18.5) are now obtained by using

$$b_n(t) = \mathrm{Tr}\left\{ \hat{B}_n \, \hat{\rho}(t) \right\}, \tag{18.14}$$

yielding a time-local system of first-order differential equations,

$$\frac{d}{dt} b_n(t) = \sum_m K_{nm}(t) b_m(t), \tag{18.15}$$

with time-dependent coefficients,

$$K_{nm}(t) = \mathrm{Tr}\left\{ \hat{B}_n \, \hat{\mathcal{K}}(t) \, \hat{B}_m \right\} = K_{nm}^{(2)}(t) + K_{nm}^{(4)}(t) + \cdots, \tag{18.16}$$

according to the set of relevant observables $\hat{B}_n$. Explicitly, the second-order term is given by

$$K_{nm}^{(2)}(t) = \int_0^t dt_1 \, \mathrm{Tr}\left\{ \hat{B}_n \hat{\mathcal{L}}(t) \hat{\mathcal{L}}(t_1) \hat{B}_m \right\}, \tag{18.17}$$

which is a two-time correlation function. Using (18.3) we obtain

$$K_{nm}^{(2)}(t) = -\int_0^t dt_1 \, \mathrm{Tr}\left\{ \hat{B}_n [\hat{V}(t), [\hat{V}(t_1), \hat{B}_m]] \right\}. \tag{18.18}$$

In the TCL equation (18.11) we have made use of the initial condition $\hat{\mathcal{P}}\hat{\rho}(0) = \hat{\rho}(0)$. For more general initial states one has to add a certain inhomogeneity to the right-hand side of the TCL equation, which involves the initial conditions through the complementary projection $\hat{\mathcal{Q}}\hat{\rho}(0) = (\hat{\mathcal{I}} - \hat{\mathcal{P}})\hat{\rho}(0)$. A general method for the treatment of such initial states within the TCL technique is described in [3]; for a recent study on their influence in weakly coupled systems, see also [31, 32].

It is important to realize that the NZ and the TCL techniques lead to equations of motion with entirely different structures, namely to a system of integro-differential equations in the NZ approach and to a system of ordinary differential equations with time-dependent coefficients in the TCL method. Therefore, in any given order the mathematical structure of the solutions and the analytical or numerical strategies to find these are very different [33]. It is difficult to formulate general conditions that allow to decide for a given model whether the NZ or the TCL approach is more efficient. The assessment of the quality of the approximation obtained generally requires the investigation of higher orders of the expansion (cf. [7]), or else the comparison with numerical simulations or with certain limiting cases that can be treated analytically. It turns out that in many cases the degree of accuracy obtained by both methods is of the same order of magnitude. In these cases the TCL approach is of course to be preferred because it is technically much simpler to deal with.

In the following we switch to an approach with completely different background founded on the investigations introduced in the second part of this book, i.e., on Hilbert space averaging. Finally, we will show that at least in leading order this approach is equivalent to the TCL master equation discussed so far.

## 18.4 Generalization of the Hilbert Space Average

A further foundational method is the Hilbert space average method (HAM) which estimates conditional quantum expectation values through an appropriate average over a constraint region in Hilbert space. It turns out that this method is closely related to the projection operator techniques and that its dynamic extension leads to equations of motion for the relevant observables which coincide with those obtained from the projection operator methods introduced above.

So far the concept introduced in the second part of this book only states the existence of an approach to equilibrium and identifies the "position" of this equilibrium. It only relies on the relative frequency of some expectation values in certain accessible regions (ARs) in Hilbert space and does not depend on the convergence of any projection operator expansion, etc. But nothing has been said about the dynamics of the approach to equilibrium. However, the concept introduced in Chap. 8 may be enlarged to include also some relaxation dynamics if one is able to compute the Hilbert space average (HA) with respect to a set of states that is not defined by the expectation values of some projectors (as done for the equilibrium situation), but by the expectation values of some arbitrary Hermitian operators $\hat{B}_n$, i.e., $\langle \psi | \hat{B}_n | \psi \rangle = b_n$. Again without any loss of generality we require the $\hat{B}_n$ to be orthonormal in the sense of (18.6). This set of expectation values defines the AR for which the average is carried out in the following. Thus, let us define here

$$AR := \{\langle \psi | \hat{B}_n | \psi \rangle = b_n\}. \tag{18.19}$$

How this special HA can be used to obtain information about the dynamics of some relevant observables will be explained below in more detail. Let us first concentrate on the computation of such an average. We write this HA of the expectation value $\langle \psi | \hat{A} | \psi \rangle$ of an arbitrary Hermitian operator $\hat{A}$ over the AR (18.19) in the following as

$$a = [\![\langle \psi | \hat{A} | \psi \rangle]\!]. \tag{18.20}$$

To repeat, this expression stands for the average of $\langle \psi | A | \psi \rangle$ over all $| \psi \rangle$ that feature $\langle \psi | \hat{B}_n | \psi \rangle = b_n$, but are uniformly distributed otherwise. Uniformly distributed means invariant with respect to all unitary transformations $\exp\{i\hat{G}\}$ that leave the respective set of expectation values unchanged, i.e.,

$$\langle \psi | e^{i\hat{G}} \hat{B}_n e^{-i\hat{G}} | \psi \rangle = \langle \psi | \hat{B}_n | \psi \rangle. \tag{18.21}$$

Thus the respective transformations may be characterized by

$$[\hat{G}, \hat{B}_n] = 0. \tag{18.22}$$

Instead of computing the so-defined HA (18.20) directly by integration, as has been done, e.g., in Chap. 8 and in [34–36], we will proceed in a slightly different way here. To those ends we change from the notion of an expectation value of a state to one of a density operator

$$a = [\![\langle \psi | \hat{A} | \psi \rangle]\!] = [\![\text{Tr}\left\{\hat{A} | \psi \rangle \langle \psi |\right\}]\!]. \tag{18.23}$$

Exchanging the average and the trace, one may rewrite

$$a = \text{Tr}\left\{\hat{A} [\![| \psi \rangle \langle \psi |]\!]\right\} = \text{Tr}\left\{\hat{A} \hat{\Omega}\right\}, \tag{18.24}$$

with

$$\hat{\Omega} = [\![| \psi \rangle \langle \psi |]\!]. \tag{18.25}$$

To compute $\hat{\Omega}$ we now exploit its invariance properties. Since the set of all $| \psi \rangle$ that "make up" $\hat{\Omega}$ [that belong to the averaging region or AR of (18.25)] is characterized by being invariant under the above transformations $\exp\{-i\hat{G}\}$, $\hat{\Omega}$ itself has to be invariant under those transformations, i.e.,

$$e^{i\hat{G}} \hat{\Omega} e^{-i\hat{G}} = \hat{\Omega} . \tag{18.26}$$

This, however, can only be fulfilled if $[\hat{G}, \hat{\Omega}] = 0$ for all possible $\hat{G}$. Due to (18.22) the most general form of $\hat{\Omega}$ which is consistent with the respective invariance properties is

$$\hat{\Omega} = \sum_n \beta_n \hat{B}_n , \tag{18.27}$$

where the coefficients $\beta_n$ are still to be determined. In principle the above sum could contain addends of higher order, i.e., products of the operators $\hat{B}_n$. But here we restrict ourselves to sets of $\hat{B}_n$s which form a group in the sense that each possible product of any two $\hat{B}_n$s only yields some other, primary $\hat{B}_n$. Within this restriction (18.27) is indeed already the most general form.

It remains to determine the coefficients $\beta_n$. From the definition of $\hat{\Omega}$ (18.25) follows that

$$\mathrm{Tr} \left\{ \hat{\Omega} \hat{B}_n \right\} = b_n . \tag{18.28}$$

By inserting (18.27) into (18.28) and exploiting (18.6) the coefficients are found to be

$$\beta_n = b_n . \tag{18.29}$$

Thus, we finally get for the Hilbert space average (18.25)

$$\hat{\Omega} = \sum_n b_n \hat{B}_n . \tag{18.30}$$

Being now equipped with the above method to evaluate such HAs, we are going to exploit these techniques to gain information about the dynamics of the $b_n$ in the following.

## 18.5 Dynamical Hilbert Space Average Method

The considerations in the following are carried out for a Hamiltonian separated into an unperturbed part $\hat{H}_0$ and a perturbation $\hat{V}$ as introduced in Sect. 18.1. Furthermore, we assume here that the considered variables $\hat{B}_n$ are constants of motion with respect to the unperturbed Hamiltonian $\hat{H}_0$, i.e., $[\hat{B}_n, \hat{H}_0] = 0$. That means that the dynamics of these observables is invariant under the transformation to the interaction picture. Certainly, a more general case can be investigated by the same procedure, too. However, one has to transform back to the Schrödinger picture at the end of the calculation to get the correct dynamical equations. For an example of such a situation see [37, 38].

To find the effective dynamics for the $b_n$ from HAM we employ the following scheme: Based on HAM we compute a guess for the most likely value of the set $b_n$ at time $t + \tau$ [i.e., $b_n(t + \tau)$], assuming that we knew the values for the $b_n$ at time $t$ [i.e., $b_n(t)$]. Once such a map $b_n(t) \rightarrow b_n(t + \tau)$ is established it can of course be iterated to produce the full time dynamics. This implies repeated guessing, since in each iteration step the guess from the step before has to be taken for granted. However, if each single guess is sufficiently reliable, i.e., the spectrum of possible outcomes is rather sharply concentrated around the most frequent one (which one guesses), even repeated guessing may yield a good "total" guess for the full time evolution. The scheme is sketched in Fig. 18.1. For more detailed information on the reliability of this scheme we refer to [39].

**Fig. 18.1** Repeated guessing scheme. To each iteration time step $\tau$ corresponds an increasing uncertainty (variance) of the guess

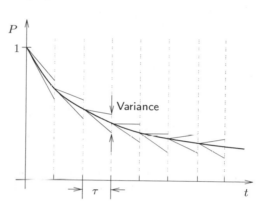

To implement the above scheme we consider the HA of an operator $\hat{B}_n$ over states which are slightly propagated forward in time, $|\psi(t + \tau)\rangle$, assuming that we knew the expectation values of $\hat{B}_n$ at time $t$. Here the superoperator that propagates density matrices from $t$ to $t + \tau$ according to (18.3) is defined by

$$\hat{\rho}(t + \tau) = \hat{\mathcal{U}}(t, \tau)\hat{\rho}(t) \, . \tag{18.31}$$

The HA of the propagated expectation value can thus be written as

$$[\![\langle\psi(t + \tau)|\hat{B}_n|\psi(t + \tau)\rangle]\!] = \mathrm{Tr}\left\{\hat{B}_n[\![|\psi(t + \tau)\rangle\langle\psi(t + \tau)|]\!]\right\} \, , \tag{18.32}$$

and with the propagator $\hat{\mathcal{U}}(t, \tau)$, this yields

$$
\begin{aligned}
[\![\langle\psi(t + \tau)|\hat{B}_n|\psi(t + \tau)\rangle]\!] &= \mathrm{Tr}\left\{\hat{B}_n\hat{\mathcal{U}}(t, \tau)[\![|\psi(t)\rangle\langle\psi(t)|]\!]\right\} \\
&= \mathrm{Tr}\left\{\hat{B}_n\hat{\mathcal{U}}(t, \tau)\,\hat{\Omega}\right\} \\
&= \sum_m b_m(t)\,\mathrm{Tr}\left\{\hat{B}_n\hat{\mathcal{U}}(t, \tau)\,\hat{B}_m\right\} \, ,
\end{aligned}
\tag{18.33}
$$

where we have used (18.30) to express the Hilbert space average $\hat{\Omega}$.

The (iterative) guess now simply consists of replacing the HA on the left-hand side of (18.33) by the actual value of the expectation value at $t + \tau$, i.e.,

$$b_n(t + \tau) \approx \sum_m b_m(t) \, \mathrm{Tr} \left\{ \hat{B}_n \, \hat{\mathcal{U}}(t, \tau) \, \hat{B}_m \right\} . \qquad (18.34)$$

However, we need an appropriate approximation for $\hat{\mathcal{U}}(t, \tau)$ since the general solution of (18.3), which is necessary due to the action of the propagator $\hat{\mathcal{U}}(t, \tau)$, amounts to the solution of the full quantum problem. Since (18.3) is a linear differential equation with time-dependent coefficients, $\hat{\mathcal{U}}(t, \tau)$ is formally given by a Dyson series

$$\hat{\mathcal{U}}(t, \tau) \approx \hat{\mathcal{I}} + \int_t^{t+\tau} d\tau_1 \hat{\mathcal{L}}(\tau_1) + \int_t^{t+\tau} d\tau_1 \int_t^{\tau_1} d\tau_2 \hat{\mathcal{L}}(\tau_1) \hat{\mathcal{L}}(\tau_2) + \cdots , \qquad (18.35)$$

which is here explicitly given to second order. From this equation an approximation to $\hat{\mathcal{U}}(t, \tau)$ for small $\tau$ may be computed and inserted into (18.34). In order to do so the following should be noted: If the considered observables commute with the unperturbed Hamiltonian as required above, then the dependence on the time variable $t$ of the expression $\mathrm{Tr} \left\{ \hat{B}_n \, \hat{\mathcal{U}}(t, \tau) \, \hat{B}_m \right\}$ vanishes and hence for the evaluation of such objects $t$ may be set to zero everywhere in (18.35). The whole iterative guessing scheme can also be applied to slightly more complicated observables [39], but here we focus on observables featuring the above properties. If furthermore $\mathrm{Tr} \left\{ \hat{B}_n \, \hat{\mathcal{L}}(\tau_1) \, \hat{B}_m \right\} = 0$, which depends on the model, but always holds if $[\hat{B}_n, \hat{B}_m] = 0$, the first term of (18.35) vanishes. The condition used here is equivalent to the disappearance of the first-order term in the expansion of the TCL generator [cf. (18.12) and the following discussion].

Using (18.35) approximated to second order within (18.34) yields

$$b_n(t + \tau) - b_n(t) \approx \sum_m b_m(t) \int_0^\tau d\tau_1 \int_0^{\tau_1} d\tau_2 \mathrm{Tr} \left\{ \hat{B}_n \, \hat{\mathcal{L}}(\tau_1) \hat{\mathcal{L}}(\tau_2) \, \hat{B}_m \right\} . \qquad (18.36)$$

Realizing that the integration over $\tau_2$ is the same correlation function (18.17) as obtained in the TCL approach we get

$$b_n(t + \tau) - b_n(t) \approx \sum_m b_m(t) \int_0^\tau d\tau_1 K_{nm}^{(2)}(\tau_1) . \qquad (18.37)$$

If the correlation function $K_{nm}^{(2)}(\tau_1)$ decays on a timescale $\tau_c$ we may write for all times larger than the correlation time, i.e., for $\tau_1 > \tau_c$,

$$K_{nm}^{(2)}(\tau_1) \approx R_{nm} = \mathrm{const.} \qquad (18.38)$$

Therefore, the integration becomes trivial and leads to a result linear in $\tau$. Thus, for $\tau \gg \tau_c$ we find

$$\frac{b_n(t+\tau) - b_n(t)}{\tau} \approx \sum_m R_{nm} \, b_m(t) , \qquad (18.39)$$

which, in the case of $1/\tau \gg R_{nm}$, may simply be approximated by

$$\frac{\mathrm{d}}{\mathrm{d}t} b_n(t) = \sum_m R_{nm} \, b_m(t) . \qquad (18.40)$$

This equation describes the reduced time evolution of the interesting quantities obtained by HAM.

## 18.6 Comparison of TCL and HAM

Equation (18.40) is essentially the same as the one that one gets by an application of the projection operator techniques in second order, using the projection operator defined in (18.10) and assuming that the relaxation dynamics is slow compared to the decay of the correlation functions. Thus, we have established the relation between apparently rather different methods of treating the relaxation dynamics in complex quantum systems, namely the projection operator techniques and the method of the Hilbert space average (HAM). Several implications emerge from these findings: First, HAM is only a guess, but as a guess it is valid for all initial states with no class of initial states being excluded a priori. Applying the projection operator techniques without considering the inhomogeneity that arises, if $\hat{\mathcal{Q}}\rho(0)$ does not vanish, one is restricted to a very specific class of initial states. The fact that to second order in both cases the same equations of motion result implies that the validity of the projection operator approach may include much more initial states than explicitly covered in the framework of these techniques. Some concrete studies that support this view are described in Chaps. 19 and 20. Ideas along these lines appear also in the context of the Mori formalism (see [17]) where the inhomogeneity is essentially classified as "noise" and not taken into account any further.

Another implication concerns the choice of a "good" projection operator $\hat{\mathcal{P}}$, i.e., of an appropriate set of relevant observables $\hat{B}_n$. Good here means that the corresponding expansion converges rapidly and that the projection contains the relevant information of interest. There is no general constructive method to find such a projection operator for a given microscopic model. In fact, the projection operator techniques only provide a formal mathematical framework and the choice of the projection operator naturally depends on what one wants to know about the system. However, the choice of an appropriate $\hat{\mathcal{P}}$ is not arbitrary if one intends to develop an efficient description of the dynamics, i.e., a description that yields accurate results even in low orders of the perturbation expansion (several examples are discussed in

[38]). The great advantage of the projection operator techniques consists in the fact that they provide a systematic expansion procedure. Hence, once a certain projection $\hat{P}$ has been chosen, one can use the perturbation expansion in order to check explicitly whether or not higher orders remain small and, thus, whether or not $\hat{P}$ enables a computationally efficient treatment of the reduced dynamics. An example is discussed in [6], where the TCL expansions corresponding to different projection operators have been compared. Finally, the existence of exact or approximate conserved quantities plays an important role in the construction of suitable projection operators as is discussed in more detail in [19].

# References

1. H.P. Breuer, J. Gemmer, Eur. Phys. J. Special Topics **151**, 1 (2007)
2. G. Uhlenbeck, *Probability and Related Topics in Physical Sciences* (Interscience Publishers, London, 1959), *Lectures in Applied Mathematics (Proc. Summer Seminar, Boulder, Colorado, 1957)*, vol. 1, chap. "The Boltzmann Equation", pp. 183–203
3. H.P. Breuer, F. Petruccione, *The Theory of Open Quantum Systems* (Oxford University Press, Oxford, 2002)
4. J. Gemmer, M. Michel, Europhys. Lett. **73**, 1 (2006)
5. M. Michel, G. Mahler, J. Gemmer, Phys. Rev. Lett. **95**, 180602 (2005)
6. H.P. Breuer, J. Gemmer, M. Michel, Phys. Rev. E **73**, 016139 (2006)
7. C. Bartsch, J. Gemmer, Phys. Rev. E **77**, 011119 (2007)
8. S. Nakajima, Prog. Theor. Phys. **20**, 948 (1958)
9. R. Zwanzig, J. Chem. Phys. **33**, 1338 (1960)
10. F. Shibata, Y. Takahashi, N. Hashitsume, J. Stat. Phys. **17**, 171 (1977)
11. S. Chaturvedi, F. Shibata, Z. Phys. B **35**, 297 (1979)
12. F. Shibata, T. Arimitsu, J. Phys. Soc. Jpn. **49**, 891 (1980)
13. C. Uchiyama, F. Shibata, Phys. Rev. E **60**, 2636 (1999)
14. H. Mori, Prog. Theor. Phys. **34**, 399 (1965)
15. H. Mori, Prog. Theor. Phys. **33**, 423 (1965)
16. D. Zubarev, *Nonequilibrium Statistical Thermodynamics*. Studies in Soviet Sciences (Consultants Bureau, New York, London, 1974). Transl. by P.J. Shepherd
17. R. Kubo, M. Toda, N. Hashitsume, *Statistical Physics II: Nonequilibrium Statistical Mechanics*, 2nd edn. No. 31 in Solid-State Sciences (Springer, Berlin, Heidelberg, New-York, 1991)
18. H. Grabert, *Projection Operator Techniques in Nonequilibrium Statistical Mechanics, Tracts in Modern Physics*, vol. 95 (Springer, New York, 1982)
19. H.P. Breuer, Phys. Rev. A **75**, 022103 (2007)
20. V. Romero-Rochin, I. Oppenheim, Physica A **155**, 52 (1989)
21. V. Romero-Rochin, A. Orsky, I. Oppenheim, Physica A **156**, 244 (1989)
22. V. Gorini, M. Verri, A. Frigerio, Physica A **161**, 357 (1989)
23. R. Kubo, J. Math. Phys. **4**, 174 (1963)
24. A. Royer, Phys. Rev. A **6**, 1741 (1972)
25. N. van Kampen, Physica **74**, 215 (1974)
26. N. van Kampen, Physica **74**, 239 (1974)
27. H.P. Breuer, D. Burgarth, F. Petruccione, Phys. Rev. B **70**, 045323 (2004)
28. H.P. Breuer, B. Kappler, F. Petruccione, Ann. Phys. (N. Y.) **291**, 36 (2001)
29. H.P. Breuer, B. Kappler, F. Petruccione, Phys. Rev. A **59**(2), 1633 (1999)
30. H.P. Breuer, D. Faller, B. Kappler, F. Petruccione, Europhys. Lett. **54**, 14 (2001)

31. S. Tasaki, K. Yuasa, P. Facchi, G. Kimura, H. Nakazato, I. Ohba, S. Pascazio, Ann. Phys. (N. Y.) **322**, 631 (2007)
32. K. Yuasa, S. Tasaki, P. Facchi, G. Kimura, H. Nakazato, I. Ohba, S. Pascazio, Ann. Phys. (N. Y.) **322**, 657 (2007)
33. A. Royer, *Aspects of Open Quantum Dynamics* (Springer, 2003), *LNP*, vol. 622, pp. 47–63
34. J. Gemmer, M. Michel, Physica E **29**, 136 (2005)
35. J. Gemmer, G. Mahler, Eur. Phys. J. B **31**, 249 (2003)
36. M. Michel, J. Gemmer, G. Mahler, Phys. Rev. E **73**, 016101 (2006)
37. H. Weimer, M. Michel, J. Gemmer, G. Mahler, Phys. Rev. E **77**, 011118 (2007)
38. M. Michel, R. Steinigeweg, H. Weimer, EPJST **151**, 13 (2007)
39. J. Gemmer, M. Michel, Eur. Phys. J B **53**, 517 (2006)

# Chapter 19
# Finite Systems as Thermostats

*...environmental induced decoherence is omnipresent
in the microscopic world. It is found, e.g., for an atom confined
in a quantum optical trap, or for electron propagation
in a mesoscopic device.*

— U. Weiss [1]

**Abstract** Having derived the formal mathematical background of projection operator techniques (TCL) and the Hilbert space average method (HAM) in the last
chapter, we will now use these techniques to analyze a system–reservoir scenario.
The above introduced partitioning scheme into a relevant part and the rest will be
used to consider the relaxation of a small quantum system, e.g., a two-level atom to
equilibrium. Contrary to the standard bath scenario the atom is coupled to a finite
environment here, i.e., another not too big quantum system. We show, by comparing
with the (numerically) exact dynamics given by the Schrödinger equation, that the
projective approach may work or fail, depending on the projector chosen.

The convergence of the projection operator technique could depend on the chosen
projection superoperator (see [2]). Thus, we show that the standard projection superoperator, frequently used for the derivation of the quantum master equation, fails to
describe the dynamical behavior of the finite bath model correctly (see Sect. 19.2
and [3, 4]). However, a better adapted projector taking correlations between system
and environment into account finally results in a convergent expansion of the TCL
generator (see Sect. 19.3) and yields a reduced dynamical description which mirrors
the full dynamics in a reasonably good way.

Let us emphasize that statistical relaxation behavior can be induced by finite
reservoirs already. Here, the correlations between system and environment, produced even if the full system is in a product state at the beginning, play an important
role. This also supports the concept of systems being driven toward equilibrium
through increasing correlations with their baths [5–9], cf. Chap. 11, rather than the
idea of system and bath remaining truly factorizable, which is often confused with
the Born approximation [1, 10].

Gemmer, J. et al.: *Finite Systems as Thermostats*. Lect. Notes Phys. **784**, 215–225 (2009)
DOI 10.1007/978-3-540-70510-9_19

## 19.1 Finite Quantum Environments

The model we analyze is characterized by a two-level system S coupled to a many-level system E consisting of two relevant bands featuring the same width and equidistant level spacing (see Fig. 19.1). So this may be viewed as an atom or spin coupled to a single molecule, a one-particle quantum dot, another atom, or simply a single harmonic oscillator. Note that the atom, unlike in typical oscillator baths or the Jaynes–Cummings model, is here not in resonance with the environment-level spacing but with the energy distance between the bands. There are two principal differences of such a finite environment-level scheme from the level scheme of, say, a standard oscillator bath:

1. The total amount of levels within a band may be finite.
2. The relevant bands of a standard bath at zero temperature would consist of only one state in the lower and infinitely many states in the upper band.

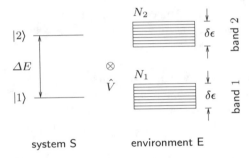

system S                     environment E

**Fig. 19.1** Two-level system coupled to a finite environment. Special case for only two bands in the environment. As a simple example we will use the same amount of levels in the lower and the upper band of the environment

As discussed in Chap. 18, the complete Schrödinger picture Hamiltonian of the model can be partitioned into an unperturbed, here local, and an interaction part

$$\hat{H} = \hat{H}_{\mathrm{loc}} + \lambda \hat{V} , \tag{19.1}$$

with $\lambda$ being the coupling strength between system and reservoir. The first part, i.e., the unperturbed part, contains the local Hamiltonian $\hat{H}_S$ of the two-level system and the local Hamiltonian $\hat{H}_E$ of the environment. Thus, the complete local Hamiltonian decomposes into a sum of two terms:

$$\hat{H}_S = \frac{\Delta E}{2} \, \hat{\sigma}_z , \tag{19.2}$$

$$\hat{H}_E = \sum_{a=1}^{2} \sum_{n_a=1}^{N_a} \left( (a-1)\Delta E + \frac{\delta \varepsilon \, n_a}{N_a} \right) |n_a\rangle\langle n_a| , \tag{19.3}$$

where $|n_a\rangle$ are energy eigenstates of the environment, $a = 1, 2$ labels the band, and $\hat{\sigma}_i$ refers to standard Pauli operators. The number of levels in the band $a$ is given by $N_a$ and the band width by $\delta\varepsilon$. Thus the environment essentially consists of two bands with equidistant level spacing, cf. Fig. 19.1. We assume the interaction to be of the following form:

$$\hat{V} = \hat{\sigma}_+ \otimes \hat{S} + \hat{\sigma}_- \otimes \hat{S}^\dagger , \tag{19.4}$$

where $\hat{\sigma}_\pm$ refer to the ladder Pauli operators. The operator $\hat{S}$ acts on the state space of the environment and we choose it to comprise only transitions from the excited band to the lower one, i.e.,

$$\hat{S} = \sum_{n_1, n_2} c_{n_1 n_2} |n_1\rangle\langle n_2| . \tag{19.5}$$

The operator $\hat{S}^\dagger$ carries out the reverse transitions. The constants $c_{n_1 n_2}$ we choose as Gaussian-distributed random numbers with zero mean and a variance according to the normalization

$$\frac{1}{N_1 N_2} \sum_{n_1, n_2} |c_{n_1 n_2}|^2 = 1 , \tag{19.6}$$

such that $\lambda$ characterizes an overall coupling strength.

Furthermore, we define band projection operators. Those operators implement projections onto the lower, respectively upper, band of the environment by

$$\hat{\Pi}_a = \sum_{n_a} |n_a\rangle\langle n_a| . \tag{19.7}$$

## 19.2 Standard Projection Superoperator

For the further analysis let us choose an initial state, where the system is excited and the environment is found to be somewhere in its lower band. Hence, the state of the environment refers to a thermal state of very low temperature. To analyze the model described above within the theory of open quantum systems (see, e.g., [10]) the following projection superoperator is frequently used:

$$\hat{\mathcal{P}}\hat{\rho}(t) = \mathrm{Tr}_E\left\{\hat{\rho}(t)\right\} \otimes \hat{\rho}_E = \hat{\rho}_S(t) \otimes \hat{\rho}_E . \tag{19.8}$$

This superoperator projects onto a product state of the reduced density operator of the system $\hat{\rho}_S$, given as a partial trace over the environment of the full density $\hat{\rho}(t)$ and an arbitrary constant state $\hat{\rho}_E$ of the environment, mostly chosen to be a thermal one. Since this state should be in accord with the low temperature thermal initial state in the environment, the reference state is chosen to be

$$\hat{\rho}_E = \hat{\Pi}_1/N_1 , \qquad (19.9)$$

in terms of the band projection operators (19.7). Thus, the projection superoperator is defined as

$$\hat{P}\hat{\rho}(t) = \hat{\rho}_S(t) \otimes \frac{1}{N_1}\hat{\Pi}_1 . \qquad (19.10)$$

To derive the dynamical equations for the reduced system density operator $\hat{\rho}_S$ in second order one plugs the projection superoperator (19.10) into the second order of the TCL generator (18.13). The time-local differential equation then reads

$$\frac{d}{dt}\hat{\rho}_S(t) = \int_0^t dt_1 \, f_2(t, t_1) \left( 2\hat{\sigma}_- \hat{\rho}_S(t)\hat{\sigma}_+ - [\hat{\sigma}_+\hat{\sigma}_-, \hat{\rho}_S]_+ \right) \otimes \hat{\rho}_E , \qquad (19.11)$$

where $[\ldots, \ldots]_+$ denotes the anti-commutator. The time-dependent function is a two-time environmental correlation function

$$f_2(t, t_1) = f(\tau) = \text{Tr}_E \left\{ \hat{S}(\tau)\hat{S}^\dagger \hat{\rho}_E \right\} , \qquad (19.12)$$

with $\tau = t - t_1$. The time argument of the operator $\hat{S}(\tau)$ refers to the transformation into the interaction picture according to (18.2).

Assuming both a random interaction $\hat{V}$ between system and environment as suggested above and a constant finite state density, the correlation function can be evaluated exploiting the same considerations which are also used in the derivation of Fermi's golden rule (see [2, 11]). This yields

$$f(\tau) = \gamma h(\tau), \qquad (19.13)$$

with the function

$$h(\tau) = \frac{\delta\epsilon}{2\pi} \frac{\sin^2(\delta\epsilon \, \tau/2)}{(\delta\epsilon \, \tau/2)^2} , \qquad (19.14)$$

and the constant

$$\gamma = 2\pi\lambda^2 \frac{N_2}{\delta\varepsilon} . \qquad (19.15)$$

The function given in (19.13) exhibits a sharp peak of width $\delta\epsilon^{-1}$ at $\tau = 0$ and may be approximated by a delta function for times $t$, which are large compared to the inverse band width, i.e., for $\delta\epsilon \, t \gg 1$, and thus the correlation function yields

$$f(\tau) \approx \gamma \, \delta(\tau). \qquad (19.16)$$

For further details on the approximate evaluation of such reservoir correlation functions see [2, 10–12].

Using (19.16) in (19.11) one gets the second-order master equation in Lindblad form:

$$\frac{d}{dt}\hat{\rho}_S = \gamma\left(\hat{\sigma}_-\hat{\rho}_S\hat{\sigma}_+ - \frac{1}{2}[\hat{\sigma}_+\hat{\sigma}_-, \hat{\rho}_S]_+\right).\tag{19.17}$$

Unfortunately, this second-order approximation leads to the wrong dynamical behavior, even and especially in a parameter range in which, as mentioned above, relaxation times are much larger than correlation times. In Fig. 19.2 we compare the dynamics of (19.17) represented by the *solid line* and the exact result from the Schrödinger equation (crosses) for the diagonal element $\rho_{22}$ of the reduced density matrix $\hat{\rho}_S$. This density matrix element represents the occupation probability of the excited state of the system. Note that due to our choice $\hbar = 1$ all times are given in units of $(\Delta E)^{-1}$. To check the convergence of the series expansion one can account for the next higher order of the TCL expansion which is the fourth order here (see Chap. 18 and for details concerning the fourth order [2, 13]). The *dashed line* in Fig. 19.2 shows a computation of the fourth order contribution as derived by H.-P. Breuer [2] As the fourth order even diverges the series expansion does not converge here. Note that for the parameters chosen the correlation time is on the order of $\tau_c \approx 2$. Hence the routinely mentioned "timescale separation" is not a sufficient criterion for a fast convergence of the projective expansion.

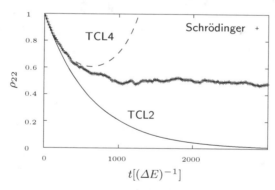

$t[(\Delta E)^{-1}]$

**Fig. 19.2** Comparison of the numerical solution of the Schrödinger equation for the model and the solution of the second-order TCL master equation (19.17) labeled as TCL2. Furthermore, the next higher order of the expansion which is the fourth order is shown as TCL4 (cf. [2]). Model parameters: $\Delta E = 1$, $N_1 = N_2 = 500$, $\delta\varepsilon = 0.5$, and $\lambda = 5 \cdot 10^{-4}$

When S has reached equilibrium the full system is in a superposition of S in the excited state, E in the lower band, and S in the ground state, E in the upper band. This is a maximum entangled state with two orthogonal addends, one of which features a bath population corresponding to $T_E \approx 0$, the other a bath population inversion, i.e., even a negative bath temperature. These findings contradict the concept of factorizability, which is often confused with the applicability of the Born approximation, but are in accord with a result from [14] claiming that an evolution toward local equilibrium is always accompanied by an increase of system–bath correlations. Thus, in spite of the system's finiteness and the reversibility of the underlying Schrödinger equation S evolves toward maximum local von Neumann entropy, which supports the concepts of Lubkin introduced in [6] and the whole approach to equilibration described in Part II.

Let us stress the fact that the choice of a suitable projection superoperator is vital for the convergence of the series expansion and to obtain a proper reduced dynamical equation for the system. In the next section we will show that there is indeed a proper projector which produces the correct dynamical behavior.

## 19.3 Correlated Projection Superoperator

In the following, let us consider the correlated projection superoperator:

$$
\begin{aligned}
\hat{\mathcal{P}}\hat{\rho}(t) &= \mathrm{Tr_E}\left\{\hat{\Pi}_1\hat{\rho}(t)\right\} \otimes \frac{1}{N_1}\hat{\Pi}_1 + \mathrm{Tr_E}\left\{\hat{\Pi}_2\hat{\rho}(t)\right\} \otimes \frac{1}{N_2}\hat{\Pi}_2 \\
&= \hat{\rho}_1(t) \otimes \frac{1}{N_1}\hat{\Pi}_1 + \hat{\rho}_2(t) \otimes \frac{1}{N_2}\hat{\Pi}_2 .
\end{aligned}
\tag{19.18}
$$

This projection belongs to the class of superoperators suggested in (18.5) (see also [2, 13, 15]). The band projection operators $\hat{\Pi}_a$ play the role of the set of arbitrary observables $\hat{B}_n$ defined in Sect. 18.2. The dynamical variables are the non-normalized matrices $\hat{\rho}_i(t)$ which are correlated with the projections onto the lower and the upper band, respectively. Thus, the reduced density matrix of the two-state system is found to be

$$
\hat{\rho}_S(t) = \mathrm{Tr_E}\left\{\hat{\mathcal{P}}\hat{\rho}(t)\right\} = \hat{\rho}_1(t) + \hat{\rho}_2(t) .
\tag{19.19}
$$

Using the correlated projection superoperator (19.18) in the TCL expansion one gets the second-order contribution defined in (18.18). Replacing the observables $\hat{B}_n$ by the band projectors yields

$$
\frac{\mathrm{d}}{\mathrm{d}t}\hat{\rho}_j(t) = \sum_i \int_0^t \mathrm{d}t_1 \, \mathrm{Tr_E}\left\{\hat{\Pi}_j[\hat{V}(t), [\hat{V}(t_1), \hat{\Pi}_i]]\right\} \frac{1}{N_i}\hat{\rho}_i(t) .
\tag{19.20}
$$

Based on the interaction operator (19.4) we finally find the dynamical equations:

$$
\frac{\mathrm{d}}{\mathrm{d}t}\hat{\rho}_1 = \gamma_1\,\hat{\sigma}_+\hat{\rho}_2\hat{\sigma}_- - \frac{\gamma_2}{2}\,[\hat{\sigma}_+\hat{\sigma}_-, \hat{\rho}_1]_+ ,
\tag{19.21}
$$

$$
\frac{\mathrm{d}}{\mathrm{d}t}\hat{\rho}_2 = \gamma_2\,\hat{\sigma}_-\hat{\rho}_1\hat{\sigma}_+ - \frac{\gamma_1}{2}\,[\hat{\sigma}_-\hat{\sigma}_+, \hat{\rho}_2]_+ ,
\tag{19.22}
$$

with the rates

$$\gamma_i = 2\pi \lambda^2 \frac{N_i}{\delta\varepsilon}. \tag{19.23}$$

Again, these rates are obtained from an approximate integration over a two-time environmental correlation function as already done in Sect. 19.2 (more details concerning the derivation can be found in [2]).

For the simple case of two bands of the same amount of levels in the environment the two decay constants are equal: $\gamma_1 = \gamma_2 = \gamma$. In this special case it is possible to derive a closed dynamical equation for the reduced density matrix of the system, $\hat{\rho}_S$, as well. Finally, this leads to the elements of the reduced density matrix:

$$\rho_{ij}(t) = \langle i|\hat{\rho}_1(t)|j\rangle + \langle i|\hat{\rho}_2(t)|j\rangle. \tag{19.24}$$

The population of the upper level of the system is then given by

$$\frac{d}{dt}\rho_{22}(t) = -2\gamma\,\rho_{22}(t) + \gamma\,\rho_{22}(0), \tag{19.25}$$

and the coherences by

$$\frac{d}{dt}\rho_{12}(t) = -\frac{\gamma}{2}\,\rho_{12}(t). \tag{19.26}$$

This is basically an exponential decay to the expected equilibrium value. However, note that in general it is not always possible to find such a closed equation for the density operator (cf. [11]). In more general cases one has to solve equations (19.21) and (19.22). Then, the reduced density operator can be reconstructed by using (19.19).

We observe that the dynamics of the populations $\rho_{22}(t)$ is on the one hand strongly "non-Markovian" because of the presence of the initial condition $\rho_{22}(0)$ on the right-hand side of (19.25). This term expresses a pronounced memory effect, namely it implies that the dynamics of the populations never forgets its initial data. Note also that the dynamics of the reduced density matrix is not in standard Lindblad form. It does, however, lead to a positive dynamical map, as can easily be verified.

However, on the other hand, the full model is also "Markovian" in the sense that bath correlations decay much faster than the system relaxes. For example, with the same amount of levels in both bands and the concrete system parameters $N_1 = N_2 = 500$, $\delta\epsilon = 0.5$, and $\lambda = 5 \cdot 10^{-4}$, the bath correlations decay on a timescale of $\tau_c \approx (\delta\epsilon)^{-1} = 2$. On the other hand the system relaxes on a timescale $\tau_r = \gamma^{-1} \approx 640$ with $\gamma$ from (19.23).

In Fig. 19.3 we compare the solution of the rate equation (19.25) with the full numerical integration of the Schrödinger equation for the model depicted in Fig. 19.1. Here, we have used the same number of levels in both bands as well as the mentioned random interaction between system and environment. The figure clearly shows that already the lowest order of the TCL expansion with the correlated projection superoperator results in a good approximation of the actual dynamics. It not only yields the correct stationary state but also reasonable predictions about the relaxation time.

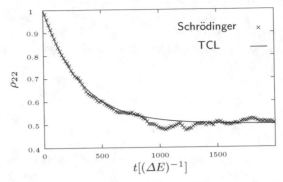

**Fig. 19.3** Comparison of the second-order TCL master equation (19.25) with a numerical solution of the Schrödinger equation. Parameters: $N_1 = N_2 = 500$, $\delta\varepsilon = 0.5$, and $\lambda = 0.001$

Since the whole system is finite there is a finite recurrence time of course. But due to the incommensurability of the full system's frequencies it appears to be very large ($> 10^8$), i.e., orders of magnitudes larger than the relaxation time of S. Furthermore, the special example at hand features, due to the artificial equidistant level spacing in the environment, a recurrence time for the bath correlations of approximately $10^4$, but that does not induce a recurrence in S.

Of course one could have used the Hilbert space average method to investigate the above given model as well. For a full analysis based on HAM see, e.g., [3, 4, 11]. As shown in Chap. 18 the result would have been the same. The projection operator technique according to the correlated projection operator produces in second order the same dynamical equations as HAM.

## 19.4 Accuracy of the Reduced Description

In this section we investigate the quality of the reduced description developed above. On one hand the dependence of the quality on the initial state is interesting because of the different treatment of initial conditions within HAM and TCL. Furthermore, the behavior of the error or the fluctuations of the exact solution around the reduced description in dependence of the system size is important for statements concerning the thermodynamic limit. To analyze the accuracy of the predictions for $\rho_{22}(t)$, we introduce the measure $D^2$, being the time-averaged quadratic deviation of HAM from the exact Schrödinger result

$$D^2 = \frac{1}{\tau} \int_0^\tau dt \left[\rho_{22}^{\mathrm{TCL}}(t) - \rho_{22}^{\mathrm{exact}}(t)\right]^2, \tag{19.27}$$

for an integration time $\tau$. Since we are not going to investigate equilibrium fluctuations here, we choose $\tau$ to be a couple of relaxation times of the system. $D$ is a measure of the deviations from the predicted behavior.

In Fig. 19.4 a histogram of the deviation $D^2$ for different initial states is shown ($\tau = 2000$). The set of respective pure initial states is characterized by a probability of 3/4 for S to be in its excited state, E in its lower band, and 1/4 for S to be in its ground state, E in its upper band, thus all of them are entangled. Within these restrictions the pure initial states are uniformly distributed in the corresponding Hilbert subspace. Due to them being correlated and pure none of these states is identically reproduced, neither by the standard projector nor by the correlated projector. Thus considering those initial states, even for the correlated projector, one would have to take the already mentioned inhomogeneity (cf. Sect. 4.8) into account. However, as Fig. 19.4 shows, the vast majority of evolutions follow the prediction from Sect. 19.3, i.e., not taking any inhomogeneity into account, quite closely. However, there is a typical fluctuation of $D = \sqrt{2} \cdot 10^{-2}$ but this is small compared to the features of the predicted behavior which is on the order of 1. These fluctuations are due to the finite size of the environment (cf. also the fluctuations in Fig. 19.3). This supports considerations in Sect. 18.6, which indicate that for the majority of initial states the inhomogeneity should only have negligible influence.

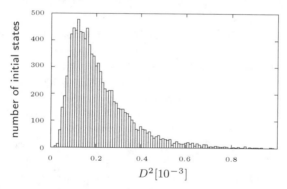

**Fig. 19.4** Deviation of the exact evolution of the excitation probability from the prediction for a set of entangled initial states

In Fig. 19.5 the dependence of $D^2$ on the number of states of E is displayed for $N = 10, \ldots, 800$. Here, we have used the same type of initial state and just one evolution for each environment size. At $N = 500$ as used in the above accuracy investigation we find the same typical fluctuation, whereas for smaller environments the typical deviation is much bigger. We find that the squared deviation scales as $1/N$ with the environment size, thus making TCL or HAM a reasonably reliable guess for many-state environments.

The claim that reduced dynamics need not be completely positive, once S and E are correlated [1, 16], in principle holds here too. But regardless of there being correlations, just a small fraction of all states from Fig. 19.4 shows significant deviations from the prediction. Thus a smooth evolution toward equilibrium, though

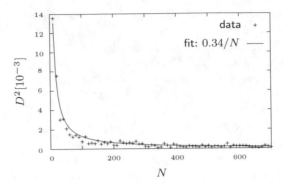

**Fig. 19.5** Deviation of the exact evolution of the excitation probability from the HAM prediction for increasing number $N$ of states in the environment

not necessarily of Lindblad type [17], can typically be expected for the reduced dynamics, regardless of the initial state being correlated or not.

Let us close with a comment on reversibility and interaction strengths. The Schrödinger equation is time reversible and indeed there is no apparent time asymmetry in Fig. 19.6. For $t > 0$ the behavior of the above system (solid line) is well described by the above rate equation scheme featuring an attractive fix point. But for $t < 0$ it would have to be described by a scheme with an repulsive fix point! Thus, any of the full model states at $t < 0$ is a paradigm for an initial state that does not yield statistical decay behavior even though the model typically generates decay. The *dashed line* shows the behavior of a model like the above one, only with $\delta\epsilon \approx 0$, and in order to keep the timescales comparable $\lambda = 10^{-4}\Delta E$. Thus, the model is no longer Markovian in the above-mentioned sense, and indeed even for $t > 0$ there is no exponential decay although the model features as many states as the above one. Nevertheless, the same equilibrium state as in the above case is reached (cf. also [18]) regardless of whether the system is propagated forward or backward in time.

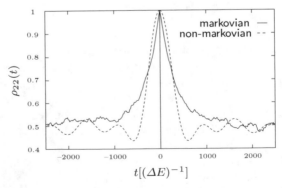

**Fig. 19.6** Evolution of the excitation probability before and after full excitation for two different sets of model parameters

# References

1. U. Weiss, *Quantum Dissipative Systems, Series in Modern Condensed Matter Physics*, vol. 10, 2nd edn. (World Scientific, Singapore, New Jersey, London, Hong Kong, 1999)
2. H.P. Breuer, J. Gemmer, M. Michel, Phys. Rev. E **73**, 016139 (2006)
3. J. Gemmer, M. Michel, Europhys. Lett. **73**, 1 (2006)
4. J. Gemmer, M. Michel, Eur. Phys. J. B **53**, 517 (2006)
5. E. Lubkin, J. Math. Phys. **19**, 1028 (1978)
6. E. Lubkin, T. Lubkin, Int. J. Theor. Phys. **32**, 933 (1993)
7. W. Zurek, J. Paz, Phys. Rev. Lett. **72**, 2508 (1994)
8. J. Gemmer, A. Otte, G. Mahler, Phys. Rev. Lett. **86**, 1927 (2001)
9. V. Scarani, M. Ziman, P. Stelmachovic, N. Gisin, V. Buzek, Phys. Rev. Lett. **88**, 097905 (2002)
10. H.P. Breuer, F. Petruccione, *The Theory of Open Quantum Systems* (Oxford University Press, Oxford, 2002)
11. M. Michel, *Nonequilibrium Aspects of Quantum Thermodynamics*. Ph.D. thesis, Universität Stuttgart, Stuttgart (2006). Also available online: http://elib.uni-stuttgart.de/opus/volltexte/2006/2803/
12. J. Gemmer, M. Michel, Physica E **29**, 136 (2005)
13. H.P. Breuer, J. Gemmer, Eur. Phys. J. Special Topics **151**, 1 (2007)
14. J. Gemmer, G. Mahler, Eur. Phys. J. B **31**, 249 (2003)
15. H.P. Breuer, Phys. Rev. A **75**, 022103 (2007)
16. P. Pechukas, Phys. Rev. Lett. **73**, 1060 (1994)
17. R. Alicki, K. Lendi, *Quantum Dynamical Semigroups and Applications* (Springer, Berlin, 2001)
18. H.P. Breuer, D. Burgarth, F. Petruccione, Phys. Rev. B **70**, 045323 (2004)

# Chapter 20
# Projective Approach to Dynamical Transport

*During the International Congress on Mathematical Physics held in London in the year 2000, J. L. Lebowitz expressed his opinion that one of the great challenges to mathematical physics in the twenty-first century is the theory of heat conductivity and other transport phenomena in macroscopic bodies.*

— R. Alicki and M. Fanes [1]

**Abstract** In this chapter we analyze transport properties of spatially structured, simplified quantum systems using the time-convolutionless (TCL) method introduced in Chap. 18. Therefore the system is initially prepared in a non-equilibrium state, i.e., a non-equilibrium distribution of energy or heat, for example. After the relaxation process, the system will be at a global equilibrium in the sense of a uniform energy distribution but not in the sense of a maximum global von Neumann entropy. From the concrete dynamics of this decay we extract information about the transport properties of the model.

## 20.1 Classification of Transport Behavior

Here we want to classify the relaxation behavior according to diffusive or ballistic transport. Diffusive behavior is to be identified simply by comparison with the heat conduction or heat diffusion equation. Fourier's law of heat conduction reads

$$J = -\gamma \nabla u = -\kappa \nabla T, \qquad \kappa \equiv \gamma\, c, \tag{20.1}$$

where $J$ is the energy or heat current, $\kappa$ the heat conductivity, $\gamma$ the "energy diffusion constant," $u$ the spatial energy density, and $c$ the heat capacity. Combining the first part of (20.1) with the energy continuity equation (3.30) yields the heat diffusion equation

$$\dot{u} = -\gamma \Delta u. \tag{20.2}$$

We are simply going to compare the dynamics of the energy density in our model to the dynamics as defined by a discrete version of this heat conduction equation, cf. (20.12). If the quantum dynamics are in good accordance with the discrete heat

Gemmer, J. et al.: *Projective Approach to Dynamical Transport*. Lect. Notes Phys. **784**, 227–240 (2009)

DOI 10.1007/978-3-540-70510-9_20

conduction equation, we will classify the transport behavior as normal or diffusive. In this case we may simply read of $\gamma$ from the quantum dynamics. It is straightforward to show from (20.2) or the discrete form (20.12) that the spatial variance of the (normalized) energy density

$$\sigma^2 \equiv \frac{M_2}{M_0} - \left(\frac{M_1}{M_0}\right)^2 \quad \text{with} \quad M_n \equiv \int x^n u(x) \mathrm{d}x \tag{20.3}$$

(or the discrete analogon) scales linear in time, i.e., $\sigma^2 \propto t$. Therefore, linear scaling of the variance is a signature of normal transport. However, from comparison with a Boltzmann equation for non-interaction particles or the Schrödinger equation for a free particle, a quadratic scaling of the variance, i.e., $\sigma^2 \propto t^2$ is routinely interpreted as a signature of ballistic transport. It is straightforward to show that boldly replacing a constant $\gamma$ in (20.2) with a function that grows linear in time, i.e., $\gamma \propto t$ yields such a variance that scales quadratically in time. Thus whenever we find our quantum dynamics in good accord with dynamics as generated by (20.2) but with a linear increasing coefficient $\gamma$, we classify the transport behavior as ballistic.

Here, the TCL projection operator technique is used to investigate the relaxation in a spatially structured model system which exhibits, dependent on the model parameters, both types of transport behavior. Furthermore, the transport behavior in the model essentially depends on the considered length scale, which is analyzed in the Fourier space of modes.

## 20.2 Single-Particle Modular Quantum System

Let us now define the modular quantum system in some detail. According to Fig. 20.1, the model consists of altogether $N$ identical subunits which are labeled by $\mu = 0, \ldots, N - 1$. These subunits are assumed to have non-degenerate ground states $|0, \mu\rangle$, with energy $\varepsilon_0 = 0$, and $n$ excited states $|i, \mu\rangle$ each, with equidistant energies $\varepsilon_i$ ($i = 1, \ldots, n$). The latter form a local band of width $\delta\varepsilon$ which is small compared to the local gap $\Delta E$. The local Hamiltonian of the $\mu$th subunit reads

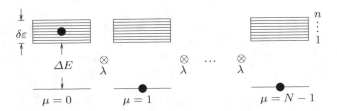

**Fig. 20.1** A chain of $N$ identical subunits: each subunit features a non-degenerate ground state, a wide energy gap $\Delta E$, and a comparatively narrow energy band $\delta\varepsilon$ which contains $n$ equidistant states. The *dots* indicate that one subunit is excited while all other subunits are in their ground state, i.e., *the dots* visualize a state from the investigated one-excitation space

$$\hat{h}_0(\mu) = \sum_{i=0}^{n} \varepsilon_i |i, \mu\rangle\langle i, \mu| \tag{20.4}$$

and, thus, the total Hamiltonian yields

$$\hat{H} = \hat{H}_0 + \hat{V} = \sum_{\mu=0}^{N-1} \left[\hat{h}_0(\mu) + \lambda\,\hat{v}(\mu, \mu+1)\right], \tag{20.5}$$

where we use periodic boundary conditions, i.e., we identify $\mu = N$ with $\mu = 0$. Hence, the model is a ring of subunits which are coupled by a next-neighbor interaction,

$$\hat{v}(\mu, \mu+1) = \sum_{i,j=1}^{n} \left[c_{ij}\,\hat{s}_i(\mu)\,\hat{s}_j^\dagger(\mu+1) + c_{ij}^*\,\hat{s}_i^\dagger(\mu)\,\hat{s}_j(\mu+1)\right]. \tag{20.6}$$

The operator $\hat{s}_i(\mu) = |0, \mu\rangle\langle i, \mu|$ acts on the state space of the $\mu$th subunit and describes a transition from the excited state $|i, \mu\rangle$ to the ground state $|0, \mu\rangle$. Inversely, $\hat{s}_i^\dagger(\mu)$ describes a transition from the ground state to the excited state. The $c_{ij}$ are Gaussian-distributed numbers as before in Chap. 19 with zero mean and form a random matrix normalized according to

$$\frac{1}{n^2} \sum_{i,j=1}^{n} |c_{ij}|^2 = 1, \tag{20.7}$$

such that $\lambda$ in (20.5) represents the overall coupling strength between adjacent sites. Note that this form of the interaction restricts the dynamics to the "one excitation subspace" if one starts in it. Note, furthermore, that the constants do not depend on the site $\mu$ here, i.e., the interaction is the same between different pairs of sites. From this it follows that the model is translationally invariant.

This model may be viewed as a simplified model for, e.g., a chain of coupled molecules or quantum dots. In this case the hopping of an excitation from one subunit to another corresponds to energy transport. However, it may as well be viewed as a tight-binding model for particles on a lattice but, according to (20.6), without particle–particle interaction in the sense of a Hubbard model. Consequently, this system may also be characterized as a "single-particle multi-band quantum wire" with random inter-band hoppings. Nevertheless, since the $c_{ij}$ in (20.6) are independent from $\mu$, these are systems without disorder in the sense of, say, an Anderson model [2].

In the following we investigate this model with the TCL method. For an analysis according to HAM, see [3–6]. A comparison to an analysis of the model by means of the Kubo formula can be found in [7], an approach based on the Boltzmann equation in [8] and some aspects of quantum chaos in [9].

## 20.3 General TCL Approach

An operator projecting onto the excited band of subsystem $\mu$ is defined by

$$\hat{\Pi}_\mu = \sum_{i=1}^{n} |i, \mu\rangle\langle i, \mu|. \tag{20.8}$$

We denote the operator which projects the $\mu$th subunit onto its band of excited states and all other subunits onto their ground state by

$$\hat{P}_\mu = |0, 0\rangle\langle 0, 0| \otimes \cdots \otimes \hat{\Pi}_\mu \otimes \cdots \otimes |0, N-1\rangle\langle 0, N-1|. \tag{20.9}$$

The total state of the system is represented by a time-dependent density matrix $\rho(t)$. Then the quantity $P_\mu(t) = \mathrm{Tr}\left\{\hat{P}_\mu \hat{\rho}(t)\right\}$ represents the probability of finding an excitation of the $\mu$th subunit, while all other subunits are in their ground state. Since the band is small compared to the gap, the local energy at site $\mu$ is approximately given by $\Delta E P_\mu$. Hence $P_\mu$ is our discrete analogon to the energy density as described in Sect. 20.1. Therefore we aim at describing the dynamical behavior of the $P_\mu(t)$ to explicitly perform the classification explained in Sect. 20.1. Since (20.6) does not incorporate particle–particle interaction in the sense of, say, a Hubbard model, and the particle number $\sum_\mu P_\mu$ is conserved as well, we choose to restrict the analysis to the one-particle space (cf. black dots in Fig. 20.1 referring to a state of this subspace). From a numerical point of view, this restriction is advantageous, since the dimension of the problem is now given by $N \times n$ and grows only linearly with $N$ and $n$.

The TCL approach introduced in Sect. 18.3 requires a set of observables, here chosen for reasons given above, to be the set of band projections $\hat{P}_\mu$ defined in (20.9). Hence, the correlated projection superoperator defined in (18.5) reads

$$\mathcal{P}\hat{\rho}(t) = \sum_\mu \mathrm{Tr}\left\{\hat{P}_\mu \hat{\rho}(t)\right\} \frac{1}{n} \hat{P}_\mu = \frac{1}{n} \sum_\mu P_\mu(t) \hat{P}_\mu \tag{20.10}$$

(see also [10–12]). Proceeding as in Chap. 19 to derive a second-order reduced dynamical equation for the observables $P_\mu$, i.e., replacing the operators $\hat{B}_n$ in (18.18) by the operators $\hat{P}_\mu$, and using both definition (20.9) and Hamiltonian (20.5) in the interaction picture, we get to second order

$$\frac{d}{dt} P_\mu = -\sum_\nu \int_0^t dt_1 \frac{1}{n} \mathrm{Tr}\left\{\hat{\Pi}_\mu[\hat{V}(t), [\hat{V}(t_1), \hat{\Pi}_\nu]]\right\} P_\nu, \tag{20.11}$$

where we have used the orthogonality property (18.6) of the relevant operators which is obviously valid for the basis $\{\hat{P}_\mu\}$ as well (see also [3, 6]). As will be demonstrated in some detail below (Sect. 20.4) this formulation may be cast into the following form:

$$\frac{d}{dt} P_\mu = \gamma(t), (P_{\mu+1} + P_{\mu-1} - 2 P_\mu). \tag{20.12}$$

This dynamical equation obviously describes the dynamics of the set of relevant quantities, i.e., the probabilities $P_\mu$. However, it is only a suitable approximation to the full dynamical behavior, if all higher order terms of the expansion (18.16) of the TCL generator are sufficiently small compared to this leading second-order term. As already mentioned (cf. Sect. 19.2), this is not only a question of the separation of timescales. Whether this is indeed the case, can only be checked by computing some of the higher order terms or use sophisticated criteria to estimate their magnitude. For more details about these convergence tests, see [10, 12–14]. On the other hand comparing the solution of the second-order equation (20.11) directly with the exact dynamics which is given by the numerical solution of the full Schrödinger equation and finding a good accordance between both descriptions is a sufficient criterion for the convergence of the expansion as well. For the model at hand we are going to proceed this way. However, for more realistic models this "check" can hardly be performed because of the high dimensionality of the problems. Thus the modular quantum system is an instructive example for a system for which a second-order TCL description captures all relevant transport dynamics. And such a second-order TCL description may also be performed for much more complicated systems.

However, the dynamical equation (20.12) may describe both diffusive and ballistic transport behaviors. If the coefficient $\gamma$ is time independent, it is basically a discrete diffusion equation; thus, as outlined in Sect. 20.1, in this case we classify the dynamics as diffusive. A time-dependent coefficient, however, would indicate non-normal or possibly ballistic behavior. In the following section we show how (20.12) follows from the TCL master equation in second-order (20.11). We investigate the time dependence of $\gamma$ essentially by analyzing a two-time correlation function.

## 20.4  Concrete TCL Description and Timescales

Plugging the concrete interaction operator (20.6) into (20.11) and using the orthogonality of the relevant observables yield

$$\frac{d}{dt} P_\mu = - \sum_\nu \int_0^t dt_1 \, f_2(t, t_1) [\delta_{\mu,\nu+1} + \delta_{\mu,\nu-1} - 2\delta_{\mu,\nu}] P_\nu. \tag{20.13}$$

Here, the possibly time-dependent coefficient is found to be an integral over a two-time correlation function (cf. [15])

$$f_2(t, t_1) = f(\tau) = \frac{1}{n} \mathrm{Tr} \left\{ \hat{V}(\tau) \hat{V} \hat{\Pi}_\mu \right\}, \tag{20.14}$$

with $\tau = t - t_1$, similar to the relaxation coefficient in Chap. 19. (Here again the time argument refers to the interaction picture.) Therefore, it is convenient to define

$$\gamma(t) = \int_0^t dt_1\, f_2(t, t_1)\,. \tag{20.15}$$

Since the model features translational invariance, the function $f(\tau)$ is independent of $\mu$. Of course, the correlation function depends on the concrete realization of the interaction matrix. Nevertheless, due to the law of large numbers, the crucial features are the same for the overwhelming majority of all possible realizations, as long as $\sqrt{n} \gg 1$. Integral (20.15) already appeared and has been analyzed in detail in Sect. 19.2. The typical form of $f(\tau)$ is depicted in Fig. 20.2. It decays like a standard correlation function on a timescale on the order of $\tau_c = 1/\delta\varepsilon$. The area under the first peak is approximately given by $\gamma_0 = 2\pi n\lambda^2/\delta\varepsilon$. However, unlike standard correlation functions, due to the equidistant energies in the local bands, $f(\tau)$ is strictly periodic with the period $\vartheta = 2\pi n/\delta\varepsilon$. Consequently, its time integral $\gamma(t)$ defined in (20.15) is a step-like function, as shown in Fig. 20.2.

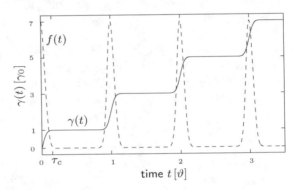

**Fig. 20.2** Typical form of the correlation function $f_2(t, t_1)$ (*dashed line*) as well as its time integral $\gamma(t)$ (*solid line*). The correlation function decays on a timescale $\tau_c$, but features complete revivals with period $\vartheta$. Consequently, $\gamma(t)$ is a step-like function. For details, see [16]

Thus, on a timescale $t$ with $\tau_c < t < \vartheta$ one finds a time-independent rate $\gamma(t) = \gamma_0$. Together with (20.13) this constant rate directly yields the discrete diffusion equation already given in (20.12). Hence, on this timescale there is indeed full agreement with a discrete diffusion equation featuring a diffusion coefficient $\gamma_0$. In Fig. 20.3 the local probability dynamics for a system which relaxes entirely on the above timescale is displayed. Compared are data from the solution of (20.12) with data from the numerical solution of the full time-dependent Schrödinger equation. Obviously there is very good agreement which again justifies the use of a second-order projective description.

However, on a timescale $t$ with $t \gg \vartheta$, one may roughly approximate $\gamma$ by a linearly increasing function $\gamma(t) \approx 2\gamma_0 t$. Due to reasons explained in Sect. 20.1 this indicates ballistic behavior. The timescale by which the relaxation will eventually be controlled is set by the typical length scale of the decaying probability distribution. This will be analyzed and explained in the next section.

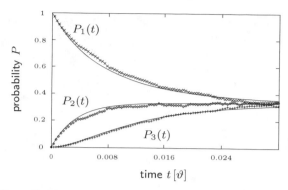

**Fig. 20.3** Probability to find the $\mu$th subunit excited. Comparison of the solution of the TCL master equation (20.12) (*solid lines*) and the solution of the exact Schrödinger equation (*crosses*). For details, see [3–6]. Model parameters: $N = 3, n = 500, \delta\varepsilon = 0.005, \lambda = 5 \times 10^{-5}$

## 20.5  TCL Approach in Fourier Space

In order to get a more detailed picture of the dynamical behavior of the system with respect to length scales, we now switch to the Fourier representation of the spatial density (cf. [16]). It is well known from Fourier's work that a diffusion equation decouples with respect to, e.g., cosine-shaped spatial density profiles (see Sect. 17.2 and [17]). Thus, starting from the diffusion equation (20.12) using the transformation

$$F_q = C_q \sum_{\mu=0}^{N-1} \cos(q\,\mu)\,P_\mu \,, \tag{20.16}$$

where $C_q$ is an appropriate normalization constant yields

$$\frac{\mathrm{d}}{\mathrm{d}t} F_q = -2\,(1 - \cos q)\,\gamma\,F_q, \tag{20.17}$$

with $q = 2\pi k/N$ and $k = 0, 1, \ldots, N/2$. Consequently, if the model indeed shows diffusive transport, the modes $F_q$ have to relax exponentially according to

$$F_q(t) = \mathrm{e}^{-2(1-\cos q)\gamma t}\,F_q(0) \,, \tag{20.18}$$

where, for diffusive transport, $\gamma$ has of course to be time independent. However, if the $F_q$ are found to relax exponentially for some regime of $q$ only, the model is said to behave diffusively on the corresponding length scale

$$l = \frac{2\pi}{q} \,. \tag{20.19}$$

Within a microscopic quantum mechanical picture the exact dynamics of the modes $F_q$ is most conveniently expressed in terms of expectation values of mode operators

$$\hat{F}_q = C_q \sum_{\mu=0}^{N-1} \cos(q\,\mu)\,\hat{P}_\mu \tag{20.20}$$

as

$$F_q(t) = \text{Tr}\left\{\hat{F}_q\,\hat{\rho}(t)\right\}, \tag{20.21}$$

where we choose $C_q = \sqrt{2/nN}$ for $q \neq 0, \pi$ and $C_0 = C_\pi = \sqrt{1/nN}$. Due to this choice the mode operators are orthonormal with respect to (18.6). Consequently, a suitable correlated projection superoperator according to the observables $\hat{F}_q$ is defined by

$$\mathcal{F}\hat{\rho}(t) = \sum_q \text{Tr}\left\{\hat{F}_q\,\hat{\rho}(t)\right\}\hat{F}_q = \sum_q F_q(t)\,\hat{F}_q. \tag{20.22}$$

The orthonormality of the mode operators again ensures $\mathcal{F}^2 = \mathcal{F}$. Thus, the TCL projection operator technique directly yields an equation of motion for modes $F_q(t)$ here (cf. [10, 16, 18]). Note that modes with different $q$ do not couple, due to the translational invariance of the model. This application of the TCL technique, i.e., replacing the operators $\hat{B}_n$ in (18.18) by the set of relevant observables $\hat{F}_q$, leads to a time-local differential equation of the form

$$\frac{d}{dt}F_q(t) = -2\,(1 - \cos q)\gamma(t)\,F_q(t), \tag{20.23}$$

in second order, with $\gamma(t)$ as defined already in (20.15) and displayed in Fig. 20.2. The Fourier mode dynamics for a mode which relaxes essentially on timescale $t$ with $\tau_c < t < \vartheta$ is shown in Fig. 20.4. Compared are data from the solution of (20.23) with data from the numerical solution of the full time-dependent Schrödinger equation. Obviously there is also very good agreement which again justifies the use of a second-order projective description.

## 20.6  Length Scale-Dependent Transport Behavior

So far, our analysis is still incomplete in two limits: When $F_q(t)$, depending on the wavelength, decays on a timescale, which is long compared to $\vartheta$ and when the decay is fast compared to $\tau_c$. That means that the above approximate integration of the correlation function breaks down. Let us discuss these two cases in the following.

We start with the long-time limit first. As will be shown below, this limit is connected to the breakdown of the reduced description for long wavelengths and

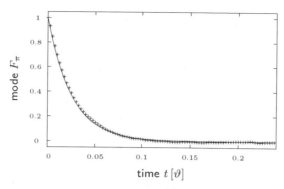

**Fig. 20.4** Mode $F_\pi(t)$ which decays on an intermediate timescale $\tau_c < t < \vartheta$. Comparison of the solution of the TCL master equation (20.23) (*solid line*) and the solution of the exact Schrödinger equation (*crosses*). For details see also [16]. Model parameters: $N = 120$, $n = 500$, $\delta\varepsilon = 0.5$, and $\lambda = 0.0005$

is therefore associated with a length scale within the system below which it behaves diffusively. According to (20.23), $F_q(t)$ decays on a timescale which is long compared to $\vartheta$, if $2(1 - \cos q)\gamma\vartheta \gg 1$ is violated. For rather small $q$ or, equivalently, for rather large wavelengths $l$ as defined in (20.19), we may approximate $2(1 - \cos q) \approx q^2 = 4\pi^2/l^2$ which eventually leads to the condition

$$\left(\frac{4\pi^2 n\lambda}{l\delta\varepsilon}\right)^? \gg 1,\qquad(20.24)$$

where we again used the estimation $\gamma_0$ for $\gamma$, which has been explained below (20.15). Consequently, if this condition is satisfied for the largest possible $l$, i.e., for $l = N$, the system behaves diffusively for all wavelengths. But if the system is large enough to allow for some $l$ such that (20.24) is violated, then diffusive behavior breaks down for long wavelengths.

Let us again use the measure $D^2$ already introduced in (19.27) to estimate the distance between the exact solution and the fully diffusive dynamics as given by (20.18) with $\gamma = \gamma_0$ for the longest possible mode $l = N$ (see also [9]). Thus, concretely, we consider

$$D_q^2 = \frac{\Gamma}{5}\int_0^{5/\Gamma}\left[F_q^{\text{diffusive}} - F_q^{\text{exact}}\right]^2,\qquad(20.25)$$

with $\Gamma = 2(1 - \cos q)\gamma_0$. In Fig. 20.5 we show this measure (gray coded) in dependence of the number of subunits $N$ and the amount of levels $n$. Dark regions belong to large deviations and thus large $D^2$, where light ones refer to small values of $D^2$ and thus a very good accordance between diffusive dynamics and the exact Schrödinger dynamics. As can be seen from the figure, there is diffusive behavior for smaller number of subunits, $N$, only if the number of levels within the subunits, $n$, is reasonably high. For more subunits, i.e., longer length scales, even more levels within each subunit are required for the system to show diffusive behavior. The "boundary" between diffusive and non-diffusive behaviors as appearing in Fig. 20.5 is in accord with (20.24). Thus we may conclude that in the thermodynamic limit, i.e., for an infinite number of subunits, but a finite number of levels within each subsystem, there must be non-diffusive behavior.

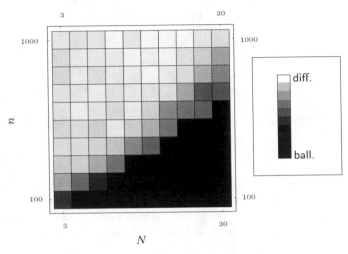

**Fig. 20.5** Deviations $D^2$, cf. (20.25), of the time evolution of the longest modes $F_q(t)$ with $q = 2\pi/N$ from the exact solution for different model parameters $N$ and $n$. The result is in accord with the claim that diffusive transport behavior is restricted to the regime defined by condition (20.24). For details see also [16]. Model parameters: $\delta\varepsilon = 0.5$, $\lambda = 0.0005$

However, it still remains to clarify what type of transport behavior is on hand, if (20.24) is violated. According to Fig. 20.2, and as already mentioned at the end of Sect. 20.4 one may approximate $\gamma(t) \approx 2\gamma_0 t$, if the timescales are much larger than $\vartheta$. To repeat, this linear scaling implies ballistic transport. The solution of (20.23) with this linear increasing $\gamma$ is then a Gaussian. And, as may be seen from Fig. 20.6a, this prediction is still in very good agreement with the numerical solution of the Schrödinger equation. Thus we may conclude that there is ballistic transport in the limit of long wavelengths. Note that in this example a second-order TCL description yields a good result even for times much larger than the recurrence time of the correlation function.

Second, $F_q(t)$ may decay on a timescale which is short compared to $\tau_c$. This will happen, if the inequality $2(1 - \cos q)\gamma\tau_c \ll 1$ is violated. For the largest possible $q$ we may approximate $2(1 - \cos q) \approx 4$. Thus, the inequality reads

$$\frac{8\pi n\lambda^2}{\delta\varepsilon^2} \ll 1 . \tag{20.26}$$

If this condition is violated, the system exhibits non-diffusive transport behavior in the limit of short wavelengths. (Note that due to the parameters chosen this transi-

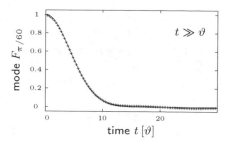

**Fig. 20.6a** Mode $F_{\pi/60}(t)$ decaying on a timescale $t \gg T$. Model parameters $N = 120$, $n = 500$, $\delta\varepsilon = 0.5$, $\lambda = 0.0005$. Comparison between exact evolution (*crosses*) and the theoretical prediction (*solid line*)

tion does not appear in Fig. 20.5.) Moreover, if we assume that the leading order contribution of the TCL generator still yields reasonable results for not too large $\lambda$, we expect a linearly growing rate $\gamma(t)$ and therefore a Gaussian decay, i.e., ballistic transport behavior again. This conclusion is in accord with Fig. 20.6b. Note that in this example a second-order TCL description yields a good result although the correlation time and the relaxation time are of comparable magnitude.

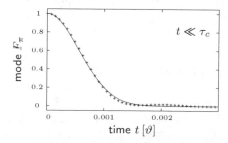

**Fig. 20.6b** Mode $F_{\pi}(t)$ decaying on a timescale $t \ll \tau_c$. Model parameters $N = 120$, $n = 500$, $\delta\varepsilon = 0.5$, $\lambda = 0.004$. Comparison between exact evolution (*crosses*) and the theoretical prediction (*solid line*)

The second transition to non-diffusive behavior in the limit of short wavelengths may be routinely interpreted as the breakdown of diffusive behavior below some length scale which is supposed to appear when the motion of, say, individual particles is resolved and length scales become short compared to mean free paths. However, the first transition to ballistic transport behavior in the limit of long wavelengths does not fit into this scheme of interpretation. Nevertheless, it may be understood in the framework of standard solid state theory, where translational invariant systems without particle–particle interactions exhibit ballistic transport behavior with quasi-particle velocities, which are determined by the dispersion relation of the bands. The analysis of the band structure of our model shows that the above transition to ballistic behavior coincides with the transition of the $E(k)$ versus $k$ diagram from being just a set of disconnected points to the standard smooth band diagram, which allows for a definition of the quasi-particle velocities. It is also possible to show that the region where condition (20.24) is violated is the only region where current eigenstates coincide with Bloch energy eigenstates. Thus, the size below which condition (20.24) holds may be interpreted as the size below which the concepts of standard solid state theory become inapplicable.

Let us finally remark that the excellent agreement of the numerical with the TCL results is really surprising in the following sense: The fact that the correlation function $f(\tau)$ features complete revivals is a hint for strong memory effects. It appears to be a widespread belief that long memory times have to be treated by means of the Nakajima–Zwanzig projection operator technique (NZ) [19, 20]. The corresponding leading order equation is obtained by replacing $F_q(t)$ with $F_q(t - \tau)$ on the right-hand side of (20.23). This leads to an integro-differential equation with the memory kernel $f(\tau)$. In the diffusive regime, $t < \vartheta$, the solutions of the TCL and the NZ technique are almost identical. But in ballistic regime, $t \gg \vartheta$, the NZ technique predicts a pure oscillatory behavior

$$F_q(t) = \cos\left(2\sqrt{1 - \cos q}\ \lambda t\right) F_q(0), \qquad (20.27)$$

which obviously contradicts the numerical solutions of the Schrödinger equation and clearly demonstrates the failure of the NZ approach in the description of the long-time dynamics.

## 20.7 Energy and Heat Transport Coefficient

We consider the model system introduced in Fig. 20.1 in the diffusive regime, however, for simplicity with $N = 2$ subunits only. Thus, the dynamics of the system is well described by the solution of the diffusion equation (20.12). The energy current between the two sites is defined by the change of the internal energy $U_\mu$,

$$J = \frac{1}{2}\left(\frac{dU_1}{dt} - \frac{dU_2}{dt}\right). \qquad (20.28)$$

Realizing that the internal energy is given by the probability to be in the excited band of the subunit times the width of the energy gap, $U_\mu = \Delta E P_\mu$, the current can be reformulated as follows:

$$J = \frac{\Delta E}{2}\left(\frac{dP_1}{dt} - \frac{dP_2}{dt}\right). \qquad (20.29)$$

Using the diffusion equation (20.12), here for two subunits only, we find

$$\begin{aligned}J &= -\gamma\, \Delta E\, (P_2 - P_1)\\ &= -\gamma\, (U_2 - U_1).\end{aligned} \qquad (20.30)$$

Hence, one gets Fourier's law (20.1) in case the dynamics of the system is given by a discrete diffusion equation. Here, the energy current is connected to the energy gradient inside the system. The proportionality is defined as the energy transport coefficient which is given by and below (20.15). Thus, the microscopic constant energy transport coefficient is given by the coupling strength between adjacent subunits and the state density of the band. Besides having derived a microscopic formula for the conductivity we are also equipped with criteria to decide under which circumstances a diffusive situation is present. The above consideration is easily extended to more than two subunits, see [3].

Typically, we would expect diffusive behavior near the global equilibrium only. Remarkably, in the present system diffusive behavior does not require an initially small energy gradient, i.e., a state near equilibrium. The initial state used for the investigation shown in Fig. 20.3, e.g., with one system excited, all others in the ground state, is far from the final expected equilibrium. Nevertheless, one can find a statistical diffusive behavior according to (20.12) here, as shown in Fig. 20.3.

So far we have considered energy diffusion through the system only. The final state should approach equipartition of energy over all subunits, i.e., a thermal equilibrium state (cf. Part I of this book). Close to this equilibrium we expect the system to be in a state, where the probability distribution of each subunit is approximately canonical, but still with a slightly different temperature for each site.

Specializing in those "local equilibrium states" and exploiting the TCL results allow for a direct connection of the local energy current between any two adjacent subunits with their local temperature difference. Since this connection is found to be linear in the temperature differences as well, one can simply extract the temperature-dependent heat conductivity finding

$$\kappa = \frac{2\pi\lambda^2 n}{\delta\epsilon} \left(\frac{\Delta E}{T}\right)^2 \frac{ne^{\frac{-\Delta E}{T}}}{\left(1 + ne^{\frac{-\Delta E}{T}}\right)^2}, \qquad (20.31)$$

as displayed in Fig. 20.7.

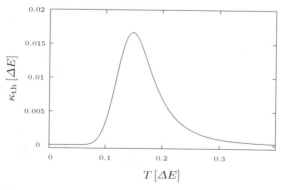

**Fig. 20.7** Heat conductivity (20.31) over temperature for a system with $n = 500$, $\delta\epsilon = 0.005$, and $\lambda = 5 \times 10^{-5}$

For our model this is in agreement with $\kappa = \gamma c$, with the specific heat $c$ of one subunit. More details of how to approach this thermal transport coefficient can be found in [3, 5, 6].

# References

1. R. Alicki, M. Fannes, *Quantum Dynamical Systems* (Oxford University Press, Oxford, 2001)
2. P.W. Anderson, Phys. Rev. **109**(5), 1492 (1958)
3. M. Michel, *Nonequilibrium Aspects of Quantum Thermodynamics*. Ph.D. thesis, Universität Stuttgart, Stuttgart (2006). Also available online: http://elib.uni-stuttgart.de/opus/volltexte/2006/2803/
4. M. Michel, J. Gemmer, G. Mahler, Int. J. Mod. Phys. B **20**, 4855 (2006)
5. M. Michel, G. Mahler, J. Gemmer, Phys. Rev. Lett. **95**, 180602 (2005)
6. M. Michel, J. Gemmer, G. Mahler, Phys. Rev. E **73**, 016101 (2006)
7. J. Gemmer, R. Steinigeweg, M. Michel, Phys. Rev. B **73**, 104302 (2006)
8. M. Kadiroglu, J. Gemmer, Phys. Rev. B **76**, 024306 (2007)
9. R. Steinigeweg, J. Gemmer, M. Michel, Europhys. Lett. **75**, 406 (2006)
10. H.P. Breuer, J. Gemmer, Eur. Phys. J. Special Topics **151**, 1 (2007)
11. H.P. Breuer, Phys. Rev. A **75**, 022103 (2007)
12. H.P. Breuer, J. Gemmer, M. Michel, Phys. Rev. E **73**, 016139 (2006)
13. C. Bartsch, P. Vidal, *Transport and Relaxation: Foundation and Perspectives* (Springer, 2007), chap. Statistical Relaxation in Closed Quantum Systems and the Van Hove-Limit. EPJST
14. C. Bartsch, J. Gemmer, Phys. Rev. E **77**, 011119 (2007)
15. H.P. Breuer, TCL for Quantum Transport Models (2006). Private communication.
16. R. Steinigeweg, H.P. Breuer, J. Gemmer, Phys. Rev. Lett. **99**(15), 150601 (2007)
17. J. Fourier, *The Analytical Theory of Heat* (Dover Publ., New York, 1955). Transl. by Alexander Freeman. Original: Théorie analytique de la chaleur (publ. 1822). Unabridged reprinting of the engl. Transl. anno 1878.
18. H.P. Breuer, F. Petruccione, *The Theory of Open Quantum Systems* (Oxford University Press, Oxford, 2002)
19. S. Nakajima, Prog. Theor. Phys. **20**, 948 (1958)
20. R. Zwanzig, J. Chem. Phys. **33**, 1338 (1960)

# Chapter 21
# Open System Approach to Transport[1]

> No subject has more extensive relations with the progress of
> industry and the natural sciences; for the action of heat is
> always present, it penetrates all bodies and spaces, it
> influences the processes of the arts, and occurs in all
> phenomena of the universe.
>
> — J. Fourier [2]

**Abstract** In this section yet another approach to transport is examined. Rather than analyzing spatial energy density dynamics as done in the previous section, we now aim at investigating a stationary non-equilibrium state. This state is induced through the local coupling of reservoirs with different temperatures at either end of the system. The system is a chain of two-level systems coupled according to the Heisenberg model. The baths are modeled according to standard open system theory (see [3–6]). The resulting equations are numerically solved using a stochastic unraveling scheme.

The Liouville–von Neumann equation, describing the time evolution of the density operator of the system, has to be extended by incoherent damping terms simulating the influence of the heat baths. How to set up the correct dynamical equation here is highly nontrivial and is based on many subtle approximation schemes. In case of an improper approach the derivation can lead to mathematically correct, but physically irrelevant dynamical equations, as discussed recently [7].

Having derived a proper quantum master equation (QME), the interpretation of the results for finite systems is relatively easy: After finding the stationary state of the dissipative dynamics all interesting quantities, such as currents and energy profiles, are simply accessible by computing the expectation value of the respective operator. However, also this approach is restricted to finite systems since a complete analytical solution for larger systems is not available. Thus, the extrapolation to infinite systems needs a careful discussion to exclude errors due to the finite size of the investigated models as well.

---

[1] This chapter is based on [1].

Gemmer, J. et al.: *Open System Approach to Transport.* Lect. Notes Phys. **784**, 241–253 (2009)
DOI 10.1007/978-3-540-70510-9_21      © Springer-Verlag Berlin Heidelberg 2009

## 21.1 Model System

The considered system consists of $N$ weakly coupled subunits described by the Hamiltonian

$$\hat{H} = \hat{H}_{\text{loc}} + \hat{H}_{\text{int}} = \sum_{\mu=1}^{N} \hat{h}^{(\mu)} + J \sum_{\mu=1}^{N-1} \hat{h}^{(\mu,\mu+1)} . \tag{21.1}$$

The first part $\hat{H}_{\text{loc}}$ of the Hamiltonian contains the local spectra of the subunits. The second part $\hat{H}_{\text{int}}$ describes the interaction between adjacent sites with the coupling strength $J$. Here, we require that the interaction is weak in the sense that the energy contained in the local part is much larger than the energy contained in the interaction part, $\langle \hat{H}_{\text{loc}} \rangle \gg \langle \hat{H}_{\text{int}} \rangle$.

We will investigate one-dimensional chains of two-level atoms or spin-1/2 particles. Both, two-level atoms and spins, are described by the same algebra, and thus, it is convenient to use the Pauli operators (see Sect. 2.2.2) as a suitable operator basis. The above-mentioned weak coupling is fulfilled by introducing a local Zeeman splitting $\frac{\Omega}{2} \hat{\sigma}_z$, where we require $\Omega$ to be much larger than the coupling constant $J$. Note that this weak internal coupling constraint is a necessary precondition for the validity of the master equation introduced below, i.e., we are not able to consider systems with $\Omega$ approaching the same magnitude as $J$.

To investigate the transport properties of these systems, they will be explicitly coupled to independent environments of different temperatures. In the following we discuss the appropriate QME (cf. Sect. 4.8 and [8]).

## 21.2 Lindblad Quantum Master Equation

In general, the derivation of the QME from a microscopic model, i.e., a system coupled to an infinitely large environment, relies on some well-known approximation schemes, the Markov [9, 10] assumption, the Born approximation, and the secular approximation [9, 11]. Recently, there has been a discussion on how to derive a suitable Lindblad [12, 13] QME for a non-equilibrium scenario [7], i.e., an equation to investigate transport in weakly coupled quantum systems. The Lindblad form of a QME defines a trace and hermiticity-preserving, completely positive dynamical map [8, 14], which thus retains all properties of the density operator at all times. In order to approach this dynamical equation the approximations are carefully carried out in a minimal invasive manner, to conserve the central non-equilibrium properties of the model. It can be shown that this non-equilibrium Lindblad QME is in very good accordance with the results of the Redfield master [8] equation (non-Lindbladian), contrary to the standard Lindblad QME in the weak coupling limit [7].

In a non-equilibrium investigation one needs two heat baths at different temperatures locally coupled to the system, i.e., the heat baths couple only to a subunit at either edge of the chain. The QME of such a situation yields

$$\frac{d\hat{\rho}}{dt} = -i[\hat{H}, \hat{\rho}] + \hat{D}_L(\hat{\rho}) + \hat{D}_R(\hat{\rho}), \tag{21.2}$$

where the dissipator $\hat{D}_L$ refers to the left and $\hat{D}_R$ to the right heat bath; they both depend on the full density operator $\hat{\rho}$ of the system, i.e., the state of the chain described by the Hamiltonian (21.1). Both dissipators also depend on the coupling strength $\lambda$ between system and bath as well as on the temperature of the bath, respectively. Besides those incoherent damping terms (21.2) includes a coherent part containing the Hamiltonian of the system.

The dissipator describing the heat bath coupled to a subunit at the edge of the system yields

$$\hat{D}_F(\hat{\rho}) = \sum_{k,l=1}^{2} (\gamma_F)_{kl} \left( \hat{F}_k \hat{\rho} \hat{F}_l^\dagger - \frac{1}{2}[\hat{F}_l^\dagger \hat{F}_k, \hat{\rho}]_+ \right), \tag{21.3}$$

with $F = L$ for the left and $F = R$ for the right heat bath. The corresponding Lindblad operators $\hat{F}_k$ are given by

$$\hat{L}_1 = \hat{\sigma}_+^{(1)} \otimes \hat{1}^{(2)} \otimes \cdots \otimes \hat{1}^{(N)}, \tag{21.4}$$

$$\hat{L}_2 = \hat{\sigma}_-^{(1)} \otimes \hat{1}^{(2)} \otimes \cdots \otimes \hat{1}^{(N)}, \tag{21.5}$$

$$\hat{R}_1 = \hat{1}^{(1)} \otimes \cdots \otimes \hat{1}^{(N-1)} \otimes \hat{\sigma}_+^{(N)}, \tag{21.6}$$

$$\hat{R}_2 = \hat{1}^{(1)} \otimes \cdots \otimes \hat{1}^{(N-1)} \otimes \hat{\sigma}_-^{(N)}, \tag{21.7}$$

with the creation and annihilation operators $\hat{\sigma}_\pm$. Here, the operators (21.4) and (21.5) belong to the left bath, and (21.6) and (21.7) to the right one. The coefficient matrices depend on the respective bath temperature $\beta_F$ and are defined as

$$\gamma_F = \begin{pmatrix} \Gamma_F(\Omega) & \sqrt{\Gamma_F(\Omega)\Gamma_F(-\Omega)} \\ \sqrt{\Gamma_F(\Omega)\Gamma_F(-\Omega)} & \Gamma_F(-\Omega) \end{pmatrix}, \tag{21.8}$$

with the rates

$$\Gamma_F(\Omega) = \frac{\lambda\Omega}{e^{\beta_F\Omega} - 1} \tag{21.9}$$

controlled by the bath coupling strength $\lambda$. For a detailed motivation for this form of the dissipators, see [8, 15].

Since (21.3) is of Lindblad form it may be simplified by diagonalizing the coefficient matrices $\gamma_F$. The complete dissipative part of (21.2) then reads

$$\hat{D}(\hat{\rho}) = \sum_{k=1}^{4} \alpha_k \left( \hat{E}_k \hat{\rho} \hat{E}_k^\dagger - \frac{1}{2}[\hat{E}_k^\dagger \hat{E}_k, \hat{\rho}]_+ \right), \tag{21.10}$$

with $\hat{E}_k$ as a linear combination of the operators $\hat{F}_k$ defined in (21.4)–(21.7) and $\alpha_k$ being non-negative numbers.

## 21.3 Observables and Fourier's Law

One of the most interesting states of a non-equilibrium scenario is the local equilibrium state, i.e., the stationary state of the QME (21.2). This state can be characterized by two central observables – the energy gradient and the energy current. Here, we use

$$h^{(\mu)} = \mathrm{Tr}\left\{\hat{h}^{(\mu)}\hat{\rho}(t)\right\} \tag{21.11}$$

as a local energy density at site $\mu$ with $\hat{\rho}(t)$ being the state of the system at time $t$. Since we are investigating internally weakly coupled subunits in the limit $\Omega \gg J$ the local energy density is reasonably well approximated by the local Hamiltonian. Therefore, we neglect contributions to the local energy by the interaction completely. Due to the smallness of $J$ these parts would be very small contributions to the above given energy density, and thus, would not dramatically change the results.

In order to obtain a current operator between two adjacent sites in the system, we consider the time evolution of the local energy operator given by the Heisenberg equation of motion for operators at site $\mu$

$$\frac{\mathrm{d}}{\mathrm{d}t}\hat{h}^{(\mu)} = \mathrm{i}[\hat{H}, \hat{h}^{(\mu)}] + \frac{\partial}{\partial t}\hat{h}^{(\mu)} . \tag{21.12}$$

Since $\hat{h}^{(\mu)}$ is not explicitly time dependent the last term vanishes. Equation (21.12) becomes after inserting (21.1)

$$\frac{\mathrm{d}}{\mathrm{d}t}\hat{h}^{(\mu)} = \mathrm{i}J\left([\hat{h}^{(\mu-1,\mu)}, \hat{h}^{(\mu)}] + [\hat{h}^{(\mu,\mu+1)}, \hat{h}^{(\mu)}]\right) . \tag{21.13}$$

Assuming the local energy to be conserved, which is justified when $\Omega \gg J$, a discretized version of the continuity equation yields

$$\frac{\mathrm{d}}{\mathrm{d}t}\hat{h}^{(\mu)} = \mathrm{div}\hat{J} = \hat{J}^{(\mu,\mu+1)} - \hat{J}^{(\mu-1,\mu)} . \tag{21.14}$$

By comparing (21.13) and (21.14) we find for the current operator

$$\hat{J}^{(\mu,\mu+1)} = \mathrm{i}J[\hat{h}^{(\mu,\mu+1)}, \hat{h}^{(\mu)}] . \tag{21.15}$$

Finally, the total energy current flowing from site $\mu$ to site $\mu + 1$ is defined as

$$J^{(\mu,\mu+1)} = \mathrm{Tr}\left\{\hat{J}^{(\mu,\mu+1)}\hat{\rho}(t)\right\} . \tag{21.16}$$

For a more detailed discussion of this heat current, see Chap. 24.

The celebrated Fourier's law (here a discrete version) states that in a proper diffusive situation the current inside the system is proportional to the gradient, i.e.,

$$J^{(\mu,\mu+1)} = -\kappa [h^{(\mu+1)} - h^{(\mu)}], \qquad (21.17)$$

here written in terms of energy current and energy gradient. If both current and gradient are equal at all sites $\mu$, and furthermore, the gradient is finite, a bulk conductivity [6, 16] follows from

$$\kappa = \frac{J^{(\mu,\mu+1)}}{h^{(\mu)} - h^{(\mu+1)}} . \qquad (21.18)$$

This is called normal or diffusive transport. On the other hand, if the gradient vanishes, $\kappa$ diverges and the transport is called ballistic. However, that does not mean that the current diverges as well. Due to the resistivity at the contact to the heat bath (in our approach $\lambda$) the current will always remain finite.

Even if we directly get a result in terms of normal or ballistic behavior for all finite systems here, a non-zero gradient in the finite system is not sufficient to deduce normal transport in the infinite one, too. The influence of the contact might dominate or long ballistic waves could be suppressed in the finite system. Thus, in order to obtain statements on the properties of the infinite system, i.e., bulk properties, it is important to investigate scaling properties as well. For normal transport behavior both gradient and current must tend to zero for infinitely large systems. Then and only then the system shows diffusive behavior. A finite current within an infinite system will always indicate ballistic transport behavior.

## 21.4 Monte Carlo Wave Function Simulation

In order to investigate the transport according to the QME requires the stationary solution $\hat{\rho}$ of (21.2). From $\hat{\rho}$ all gradients and currents can be computed with (21.11) and (21.16). Unfortunately, (21.2) is an $n^2$-dimensional system of linear differential equations if $n$ is the dimension of the Hilbert space. To find the stationary state of this equation one has to diagonalize a $n^2 \times n^2$ matrix, which is numerically restricted by the available memory.

A very powerful technique to find the stationary state without diagonalizing the Liouvillian is based on the stochastic unraveling [8] of the QME. The basic idea is to depart from a statistical treatment by means of density operators and turn to a description in terms of stochastic wave functions. In fact, any QME of Lindblad form can equivalently be formulated in terms of a stochastic Schrödinger equation (SSE) for the wave function $|\psi(t)\rangle$

$$d|\psi(t)\rangle = -i\,\hat{G}(|\psi(t)\rangle)\,|\psi(t)\rangle\,dt$$

$$+ \sum_k \left( \frac{\hat{E}_k|\psi(t)\rangle}{\|\hat{E}_k|\psi(t)\rangle\|} - |\psi(t)\rangle \right) dn_k \,, \tag{21.19}$$

which describes a piecewise deterministic process in Hilbert space. The first term on the right-hand side of (21.19) describes the deterministic evolution generated by the nonlinear operator

$$\hat{G}(|\psi(t)\rangle) = \hat{H}_{\text{eff}} + \frac{i}{2} \sum_k \alpha_k \|\hat{E}_k|\psi(t)\rangle\|^2 \,, \tag{21.20}$$

where we have introduced the non-Hermitian, effective Hamiltonian

$$\hat{H}_{\text{eff}} = \hat{H} - \frac{i}{2} \sum_{k=1}^{4} \alpha_k \hat{E}_k^\dagger \hat{E}_k \,. \tag{21.21}$$

The second term in (21.19) contains the Poisson increments $dn_k \in \{0, 1\}$ which obey the following statistical properties:

$$\langle dn_k \rangle = \|\hat{E}_k|\psi(t)\rangle\|^2 \, dt \,, \tag{21.22}$$

$$dn_k \, dn_l = \delta_{kl} \, dn_k \,. \tag{21.23}$$

The stochastic process, defined by the SSE (21.19), can be conveniently simulated by the following prescription. Starting from a normalized state, the first step of the unraveling procedure is to integrate the time-dependent Schrödinger equation according to the effective Hamiltonian defined in (21.21). Since it is not Hermitian, the normalization of the state $|\psi(t)\rangle$ decreases until $\langle \psi(t)|\psi(t)\rangle = \eta$, with $\eta$ being a random number drawn from a uniform distribution on the interval $\{0, 1\}$ at the beginning of the step. Subsequently, a jump $k$ takes place according to the probability

$$p_k = \frac{\alpha_k \|\hat{E}_k|\psi(t)\rangle\|^2}{\sum_k \alpha_k \|\hat{E}_k|\psi(t)\rangle\|^2} \,. \tag{21.24}$$

Having identified the jump $k$, the state $|\psi(t)\rangle$ is replaced by the normalized state

$$|\psi(t)\rangle \rightarrow \frac{\hat{E}_k|\psi(t)\rangle}{\|\hat{E}_k|\psi(t)\rangle\|} \,. \tag{21.25}$$

Then the algorithm starts from the beginning again with a deterministic evolution step. This procedure leads to a specific realization $r$ of the stochastic process. Let $R$ be the total number of such individual realizations. Then averaging over $R \rightarrow \infty$ realizations the time evolution of (21.2) is reproduced.

The expectation value of an observable $\hat{A}$ at time $t$ can be estimated through

$$\text{Tr}\left\{\hat{A}\,\hat{\rho}(t)\right\} \approx \frac{1}{R}\sum_{r=1}^{R}\langle\psi^r(t)|\hat{A}|\psi^r(t)\rangle \tag{21.26}$$

for a finite ensemble of $R$ realizations to arbitrary precision. This is of considerable practical importance, as one deals here with wave functions with $\mathcal{O}(n)$ elements instead of density operators with $\mathcal{O}(n^2)$ elements. Furthermore, if one is interested in the stationary state, ensemble averages can be replaced by time averages [17, 18] and one single realization suffices to determine the stationary expectation value

$$\text{Tr}\left\{\hat{A}\,\hat{\rho}\right\} \approx A_r = \frac{1}{T+1}\sum_{k=0}^{T}\langle\psi^r(t_k)|\hat{A}|\psi^r(t_k)\rangle, \tag{21.27}$$

with $t_k = t_0 + k\,\Delta t$. It turns out that introducing this uniform time discretization and allowing for jumps to occur at multiples of $\Delta t$ only has several technical advantages. However, one has to bear in mind that this introduces an error of order $\mathcal{O}(\Delta t)$ [19]. A further problem is that for $\Delta t \to 0$, the total number of time steps $T$ has to be increased, in order to retain a sufficient number of jumps in the average (21.27). There is a trade-off between this overall increase of accuracy and the cost of tedious computations. Nevertheless, in practice the time-averaging procedure proves highly efficient, and results of sufficient accuracy could always be produced. It is further advisable to discard the initial time evolution in the average in order to obtain reliable results, i.e., by choosing $t_0 \gg \Delta t$.

In order to get the standard deviation as a measure for the statistical error as well, one computes the stationary expectation value for $R$ realizations,

$$\bar{A} = \frac{1}{R}\sum_{r=1}^{R} A_r, \tag{21.28}$$

and the standard deviation of the average,

$$\sigma^2 = \frac{1}{R(R-1)}\sum_{r=1}^{R}(A_r - \bar{A})^2. \tag{21.29}$$

The errors are influenced by the chosen sampling interval $\Delta t$, the neglected steps at the beginning $t_0$, and the total number of time steps $T$ being averaged over. For all numerical results below we have chosen the parameters $\Delta t = 1$, $t_0 = 10^4$, and $T$ between $10^5$ and $10^6$. For these settings the errors are already surprisingly small.

## 21.5  Chain of Two-Level Atoms

First, we consider a chain of two-level atoms or spin-1/2 particles as depicted in
Fig. 21.1. In this case the local part of the Hamiltonian is just given by the mentioned
Zeeman splitting of each individual spin

$$\hat{h}^{(\mu)} = \frac{\Omega_\mu}{2}\,\hat{\sigma}_z^{(\mu)}\,, \tag{21.30}$$

with a splitting $\Omega_\mu$, which may differ from site to site. $\Omega_\mu$ has to be large compared
to the coupling constant $J$ to remain in the weak coupling limit. The subunits are
coupled by a generalized Heisenberg interaction

$$\hat{h}^{(\mu,\mu+1)}$$
$$= \hat{\sigma}_x^{(\mu)} \otimes \hat{\sigma}_x^{(\mu+1)} + \hat{\sigma}_y^{(\mu)} \otimes \hat{\sigma}_y^{(\mu+1)} + \Delta\hat{\sigma}_z^{(\mu)} \otimes \hat{\sigma}_z^{(\mu+1)}\,. \tag{21.31}$$

For $\Delta \neq 1$ the chain is called anisotropic, while for $\Delta = 0$ the present model is
equivalent to the XY model (Förster coupling). In this case plugging (21.30) and
the interaction given by (21.31) into (21.15), the current operator yields

$$\hat{J}^{(\mu,\mu+1)} = iJ\Omega_\mu\big[\hat{\sigma}_+^{(\mu)}\hat{\sigma}_-^{(\mu+1)} - \hat{\sigma}_-^{(\mu)}\hat{\sigma}_+^{(\mu+1)}\big]\,. \tag{21.32}$$

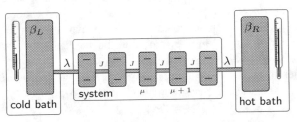

**Fig. 21.1** Chain of two-level atoms or spin-1/2 particles coupled to heat baths of different
temperatures

The above system is coupled to two heat baths of different temperatures. The left
bath is set to the inverse temperature $\beta_L = 0.5$ and the hotter one at the right-hand
side is at $\beta_R = 0.25$. Both baths couple with the same coupling strength $\lambda = 0.01$
to the system.

Having obtained the stationary state of (21.2) by the method presented in
Sect. 21.4 one can compute both the stationary energy profile within the system and
the current flowing through the system. In Fig. 21.2 the internal gradient is shown
for an isotropic chain $\Delta = 1$ of $N = 12, 13, \ldots, 16$ spins for the same constant local
field $\Omega = 1$ and coupling strength $J = 0.01$. To show that the gradient is equivalent
for the different system sizes we have normalized the chain length in Fig. 21.2 to 1.
The fit (line in Fig. 21.2) is carried out for system size $N = 16$ excluding the sites 1
and 16 to avoid the strong influences of the contacts. All chains $N = 12, \ldots, 16$ show

the same gradient. However, the energy difference between adjacent sites decreases for growing system sizes. The change in the internal gradient is shown in the upper diagram of Fig. 21.3. Here the gradient is plotted over the reciprocal chain length. The error bars refer to an average over the energy differences in all adjacent pairs of sites, where the first and the last pairs have again been neglected, as already done in Fig. 21.2. As can be seen from Fig. 21.3 the gradient decreases for larger systems until it approaches zero for an infinite chain, which is in accordance with the expected behavior in the thermodynamical limit.

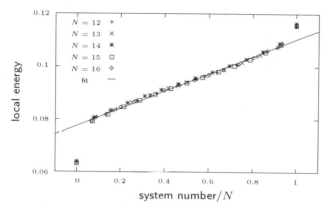

**Fig. 21.2** Local energies in a Heisenberg chain with length $N = 12 - 16$. The system number is normalized by the chain length. The fit is carried out for chain length $N = 16$ excluding sites 1 and 16. System parameters: $J = 0.01$, $\Delta = 1$, $\Omega = 1$, $\lambda = 0.01$, $\beta_L = 0.5$, $\beta_R = 0.25$

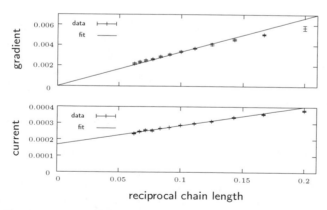

**Fig. 21.3** Scaling properties of the Heisenberg chain $N = 5 - 16$. *Lines* are fits carried out for chain length $N = 6 - 16$. System parameters: $J = 0.01$, $\Delta = 1$, $\Omega = 1$, $\lambda = 0.01$, $\beta_L = 0.5$, $\beta_R = 0.25$

The lower part of Fig. 21.3 shows the scaling behavior of the current through the system. Here, error bars refer to the stochastic algorithm given by the square root of (21.29). The current decreases similarly as the gradient; however, the extrapolation for the infinitely long chain does not approach zero. A finite current for an infinite system is a typical characteristic for ballistic transport behavior. According to the data shown in Fig. 21.3 one should eventually expect to find ballistic transport in the Heisenberg chain.

Figure 21.4 shows the local energy profile within the XY model ($\Delta = 0$). In comparison to Fig. 21.2 the profile is flat. According to Fourier's law (21.17) this could be interpreted as ballistic behavior in the investigated finite models of different lengths (cf. discussion in Sect. 21.3). The local current between sites $\mu$ and $\mu + 1$ remains finite, although the gradient within the system vanishes. This local current is constant for all investigated system sizes and we find for the chosen parameters $(5.33 \pm 0.05) \times 10^{-4}$. The results concerning the Heisenberg chain and the XY model are in accordance with some earlier results (for smaller systems) based on the full diagonalization of the Liouvillian [3, 20].

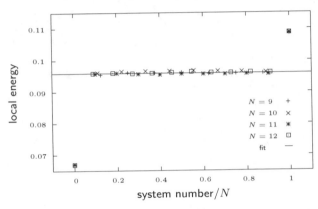

**Fig. 21.4** Local energies in an XY chain with length $N = 9-12$. The system number is normalized by the chain length. The fit is carried out for chain length $N = 12$ excluding sites 1 and 12. System parameters: $J = 0.01$, $\Delta = 0$, $\Omega = 1$, $\lambda = 0.01$, $\beta_L = 0.5$, $\beta_R = 0.25$

In Fig. 21.5 we investigate the dependence of the extrapolated value of the current and the energy gradient of an infinitely long chain on the coupling strength $\lambda$ at the contact. In order to get comparable data and errors, all other parameters are kept constant. A decrease in the external coupling strength is combined with a decrease in the global decay time of the system and a drastic change of the jump probabilities (21.24) as well. To gain a proper expectation value from (21.27) with a rather small error it is crucial that the sampling time step $\Delta t$ of the continuous stochastic trajectory is chosen such that a suitable amount of both coherent dynamics and stochastic jumps enter the average. That means, if $\Delta t$ is too large, so that typically after each coherent step already a jump follows, the result of (21.27) will deteriorate. Thus, changing the external coupling strength would also require an adaption of the sampling parameter $\Delta t$. Furthermore, in case of a larger decay time of the system also the parameter $t_0$ (initial neglect of data points) has to be increased. Thus, having fixed these parameters to get comparable data we are restricted to a small change of the external coupling strength only. For the finite system an increase in the external coupling strength λ means that a larger current is injected into the system, as follows from Fig. 21.5. The resistance of the contact is decreased. Finally, this also results in a larger gradient within the system. Nevertheless, Fig. 21.5 shows that even if the results for finite systems change drastically (especially for very small system sizes) the extrapolation for the infinite chain remains the same within the accuracy of the fit.

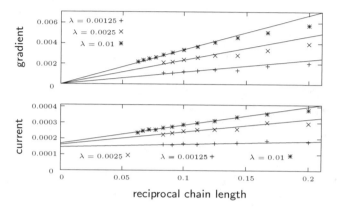

**Fig. 21.5** Dependence of the extrapolated value for the current of an infinitely long Heisenberg chain on the bath coupling strength λ

Figure 21.6 shows the scaling behavior of the current for different values of the anisotropy $\Delta$ and Fig. 21.7 shows the scaling behavior of the gradients. From the linear fits in Fig. 21.6 one could extrapolate the current within an infinitely long chain. This current is shown in Fig. 21.8 in dependence on $\Delta$ ($\Delta = 1$ refers to the Heisenberg chain and $\Delta = 0$ to the XY model). Near the anisotropy $\Delta = 1.6$ the current within the infinite system seems to vanish (cf. Fig. 21.6), i.e., the analysis at hand indicates normal transport behavior. Whether this is obtained for increasing $\Delta$ as well cannot be decided from the present analysis, but it seems probable that it remains diffusive also for higher values of $\Delta$.

Having found diffusive behavior according to the Kubo formula (i.e., a vanishing Drude peak, see, e.g., [21]) it seems nevertheless unclear how to extract the dc-conductivity from the behavior of the finite system. Contrary to the Kubo investigation a dc-conductivity directly follows from (21.18), in the present analysis, if we assume for the moment that the linear scaling of current and gradient as found in Figs. 21.6 and 21.7 is also valid for larger systems. According to the small errors found in the above investigation, this assumption seems to be plausible. Thus, we are able to compute the conductivity of the infinite system for $\Delta = 1.6$ using (21.18) by dividing the slope of the current by the slope of the energy gradient directly, finding $\kappa_\infty = [2.34 \pm 0.08] \cdot 10^{-2}$. Here, the error follows from the uncertainty of the linear regression, which is weighted already by the errors of the data points.

Unfortunately, the models which can be investigated according to the suggested method are also restricted in size. The main restriction here is not the size of mem-

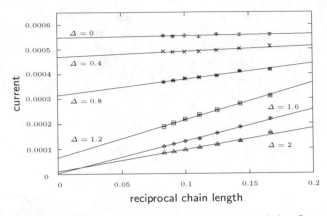

**Fig. 21.6** Scaling behavior of the current in anisotropic Heisenberg chains. System parameters: $J = 0.01, \Omega = 1, \lambda = 0.01, \beta_L = 0.5, \beta_R = 0.25$

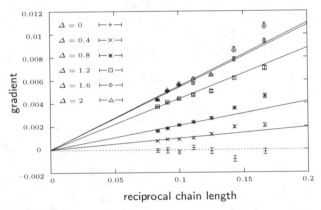

**Fig. 21.7** Scaling behavior of the gradient in anisotropic Heisenberg chains. System parameters: $J = 0.01, \Omega = 1, \lambda = 0.01, \beta_L = 0.5, \beta_R = 0.25$

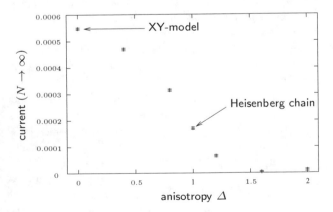

**Fig. 21.8** Extrapolated current for the infinitely long chain in dependence on the anisotropy $\Delta$

ory, but the time one accepts to wait for the data. For the present technique the computing time scales exponentially in the system size. Thus, investigations on how disorder (random offset in the local field, random couplings) would change the above results are not available yet.

# References

1. M. Michel, H. Wichterich, J. Gemmer, O. Hess, **77**, 104303 (2008)
2. J. Fourier, *The Analytical Theory of Heat* (Dover Publ., New York, 1955). Transl. by Alexander Freeman. Original: Théorie analytique de la chaleur (publ. 1822). Unabridged reprinting of the Engl. Transl. anno 1878.
3. K. Saito, Europhys. Lett. **61**, 34 (2003)
4. K. Saito, S. Takesue, S. Miyashita, Phys. Rev. E **61**, 2397 (2000)
5. M. Michel, M. Hartmann, J. Gemmer, G. Mahler, Eur. Phys. J. B **34**, 325 (2003)
6. M. Michel, *Nonequilibrium Aspects of Quantum Thermodynamics*. Ph.D. thesis, Universität Stuttgart, Stuttgart (2006). Also available online: http://elib.uni-stuttgart.de/opus/volltexte/2006/2803/
7. H. Wichterich, M.J. Henrich, H.P. Breuer, J. Gemmer, M. Michel, Phys. Rev. E **76**, 031115 (2007)
8. H.P. Breuer, F. Petruccione, *The Theory of Open Quantum Systems* (Oxford University Press, Oxford, 2002)
9. E.B. Davies, Commun. Math. Phys. **39**, 91 (1974)
10. E.B. Davies, *Quantum Theory of Open Systems* (Academic Press, London, 1976)
11. R. Dümcke, H. Spohn, Z. Phys. B **34**, 419 (1979)
12. V. Gorrini, A. Kossakowski, E. Sudarshan, J. Math. Phys. **17**, 821 (1976)
13. G. Lindblad, Commun. Math. Phys. **48**, 119 (1976)
14. R. Alicki, K. Lendi, *Quantum Dynamical Semigroups and Applications* (Springer, Berlin, 2001)
15. V. May, O. Kühn, *Charge and Energy Transfer Dynamics in Molecular Systems* (Wiley-VCH, Berlin, 2003)
16. M. Michel, J. Gemmer, G. Mahler, Int. J. Mod. Phys. B **20**, 4855 (2006)
17. K. Mølmer, Y. Castin, Quantum Semiclass. Opt. **8**, 49 (1996)
18. C. Mejia-Monasterio, H. Wichetrich, EPJST **151**, 113 (2007)
19. J. Steinbach, B.M. Garraway, P.L. Knight, Phys. Rev. A **51**(4), 3302 (1995)
20. M. Michel, J. Gemmer, G. Mahler, Eur. Phys. J. B **42**, 555 (2004)
21. F. Heidrich-Meisner, A. Honecker, W. Brenig, Phys. Rev. B **71**(18), 184415 (2005)

# Part IV
# Applications and Models

# Chapter 22
# Purity and Local Entropy in Product Hilbert Space

> *One natural way to get entropy, even for a system in a pure*
> *quantum state, is to make the coarse graining of dividing the*
> *system into subsystems and ignoring their correlations.*
> — D. N. Page [1]

**Abstract** A property might be called "typical" for some ensemble of items, if this property will show up with high probability for any randomly selected ensemble member (cf. Chap. 6). This requires a statistical characterization of the ensemble, i.e., a distribution function. For pure states in Hilbert-space of finite dimension $n_{\text{tot}}$ such a distribution can be derived by the condition of unitary invariance. As a pertinent example we will show here that for a bipartite Hilbert space of dimension $n_{\text{tot}} = n^{\text{g}} \cdot n^{\text{c}}$ the maximum local entropy (minimum local purity) within subsystem g (dimension $n^{\text{g}}$) becomes typical, provided $n^{\text{g}} \ll n^{\text{c}}$. Note that this observation is a consequence of the tensor space and virtually independent of the respective physical system.

## 22.1 Unitary Invariant Distribution of Pure States

For simplicity we start by considering a two-dimensional Hilbert space. According to (7.2), any normalized state vector $|\psi\rangle$ can be represented by the real and imaginary parts $\{\eta_i, \xi_i\}$, basis $|i\rangle$, of the complex amplitudes, and fulfill condition (7.3). In spite of this constraint let us assume for the moment that all these parameters are independent. We are looking for the probability distribution,

$$W(\eta_1, \eta_2, \xi_1, \xi_2) = W(\eta_1)W(\eta_2)W(\xi_1)W(\xi_2), \tag{22.1}$$

which is invariant under the unitary transformation

$$\hat{U} = \begin{pmatrix} 1 + i\varepsilon_3 & i\varepsilon_1 - \varepsilon_2 \\ i\varepsilon_1 + \varepsilon_2 & 1 - i\varepsilon_3 \end{pmatrix}. \tag{22.2}$$

This transformation leads to

Gemmer, J. et al.: *Purity and Local Entropy in Product Hilbert Space.* Lect. Notes Phys. **784**, 257–261 (2009)
DOI 10.1007/978-3-540-70510-9_22

$$|\psi'\rangle = \hat{U}|\psi\rangle = \sum_{i=1}^{2} \left(\eta_i' + i\xi_i'\right)|i\rangle, \tag{22.3}$$

with the coordinate transformation

$$\eta_1' = \eta_1 - \xi_2\varepsilon_1 - \eta_2\varepsilon_2 - \xi_1\varepsilon_3 \,, \tag{22.4}$$
$$\eta_2' = \eta_2 - \xi_1\varepsilon_1 + \eta_1\varepsilon_2 + \xi_2\varepsilon_3 \,, \tag{22.5}$$
$$\xi_1' = \xi_1 + \eta_2\varepsilon_1 - \xi_2\varepsilon_2 + \eta_1\varepsilon_3 \,, \tag{22.6}$$
$$\xi_2' = \xi_2 + \eta_1\varepsilon_1 + \xi_1\varepsilon_2 - \eta_2\varepsilon_3 \,. \tag{22.7}$$

It suffices to consider infinitesimal changes in the transformed probability distribution of a single coordinate:

$$W(\eta_i') = W(\eta_i) + (\eta_i' - \eta_i)\frac{\partial W(\eta_i)}{\partial \eta_i} \tag{22.8}$$

and in the same way for the distribution of the imaginary parts of the coordinates $W(\xi_i')$. Keeping only terms of first order in $\varepsilon$, we obtain for the completely transformed probability distribution

$$\begin{aligned}
&W(\eta_1')\,W(\eta_2')\,W(\xi_1')\,W(\xi_2')\\
&\approx W(\eta_2)\,W(\xi_1)\,W(\xi_2)\frac{\partial W(\eta_1)}{\partial \eta_1}\,(-\xi_2\varepsilon_1 - \eta_2\varepsilon_2 - \xi_1\varepsilon_3)\\
&+ W(\eta_1)\,W(\xi_1)\,W(\xi_2)\frac{\partial W(\eta_2)}{\partial \eta_2}\,(-\xi_1\varepsilon_1 + \eta_1\varepsilon_2 + \xi_2\varepsilon_3)\\
&+ W(\eta_1)\,W(\eta_2)\,W(\xi_2)\frac{\partial W(\xi_1)}{\partial \xi_1}\,(\eta_2\varepsilon_1 - \xi_2\varepsilon_2 + \eta_1\varepsilon_3)\\
&+ W(\eta_1)\,W(\eta_2)\,W(\xi_1)\frac{\partial W(\xi_2)}{\partial \xi_2}\,(\eta_1\varepsilon_1 + \xi_1\varepsilon_2 + \eta_2\varepsilon_3)\,. \tag{22.9}
\end{aligned}$$

Postulating

$$\frac{\partial W(\eta_i)}{\partial \eta_i} \approx -\eta_i\,W(\eta_i)\,, \qquad \frac{\partial W(\xi_i)}{\partial \xi_i} \approx -\xi_i\,W(\xi_i)\,, \tag{22.10}$$

leads to

$$W(\eta_1')\,W(\eta_2')\,W(\xi_1')\,W(\xi_2') \approx \text{const.}\cdot W(\eta_1)\,W(\eta_2)\,W(\xi_1)\,W(\xi_2)\,, \tag{22.11}$$

as required. Equation (22.10) implies the normalized solution

$$W(\eta_i) = \frac{1}{\sqrt{2\pi}\sigma}\,e^{\eta_i^2/2\sigma^2}\,, \qquad W(\xi_i) = \frac{1}{\sqrt{2\pi}\sigma}\,e^{\xi_i^2/2\sigma^2}\,. \tag{22.12}$$

As long as the complete probability distribution is a Gaussian distribution of the single coordinates, it is invariant under unitary transformations.

Generalizing this result for two-dimensional Hilbert spaces to any finite Hilbert space of dimension $n_{\text{tot}}$, we thus end up with the Gaussian

$$W(\{\eta_i, \xi_i\}) = \left(\frac{1}{\sqrt{2\pi}\sigma}\right)^{2n_{\text{tot}}} \exp\left(-\frac{1}{2\sigma^2} \sum_{i=1}^{n_{\text{tot}}} \left(\eta_i^2 + \xi_i^2\right)\right). \tag{22.13}$$

The normalization condition for the wave function, though, requires that the sum of the squares of the coordinates is 1 (see (7.3)), i.e., the parameters are not independent, contrary to our assumption. However, for large $n_{\text{tot}}$ the central limit theorem tells us that $W(\{\eta_i, \xi_i\})$ is indeed approximate a Gaussian, provided we choose [2]

$$\sigma = \frac{1}{\sqrt{2n_{\text{tot}}}}. \tag{22.14}$$

The above unitary invariant distribution holds for an $n_{\text{tot}}$-dimensional Hilbert space without further constraints. It characterizes an ensemble in Hilbert space, from which to pick "typical" pure states.


## 22.2 Application

The results of the preceding section are now applied to a bipartite system of dimension $n_{\text{tot}} = n^{\text{g}} \cdot n^{\text{c}}$, with $n^{\text{g}} \leq n^{\text{c}}$. Let $f = f(|\psi\rangle) = f(\{\eta_i, \xi_i\})$ be some function of the state vector. Then we can define its Hilbert space average $[\![f]\!]$ and its Hilbert space distribution $\{f\}$, respectively, as

$$[\![f]\!] = \int W(\{\eta_i, \xi_i\}) f(\{\eta_i, \xi_i\}) \prod_{i=1}^{n_{\text{tot}}} \mathrm{d}\eta_i \mathrm{d}\xi_i, \tag{22.15}$$

$$\{f\} = \int \delta(f(\{\eta_i, \xi_i\}) - f) W(\{\eta_i, \xi_i\}) \prod_{i=1}^{n_{\text{tot}}} \mathrm{d}\eta_i \mathrm{d}\xi_i. \tag{22.16}$$

Here we restrict ourselves to the local purity $P^{\text{g}}$ and local entropy $S^{\text{g}}$. The resulting distribution $\{P^{\text{g}}\}$ is shown in Fig. 22.1 for $n^{\text{g}} = 2$ and varying $n^{\text{c}}$. We see that this distribution tends to peak at the minimum value $P^{\text{g}} = 1/n^{\text{g}} = 1/2$. Its average is given by (see [1])

$$[\![P^{\text{g}}]\!] = \frac{n^{\text{g}} + n^{\text{c}}}{n^{\text{g}}n^{\text{c}} + 1}. \tag{22.17}$$

In Fig. 22.2 we show the Hilbert space average of $S^{\text{g}}$ for $n^{\text{g}} = 2$ as a function of $n^{\text{c}}$. Again we see that $S^{\text{g}}$ rapidly approaches its maximum value $S^{\text{g}}_{\text{max}}$ for large embedding $n^c$. For $1 \ll n^g \leq n^c$ one obtains [1]

$$[\![ S^g ]\!] \approx \ln n^g - \frac{n^g}{2n^c} . \tag{22.18}$$

Both results indicate that for $n^g \ll n^c$ a typical state of subsystem g is totally
"mixed": all its local properties have maximum uncertainty.

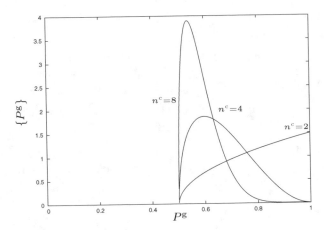

**Fig. 22.1** Purity distribution $\{ P^g \}$ for $n^g = 2$ and several container dimensions $n^c$

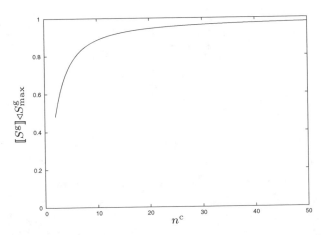

**Fig. 22.2** Average entropy $[\![ S^g ]\!] / S^g_{\max}$ of subsystem g ($n^g = 2$), depending on the dimension $n^c$ of
the environment

These findings may appear to anticipate the second law in a suprisingly simple
and fundamental way. However, while one cannot deny that the second law and
the above-discussed typical state features have a common source, the former does

not follow from the latter. Detailed physical explorations cannot be avoided, cf. Sects. 7, 8.

## References

1. D.N. Page, Phys. Rev. Lett. **71**(9), 1291 (1993)
2. A. Otte, *Separabilität in Quantennetzwerken*. Dissertation, Universität Stuttgart (2001)

# Chapter 23
# Observability of Intensive Variables

> *"Temperature fluctuation" is an oxymoron because the*
> *consistent and consensual definition of temperature admits no*
> *fluctuation.*
>
> — Ch. Kittel [1]

**Abstract** The status of the intensive variables like temperature or pressure needs further exploration; two interrelated questions are addressed in this chapter: First we investigate under what conditions temperature can be defined locally. We then discuss basic procedures by which pressure and temperature, though not observables by themselves, could be inferred from Experiment.

The distinction between intensive and extensive variables is intimately related to the composition/decomposition of systems: The intensive variables are contact variables, which should be the same for systems in thermal equilibrium. However, this composition/decomposition has originally been introduced on the macroscopic domain, leaving open the question on what scale temperature (or pressure) can be defined locally.

## 23.1 On the Existence of Local Temperatures[1]

As in Sect. 16.4 we consider a chain of spins with nearest neighbor interactions. However, instead of various values of the coupling strength we consider groups of different numbers of adjoining spins (see Fig. 23.1 and [2, 3]). The idea behind this approach is the following. If $N$ adjoining spins form a group, the energy of

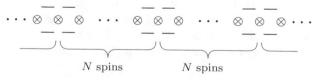

**Fig. 23.1** $N^G$ groups of $N$ adjoining spins each are formed

---

[1] Based on [2–5] by Hartmann et al.

Gemmer, J. et al.: *Observability of Intensive Variables*. Lect. Notes Phys. **784**, 263–274 (2009)
DOI 10.1007/978-3-540-70510-9_23 © Springer-Verlag Berlin Heidelberg 2009

the group is $N$ times the average energy per spin and is thus expected to grow proportionally to $N$ as the size of the group is increased. Since the spins only interact with their nearest neighbors, two adjacent groups only interact via the two spins at the respective boundaries. As a consequence, the effective coupling between two groups is independent of the group size and thus becomes less relevant compared to the energy contained in the groups as the group size increases.

Since we want to analyze the existence of a quantity usually assigned to equilibrium states, we should consider equilibrium scenarios or at least situations close to equilibrium. Therefore, we assume that our entire spin chain is in a thermal equilibrium state. One can imagine that it may have approached this state via interactions with its surrounding as described in Chap. 9, although, for the consideration here, such details will be irrelevant.

Before we can address the question of local temperatures, we have to clarify what we mean when we say that temperature exists or does not exist. The spectral temperature defined in Chap. 13 always exists, by definition, but it usually does not have all the properties temperature is supposed to have in thermal equilibrium.

We adopt here the convention that local temperature exists if the respective (local) reduced density matrix is close to a canonical one. Then, the spectral temperature, which in this case coincides with the standard temperature, fully characterizes the distribution, i.e., the diagonal elements of the corresponding density matrix. If, furthermore, the local temperatures coincide with the global one, temperature is even an intensive quantity in the scenario at hand (i.e., does not change with system size).

We thus consider a very long chain of very many spins in a global thermal equilibrium state (10.36), divide the chain into $N^G$ groups of $N$ adjoining spins each (see Fig. 23.1), and test whether the local probability distribution also has the canonical form (10.36).

### 23.1.1 Model

We start by defining the Hamiltonian of our spin chain in the form,

$$\hat{H} = \sum_{\mu} \left( \hat{H}_{\text{loc}}^{(\mu)} + \hat{H}_{\text{int}}^{(\mu,\mu+1)} \right), \tag{23.1}$$

where the index $\mu$ labels the elementary subsystems. The first term is the local Hamiltonian of subsystem $\mu$ and the second one describes the interaction between subsystem $\mu$ and $\mu+1$. We assume periodic boundary conditions. Since this section applies to all models with the structure (23.1), we do not further specify the terms in the Hamiltonian before we apply the results to the concrete spin chain model in Sect. 23.1.4.

We now form $N^G$ groups of $N$ subsystems each with index $\nu = 1, \ldots, N^G$ specifying the respective group, and $\mu = 1, \ldots, N$ numbers the elementary subsystems within such a group

$$\mu \mapsto (\nu - 1)N + \mu. \tag{23.2}$$

According to the formation of groups the total Hamiltonian splits up into two parts,

$$\hat{H} = \hat{H}_0 + \hat{I} ,$$ (23.3)

where $\hat{H}_0$ is the sum of the Hamiltonians of the isolated groups,

$$\hat{H}_0 = \sum_{\nu=1}^{N^G} \hat{H}_{\text{group}}^{(\nu)} \quad \text{with}$$ (23.4)

$$\hat{H}_{\text{group}}^{(\nu)} = \sum_{\mu=1}^{N} \hat{H}_{\text{loc}}^{((\nu-1)N+\mu)} + \sum_{\mu=1}^{N-1} \hat{H}_{\text{int}}^{((\nu-1)N+\mu,(\nu-1)N+\mu+1)},$$

and $\hat{I}$ contains the interaction terms of each group with its neighbor group only

$$\hat{I} = \sum_{\nu=1}^{N^G} \hat{H}_{\text{int}}^{(\nu N, \nu N+1)} .$$ (23.5)

We label the eigenstates of the total Hamiltonian $\hat{H}$ and their energies with Greek letters ($\varphi, \psi$) and eigenstates and energies of the group Hamiltonian $\hat{H}_0$ with Latin letters ($a, b$)

$$\hat{H}|\varphi\rangle = E_\varphi|\varphi\rangle , \qquad \hat{H}_0|a\rangle = E_a|a\rangle .$$ (23.6)

Here, the states $|a\rangle$ are products of group eigenstates defined as

$$\hat{H}_{\text{group}}^{(\nu)}|a_\nu\rangle = E_\nu|a_\nu\rangle , \qquad |a\rangle = \prod_{\nu=1}^{N^G} |a_\nu\rangle .$$ (23.7)

$E_\nu$ is the energy of one subgroup only and $E_a = \sum_{\nu=1}^{N^G} E_\nu$.

### 23.1.2 Global Thermal State in the Product Basis

We assume that the total system is in a thermal state with a density matrix $\hat{\rho}$, which reads in the eigenbasis of $\hat{H}$

$$\langle\varphi|\hat{\rho}|\psi\rangle = \frac{e^{-\beta E_\varphi}}{Z} \delta_{\varphi\psi} .$$ (23.8)

Here, $Z$ is the partition function and $\beta = 1/(k_B T)$ the inverse temperature. Transforming the density matrix (23.8) into the eigenbasis of $\hat{H}_0$ we obtain

$$\langle a|\hat{\rho}|a\rangle = \int_{E_0}^{E_1} dE\, W_a(E) \frac{e^{-\beta E}}{Z}$$ (23.9)

for the diagonal elements in the new basis. Here, the sum over all states $|a\rangle$ has been replaced by an integral over the energy. $E_0$ is the energy of the ground state and $E_1$ the upper limit of the spectrum. The density of conditional probabilities $W_a(E)$ is given by

$$W_a(E) = \frac{1}{\Delta E} \sum_{|\varphi\rangle : E \leq E_\varphi < E + \Delta E} |\langle a|\varphi\rangle|^2 , \qquad (23.10)$$

where $\Delta E$ is small and the sum runs over all states $|\varphi\rangle$ with eigenvalues $E_\varphi$ in the interval $[E, E + \Delta E]$.

To compute the integral of (23.9) we need to know the density of the conditional probabilities $W_a(E)$. For a very large number of groups, $N^G \gg 1$, it may be approximated by a Gaussian normal distribution (for a rigorous proof of this statement, which is a quantum analog of the central limit theorem, and further applications, see [3] and [6]),

$$\lim_{N^G \to \infty} W_a(E) = \frac{1}{\sqrt{2\pi}\,\Delta_a} \exp\left(-\frac{(E - E_a - \varepsilon_a)^2}{2\,\Delta_a^2}\right) , \qquad (23.11)$$

where $\varepsilon_a$ and $\Delta_a$ are defined by

$$\varepsilon_a = \langle a|\hat{H}|a\rangle - \langle a|\hat{H}_0|a\rangle , \qquad (23.12)$$

$$\Delta_a^2 = \langle a|\hat{H}^2|a\rangle - \langle a|\hat{H}|a\rangle^2 . \qquad (23.13)$$

The quantity $\varepsilon_a$ has a classical counterpart, while $\Delta_a^2$ is purely quantum mechanical. It appears because the commutator $[\hat{H}, \hat{H}_0]$ is non-zero, and the distribution $W_a(E)$ therefore has non-zero width. Equation (23.9) can now be computed for $N^G \gg 1$

$$\langle a|\hat{\rho}|a\rangle = \frac{1}{Z} \exp\left(-\beta\, y_a + \frac{\beta^2 \Delta_a^2}{2}\right) \times$$
$$\times \frac{1}{2}\left(\mathrm{erfc}\left(\frac{E_0 - y_a + \beta\Delta_a^2}{\sqrt{2}\Delta_a}\right) - \mathrm{erfc}\left(\frac{E_1 - y_a + \beta\Delta_a^2}{\sqrt{2}\Delta_a}\right)\right) , \qquad (23.14)$$

where $y_a = E_a + \varepsilon_a$ and $\mathrm{erfc}(\dots)$ is the conjugate Gaussian error function. The second term only appears if the energy is bounded and the integration extends from the energy of the ground state $E_0$ to the upper limit of the spectrum $E_1$.

The off-diagonal elements $\langle a|\hat{\rho}|b\rangle$ vanish for $|E_a - E_b| > \Delta_a + \Delta_b$ because the overlap of the two distributions of conditional probabilities becomes negligible. For $|E_a - E_b| < \Delta_a + \Delta_b$, the transformation involves an integral over frequencies and thus these terms are significantly smaller than the entries on the diagonal part.

### 23.1.3  Conditions for Local Thermal States

We now test under what conditions the diagonal elements of the (local) reduced density matrices are also canonically distributed with some local inverse temperature $\beta_{\text{loc}}^{(\nu)}$ for each subgroup $\nu = 1, \ldots, N^G$. Since the trace of a matrix is invariant under basis transformations, it is sufficient to verify that they show the correct energy dependence. If we assume periodic boundary conditions, all reduced density matrices are equal ($\beta_{\text{loc}}^{(\nu)} = \beta_{\text{loc}}$ for all $\nu$) and the products of their diagonal elements are of the form $\langle a|\hat{\rho}|a\rangle \propto \exp(-\beta_{\text{loc}} E_a)$. We thus have to verify that the logarithm of the right-hand side of (23.14) is a linear function of the energy $E_a$,

$$\ln\left(\langle a|\hat{\rho}|a\rangle\right) \approx -\beta_{\text{loc}} E_a + c \,, \tag{23.15}$$

where $\beta_{\text{loc}}$ and $c$ are real constants. Note that (23.15) does not imply that the occupation probabilities of an eigenstate $|\varphi\rangle$ with energy $E_\varphi$ and a product state with the same energy $E_a \approx E_\varphi$ are equal. Even if $\beta_{\text{loc}}$ and $\beta$ are equal with very good accuracy, but not exactly the same, occupation probabilities may differ by several orders of magnitude, provided the energy range is large enough.

Since we consider the limit $N^G \to \infty$, we approximate the conjugate error functions of (23.14) by their asymptotic expansions (cf. [7]). This is possible because $y_a$ and $\Delta_a^2$ are sums of $N^G$ terms and the arguments of the error functions grow proportionally to $\sqrt{N^G}$. Inserting the asymptotic expansions into (23.14) shows that (23.15) can only be true if

$$\frac{E_a + \varepsilon_a - E_0}{\sqrt{N^G}\,\Delta_a} > \beta\,\frac{\Delta_a^2}{\sqrt{N^G}\,\Delta_a} \tag{23.16}$$

(for a more detailed consideration see [4]). In this case, (23.14) may be taken to read

$$\langle a|\hat{\rho}|a\rangle = \frac{1}{Z}\,\exp\left(-\beta\left(E_a + \varepsilon_a - \frac{\beta\Delta_a^2}{2}\right)\right) \,, \tag{23.17}$$

where we have used that $y_a = E_a + \varepsilon_a$. To ensure that the criterion (23.15) is met, $\varepsilon_a$ and $\Delta_a^2$ have to be of the form

$$-\varepsilon_a + \frac{\beta}{2}\,\Delta_a^2 \approx c_1 E_a + c_2 \,, \tag{23.18}$$

as follows by inserting (23.17) into (23.15). In (23.18), $c_1$ and $c_2$ are arbitrary, real constants. Note that $\varepsilon_a$ and $\Delta_a^2$ need not be functions of $E_a$ and therefore in general cannot be expanded in a Taylor series. In addition, the temperature becomes intensive, if the constant $c_1$ vanishes, in

$$\beta_{\text{loc}} = \beta(1 - c_1) \,, \tag{23.19}$$

which follows by introducing the approximated diagonal matrix element (23.17) into condition (23.14) and using (23.18). If $c_1$ does not vanish, temperature would not be intensive, although it might exist locally.

Conditions (23.16) and (23.18) determine when temperature can exist locally. These conditions depend on the global inverse temperature $\beta$. In both equations the inverse temperature $\beta$ appears together with the squared width $\Delta_a^2$, which is always larger than zero as long as the commutator $[\hat{H}, \hat{H}_0]$ does not vanish. The temperature dependence of the criteria for the existence of local temperatures is thus a pure quantum effect, which appears in almost all models of interest since these show $[\hat{H}, \hat{H}_0] \neq 0$. As can already be deduced from (23.16) and (23.18), the temperature dependence is most relevant at low temperatures, where minimal length scales for the existence of temperature may even become macroscopic, as we will see below (cf. [8–11]).

It is sufficient to satisfy the conditions (23.16) and (23.18) for an adequate energy range $E_{min} \leq E_a \leq E_{max}$ only. For large systems with a modular structure, the density of states is a rapidly growing function of energy, as explained in Sect. 12.2. If the total system is in a thermal state, occupation probabilities decay exponentially with energy. The product of these two functions is thus sharply peaked at the expectation value of the energy $\overline{E}$ of the total system $\overline{E} + E_0 = \text{Tr}\{\hat{H}\hat{\rho}\}$, with $E_0$ being the ground state energy. The energy range thus needs to be centered at this peak and to be large enough. On the other hand it must not be larger than the range of values $E_a$ can take on. Therefore, a pertinent and "safe" choice for $E_{min}$ and $E_{max}$ is

$$E_{min} = \max\left([E_a]_{min},\ \alpha^{-1}\overline{E} + E_0\right), \tag{23.20}$$

$$E_{max} = \min\left([E_a]_{max},\ \alpha\overline{E} + E_0\right), \tag{23.21}$$

where $\alpha \gg 1$ and $\overline{E}$ will in general depend on the global temperature. $[x]_{min}$ and $[x]_{max}$, respectively, denote the minimal and maximal values that $x$ can take on.

The expectation value of the total energy, $\overline{E}$, in general depends on the global temperature. This temperature dependence, which is characteristic of the model (i.e., of the material) at hand, also enters into the criteria (23.16) and (23.18) via the definition of the relevant energy range in (23.20) and (23.21).

Figure 23.2 compares the logarithm of (23.14) for a model, where $\varepsilon_a = 0$ and $\Delta_a^2 = E_a^2/N^2$ with $N = 1000$ and the logarithm of a canonical distribution with the same $\beta$. The actual density matrix is more mixed than the canonical one. In the interval between the two vertical lines, both criteria (23.16) and (23.18) are satisfied. For $E < E_{low}$ (23.16) is violated and (23.18) for $E > E_{high}$. To allow for a description by means of a canonical density matrix, the group size $N$ needs to be chosen such that $E_{low} < E_{min}$ and $E_{high} > E_{max}$.

For the existence of a local temperature the two conditions (23.16) and (23.18), which constitute the general result of our approach, must both be satisfied. Or, to put it differently, the two criteria determine, for a concrete situation, the minimal number of group members $N$ (a minimal length scale) on which temperature can be defined. These fundamental criteria will now be applied to a concrete example.

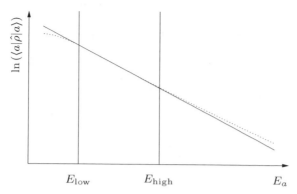

**Fig. 23.2** $\ln(\langle a|\hat{\rho}|a\rangle)$ for a canonical density matrix $\rho$ (*dashed line*) and for $\hat{\rho}$ as in (23.14) (*solid line*) with $\varepsilon_a = 0$ and $\Delta_a^2 = E_a^2/N^2$ ($N = 1000$) [4]

### 23.1.4 Spin Chain in a Transverse Field

As a concrete application, we consider a spin chain in a transverse field. For this model the Hamiltonian reads

$$\hat{H}_{\text{loc}}^{(\mu)} = -\Delta E\,\hat{\sigma}_3^{(\mu)}$$
$$\hat{H}_{\text{int}}^{(\mu,\mu+1)} = -\frac{\lambda}{2}\left(\hat{\sigma}_1^{(\mu)}\otimes\hat{\sigma}_1^{(\mu+1)} + \hat{\sigma}_2^{(\mu)}\otimes\hat{\sigma}_2^{(\mu+1)}\right). \qquad (23.22)$$

Here, $2\Delta E$ is the Zeeman splitting (we will always assume $\Delta E > 0$) and $\lambda$ the strength of the spin–spin coupling, called Förster coupling.

If one partitions the chain into $N^G$ groups of $N$ subsystems each, the groups may be diagonalized via a Jordan–Wigner and a Fourier transformation [2]. With this procedure, the quantities $E_a$, $\varepsilon_a$, and $\Delta_a$ can be expressed in terms of fermionic occupation numbers of the group eigenstates, which in turn allows us to calculate the minimal number of spins $N_{\text{min}}$ in each group (the minimal length scale) from conditions (23.16) and (23.18).

The technical details of this calculation, which involve a few further approximations, are not relevant for the understanding of the relevant physics. We thus refer the interested reader to Appendix E for a detailed discussion.

Figure 23.3 shows $N_{\text{min}}$ as a function of temperature $T$ in units of half the Zeeman splitting for weak coupling, $\lambda = 0.1\Delta E$. Here, condition (23.16) is approximated by (E.7) and condition (23.18) by (E.10) (see Appendix E).

The strong coupling case ($\lambda = 10\Delta E$) is shown in Fig. 23.4, where condition (23.16) is approximated by (E.9) and condition (23.18) by (E.10) (see Appendix E).

Apparently, criterion (23.18) and (E.10), respectively, becomes relevant only at high temperatures $T$ and strong coupling $\lambda > \Delta E$. In addition, temperature is always an intensive quantity for the spin chain considered here, irrespective of the coupling strength (see Appendix E for details).

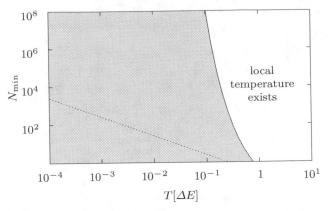

**Fig. 23.3** Weak coupling ($\lambda = 0.1\,\Delta E$). Log–log plot of $N_{\min}$ over $T$: the criteria (E.7) (*solid line*) and (E.10) (*dashed line*) define lower limits above which a local temperature is defined [4]

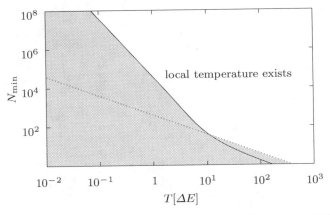

**Fig. 23.4** Strong coupling ($\lambda = 10\,\Delta E$). Log–log plot of $N_{\min}$ over $T$: the criteria (E.9) (*solid line*) and from (E.10) (*dashed line*) define a limit above which a local temperature is defined [4]

The model shows that for spin chains at high temperatures local temperatures may exist even for single spins, while the minimal group size becomes increasingly large at low temperatures. These characteristics are quite plausible if the transformation behavior of matrices under a change of basis is taken into account. At very high global temperatures, the total density matrix is almost completely mixed, i.e., proportional to the identity matrix, and thus does not change under basis transformations. There are, therefore, global temperatures that are high enough, so that local temperatures exist even for single spins.

On the other hand, if the global temperature is very low, the system is essentially in its ground state. This state is typically a superposition of product states and therefore strongly correlated [8, 9]. It becomes thus impossible to assign local temperatures to the ground state for any partition.

## 23.2 Measurement Schemes

The intensive thermodynamic variables $x = \{T, p\}$ are state parameters, not observables. In quantum mechanics they are not associated with operators. For a stationary state $x$ is time independent, by definition.

Measurements of $x$ are thus necessarily indirect. Basically there appear to be two different schemes:

(1) Process-based: Here one considers a thermodynamic process as demanded by the thermodynamic definition of $x$:

$$T = \left(\frac{\partial U}{\partial S}\right)_V = T(S, V) , \tag{23.23}$$

$$P = -\left(\frac{\partial U}{\partial V}\right)_S = P(S, V) . \tag{23.24}$$

We would thus have to perform an isochor (adiabat) to construct the tangent at the required point (S,V). We will return to this scheme in Chap. 25.

(2) Observable-based: Here one exploits the parameter dependence of some observable $A$, which can be accessed by a measurement device. If one knows the functional dependence $A(x)$, $x$ can be inferred from a measurement of $A$. In this way, the measurement of $x$ has been converted to a "conventional" measurement of $A$. In the quantum domain the measurement of $A$ is subject to the limitations imposed by quantum physics such as quantum uncertainty.

In the following we will discuss two quantum examples for scheme (2). We leave out the problem of fluctuations [1, 12, 13].

### 23.2.1 Pressure

In thermodynamics, pressure is an intensive work variable defined as (cf. Chap. 14)

$$P = -\left(\frac{\partial U}{\partial V}\right)_S , \tag{23.25}$$

where $U$ is the internal energy of the system under consideration (not of the total system) and the volume $V$ is the extensive work variable conjugate to $P$. In the one-dimensional case, $P$ has the dimension of force:

$$F = -\left(\frac{\partial U}{\partial L}\right)_S . \tag{23.26}$$

In any case, the mechanical effect will consist of a deformation (compression) of an additional degree of freedom entering $U$. Here we intend to show that the conjugation between the extensive work variable and the pressure can indeed be confirmed

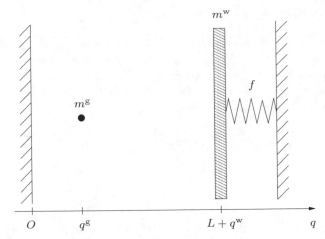

**Fig. 23.5** One-dimensional model for a confined "gas particle," $m^g$, interacting with a harmonic oscillator

down to the nano limit. For this purpose we first study an isolated bipartite system, i.e., without thermalizing environment.

In our present one-dimensional model introduced by Borowski et al. [14] one subsystem is a single particle (g) in a box, where, however, one wall (w) is now replaced by a movable piston connected to a spring, the central part of a manometer (see Fig. 23.5). The fact that the manometer is essentially mechanical is often interpreted to mean that the pressure $P$ is mechanical, too. However, this is not the case, as can be inferred from the condition $S = $ const. in (23.25). On the other hand, pressure, like temperature, can under appropriate conditions be defined for a single (embedded) subsystem. Classically this would hardly be considered meaningful. The respective two-particle Hamiltonian is given by

$$\hat{H}(q^g, q^w) = -\frac{\hbar^2}{2m^g} \frac{\partial^2}{\partial (q^g)^2} - \frac{\hbar^2}{2m^w} \frac{\partial^2}{\partial (q^w)^2} + \frac{f}{2} (q^w)^2 + \hat{V}(q^g, q^w) , \quad (23.27)$$

with the potential

$$\hat{V}(q^g, q^w) = \begin{cases} \infty & q^g < 0 \\ \infty & \text{for} \quad q^g > L + q^w \\ 0 & 0 \leq q^g \leq L + q^w \end{cases} . \quad (23.28)$$

The total potential term, which is obviously not accessible from a perturbative approach,

$$\hat{V}'(q^g, q^w) = \hat{V}(q^g, q^w) + \frac{f}{2} (q^w)^2 \quad (23.29)$$

can be shown to separate under the coordinate transformation

$$y^g = \frac{q^g}{q^w + L} L , \quad y^w = q^w ,$$
(23.30)

in the limit of $y_w \ll L$. Of course, in order to make use of this convenient feature, one also has to transform the kinetic parts to the new coordinates. The unperturbed part (particle of mass $m^g$ in a box with infinite walls at $y^g = 0$ and $y^g = L$ and a particle of mass $m^w$ in a harmonic potential) is then given by

$$\hat{H}_0 = -\frac{\hbar^2}{2m^g} \frac{\partial^2}{\partial(y^g)^2} - \frac{\hbar^2}{2m^w} \frac{\partial^2}{\partial(y^w)^2} + \hat{V}'(y^g, y^w) .$$
(23.31)

While the transformation of the kinetic energies produces an additional coupling term, a more careful analysis shows that this coupling term can now be dealt with by means of standard perturbation theory. The corresponding energy spectrum of $\hat{H}_0$ is

$$E_{j^g j^w} = E_{j^g} + E_{j^w} = \left(\frac{\pi \hbar j^g}{\sqrt{2m^g}L}\right)^2 + \hbar\sqrt{\frac{f}{m^w}} \left(j^w + \frac{1}{2}\right) ,$$
(23.32)

where $j^g$ and $j^w$ are quantum numbers. Due to the particle–particle interaction one finds for the perturbed energy eigenstate with $j^w = 0$ approximately

$$\langle q^w \rangle_{j^g} = \left(\frac{\pi \hbar j^g}{L\sqrt{m^g L f}}\right)^2 .$$
(23.33)

For such an elongation (with the particle in state $j^g$) the corresponding force ("pressure") should be

$$F_{j^g} = f \langle q^w \rangle_{j^g} .$$
(23.34)

This is identical with the definition

$$-\left(\frac{\partial E_{j^g}}{\partial L}\right)_S = F_{j^g} ,$$
(23.35)

so that in this case we confirm the functional dependence

$$q^w(F) = \frac{1}{f} F .$$
(23.36)

The interaction between the particle and the oscillator can be shown to lead to negligible entanglement, i.e., the local entropies remain zero. The whole system ("gas" with movable wall) can be used as a manometer for some other system with which an appropriate mechanical contact exists, i.e., which can "exchange" volume. From a measurement of $q^w$ (a property of the total system) we can thus infer the "pressure" $F$, applicable to that system. Of course, this model study should be generalized to include thermal states.

Note that $q^w$ is taken to be the average position of the single quantum oscillator; its measurement would thus require a sequence of conventional quantum measurements. These can be performed independently only, if the particle in the box relaxes back to its equilibrium state before the next measurement occurs. Also from this point of view an embedding is needed. In the extreme quantum limit, the respective variance can easily be of the same order as the mean value (i.e., $q^w$ shows large fluctuations).

## 23.2.2 Temperature

Insofar as the pressure also depends on temperature, the system of Sect. 23.2.1 could also serve as a thermometer.

Alternatively, we may consider a set of non-interacting spins (two-level systems) with finite Zeeman splitting in contact with some appropriate environment (or bath). Its average magnetization $M$ (or mean energy) constitutes a unique relation between $M$ and $T$ (which can be calculated, e.g., based on the canonical-level occupation probabilities). From a measurement of $M$ we may thus infer $T$: The measurement of $T$ is converted to a conventional measurement of $M$. It goes without saying that such a system may operate as a thermometer for any other system, which has been brought in thermal contact, i.e., which can exchange heat. In general, the actual measurement of $M$ will call for another specific technical design [13].

# References

1. C. Kittel, Phys. Today, 93 (1988)
2. M. Hartmann, J. Gemmer, G. Mahler, O. Hess, Euro. Phys. Lett. **65**, 613 (2004)
3. M. Hartmann, G. Mahler, O. Hess, Lett. Math. Phys. **68**, 103 (2004)
4. M. Hartmann, G. Mahler, O. Hess, Phys. Rev. Lett. **93**, 080402 (2004)
5. M. Hartmann, G. Mahler, O. Hess, Phys. Rev. E **70**, 066148 (2004)
6. M. Hartmann, G. Mahler, O. Hess, J. Stat. Phys. **19**, 1139 (2005)
7. M. Abramowitz, I. Stegun, *Handbook of Mathematical Functions*, 9th edn. (Dover Publ., New York, 1970)
8. X. Wang, Phys. Rev. A **66**, 044305 (2002)
9. A.N. Jordan, M. Büttiker, Phys. Rev. Lett. **92**(24), 247901 (2004)
10. A.E. Allahverdyan, T.M. Nieuwenhuizen, Phys. Rev. B **66**, 115309 (2002)
11. T.M. Nieuwenhuizen, A.E. Allahverdyan, Phys. Rev. E **66**, 036102 (2002)
12. B. Diu, C. Guthmann, D. Lederer, B. Roulet, *Grundlagen der Statistischen Physik* (de Gruyter Berlin New York, 1994)
13. T.C.P. Chui, D.R. Swanson, M.J. Adriaans, J.A. Nissen, J.A. Lipa, Phys. Rev. Lett. **69**, 3005 (1992)
14. P. Borowski, J. Gemmer, G. Mahler, Europhys. Lett. **62**, 629 (2003)

# Chapter 24
# Observability of Extensive Variables

*In the same way that classical mechanics proved inadequate to describe energy exchange between radiation and matter at the atomistic level, one could imagine that current theories describing thermal exchange processes occurring at very small length-scales or short times should accordingly be revised.*

— F. Ritort [1]

**Abstract** The observability of extensive thermodynamic variables is most conveniently related to thermodynamic processes: One can study the change of internal energy subject to different constraints. This leads, e.g., to the concept of work and heat, the later being related to a change of entropy. Fluctuations of work become important in the limit of small systems, i.e., in the domain of "nanothermodynamics," and in connection with non-equilibrium.

## 24.1 Work and Heat[1]

The splitting of energy change into work and heat is related to the availability of tools by which either of these two modes of change can selectively be implemented. In classical thermodynamics this is routinely done based on work reservoirs and heat reservoirs, respectively (cf. thermodynamic machines). In general, however, and, in particular, in the quantum domain such an idealized classification of the environment becomes questionable.

### 24.1.1 LEMBAS Scenario

Here we visualize a two-step scenario, in which a system $A$ has first been prepared in a thermal state (via embedding into some appropriate large environment), then the environment is removed and eventually replaced by another subsystem $B$. The latter could be some different environment, a measurement apparatus, or designed to induce thermodynamic processes on $A$, usually still close to equilibrium. Work

---

[1] Based on [2].

Gemmer, J. et al.: *Observability of Extensive Variables*. Lect. Notes Phys. **784**, 275–289 (2009)
DOI 10.1007/978-3-540-70510-9_24

and heat have meaning only in the context of such thermodynamic processes (cf. Chap. 25): One attempts to characterize how the change of the internal energy $U_A$ is being achieved. Here $U_A$ is a state function; work and heat depend on the details of the process performed. The process (or path) dependence of heat and work is constitutive for cyclic thermodynamic heat engines.

While there is no problem to define heat and work in classical thermodynamics, their classification in the quantum regime is not that obvious. A common approach is based on the change of the total energy expectation value

$$dU_A = \text{Tr}\left\{\hat{\rho}_A d\hat{H}_A + \hat{H}_A d\hat{\rho}_A\right\},\tag{24.1}$$

defining the first term (change of Hamiltonian) as work đ$W_A$ and the second (change of state) as heat đ$Q_A$. Taking $\hat{\rho}_A$ to be close to thermal equilibrium, this formulation gives an interpretation of the differential form of the first law. The rationale behind this identification is that the change of Hamilton-parameters (like the confining volume for a gas) may be called "mechanical," while the change of state may be associated with a change of entropy, i.e., heat.

However, this definition lacks a clear operational meaning and may even yield unphysical results, as we will see later. This is why we turn to a different approach incorporating two features: (i) a concrete physical embedding of $A$ causing the change of local energy $E_A$ and (ii) a prescription of how to specify (measure) $E_A$ in the presence of that very embedding (applicable also beyond the weak coupling limit). The change of local energy for a considered subsystem should thus be described within a local effective measurement basis (LEMBAS). The change of local energy can then systematically be split into a part associated with a change of entropy and another part which is not. Like in classical thermodynamics we call the first part heat and the second part work. However, in contrast to the classical definitions, work and heat are no longer absolute quantities but depend on the chosen measurement basis.

We consider subsystem $A$ together with its control $B$ as an autonomous (i.e., time-independent) bipartite quantum system with Hamiltonian

$$\hat{H} = \hat{H}_A + \hat{H}_B + \hat{H}_{AB},\tag{24.2}$$

where $\hat{H}_A$ and $\hat{H}_B$ are the local Hamiltonians of subsystems $A$ and $B$, respectively, and $\hat{H}_{AB}$ describes the interaction between both subsystems. The internal energy $U = \text{Tr}\left\{\hat{H}\hat{\rho}\right\}$ is thus, at most, approximately additive. The interaction will typically gain in importance for small subsystems. The dynamics of the total system is given by the Liouville–von Neumann equation

$$\frac{\partial}{\partial t}\hat{\rho} = -i[\hat{H}, \hat{\rho}].\tag{24.3}$$

$\hbar$ is set to 1 here. There is no external driving.

As we are interested in the properties of only one subsystem, $A$, we obtain its reduced dynamics by tracing out system $B$

$$\frac{\partial}{\partial t}\hat{\rho}_A = -i\mathrm{Tr}_B\left\{[\hat{H}_A + \hat{H}_B + \hat{H}_{AB}, \hat{\rho}]\right\} . \tag{24.4}$$

With respect to a proper definition of work and heat, we will have to split this effective dynamics into a "coherent" and an "incoherent" part. We proceed as follows.

By using some theorems on partial traces, it is possible to show that terms involving $\hat{H}_B$ vanish. For dealing with the other terms we split the density operator as

$$\hat{\rho} = \hat{\rho}_A \otimes \hat{\rho}_B + \hat{C}_{AB} , \tag{24.5}$$

where $\hat{\rho}_{A,B}$ are the reduced density operators for $A$ and $B$, respectively, and $\hat{C}_{AB}$ describes the correlations between both subsystems. Note that $\hat{C}_{AB} \neq 0$ does not necessarily imply entanglement, it rather means that the entropy becomes non-extensive. Again a partial trace theorem can be used to show that $\mathrm{Tr}_B\left\{[\hat{H}_A, \hat{C}_{AB}]\right\}$ vanishes. For the interaction part we get two terms: The first one is given by

$$\mathrm{Tr}_B\left\{[\hat{H}_{AB}, \hat{\rho}_A \otimes \hat{\rho}_B]\right\} = [\hat{H}_A^{\mathrm{eff}}, \hat{\rho}_A] , \tag{24.6}$$

with

$$\hat{H}_A^{\mathrm{eff}} := \mathrm{Tr}_B\left\{\hat{H}_{AB}(\hat{1}_A \otimes \hat{\rho}_B)\right\} . \tag{24.7}$$

The second term can be identified as an incoherent part

$$-i\mathrm{Tr}_B\left\{[\hat{H}_{AB}, \hat{C}_{AB}]\right\} =: \mathcal{L}_{\mathrm{inc}} , \tag{24.8}$$

since this term causes a change in the local von Neumann entropy $S_A$:

$$\dot{S}_A = -\mathrm{Tr}\left\{-i[\hat{H}_{AB}, \hat{C}_{AB}]\log \hat{\rho}_A \otimes \hat{1}_B\right\} \neq 0 . \tag{24.9}$$

With the definitions (24.7) and (24.8) the Liouville–von Neumann equation for subsystem $A$ (24.4) can thus be rewritten as

$$\frac{\partial}{\partial t}\hat{\rho}_A = -i[\hat{H}_A + \hat{H}_A^{\mathrm{eff}}, \hat{\rho}_A] + \mathcal{L}_{\mathrm{inc}}(\hat{\rho}) , \tag{24.10}$$

where the first term on the right-hand side describes local unitary dynamics and the second term incoherent processes. Note that while this equation is reminiscent of master equations, it is not closed here: it still refers to the total state $\hat{\rho}(t)$, which evolves unitarily. Equation (24.10) describes the dynamical impact on $A$ by its quantum embedding, irrespective of the size/dimension of the participating subsystems.

Since we are interested in the change of the internal energy of $A$, we now choose the energy basis of the isolated subsystem $A$ as our measurement basis. However, a measurement of energy within this basis will not only yield some eigenvalue of the Hamiltonian $\hat{H}_A$ but also parts of $\hat{H}_A^{\text{eff}}$ which commute with $\hat{H}_A$. Hence, we have to split $\hat{H}_A^{\text{eff}}$ into two parts: one which commutes with $\hat{H}_A$ and one which does not,

$$[\hat{H}_1^{\text{eff}}, \hat{H}_A] = 0 , \quad [\hat{H}_2^{\text{eff}}, \hat{H}_A] \neq 0 . \tag{24.11}$$

By expanding $\hat{H}_A^{\text{eff}}$ in the transition operator basis defined by the energy eigenstates $\{|j\rangle\}$ of $\hat{H}_A$ one can see that these parts are given by

$$\hat{H}_1^{\text{eff}} = \sum_j (\hat{H}_A^{\text{eff}})_{jj} |j\rangle\langle j| , \quad \hat{H}_2^{\text{eff}} = \hat{H}_A^{\text{eff}} - \hat{H}_1^{\text{eff}} . \tag{24.12}$$

As stated before, the part $\hat{H}_1^{\text{eff}}$ affects the measurement of the local energy. Thus, the effective Hamiltonian for our local description reads

$$\hat{H}_A' = \hat{H}_A + \hat{H}_1^{\text{eff}} . \tag{24.13}$$

The ensemble average over such local measurements $E_A'$ would give the respective internal energy $U_A$. Correspondingly, the change of internal energy in $A$ is given by

$$dU_A = \frac{d}{dt} \text{Tr}_A \left\{ \hat{H}_A' \hat{\rho}_A \right\} dt = \text{Tr}_A \left\{ \dot{\hat{H}}_A' \hat{\rho}_A + \hat{H}_A' \dot{\hat{\rho}}_A \right\} dt . \tag{24.14}$$

With $\hat{H}_A$ being time independent and using (24.10) as well as the cyclicity of the trace leads to

$$dU_A = \text{Tr}_A \left\{ \dot{\hat{H}}_A^{\text{eff}} \hat{\rho}_A - i[\hat{H}_A', \hat{H}_2^{\text{eff}}]\hat{\rho}_A + \hat{H}_A' \mathcal{L}_{\text{inc}}(\hat{\rho}) \right\} dt . \tag{24.15}$$

Bearing in mind that only the dynamics due to $\mathcal{L}_{\text{inc}}(\hat{\rho})$ changes the entropy, we can identify work and heat as follows:

$$đW_A = \text{Tr}_A \left\{ \dot{\hat{H}}_A^{\text{eff}} \hat{\rho}_A - i[\hat{H}_A', \hat{H}_2^{\text{eff}}]\hat{\rho}_A \right\} dt , \tag{24.16}$$

$$đQ_A = \text{Tr}_A \left\{ \hat{H}_A' \mathcal{L}_{\text{inc}}(\hat{\rho}) \right\} dt . \tag{24.17}$$

$đW_A$ and $đQ_A$ characterize modes of change of $dU_A$; in general, these terms are not observable separably. $dU_A = 0$ implies that there is no net energy exchange between $A$ and $B$.

The appearance of heat – within an entirely unitary evolution of the total system – is remarkable, but consistent with the ideas of quantum thermodynamics: Heat can be transferred to subsystem $A$ provided the total system does not remain in a product state (cf. Eqs. (24.8), (24.5)). The transfer of work requires a time-dependent Hamiltonian $\hat{H}_A^{\text{eff}}$ (i.e. a non-stationary $\hat{\rho}_B$) and/or a non-zero $\hat{H}_2^{\text{eff}}$. One can easily check that (24.16) and (24.17) are equivalent to the definitions based on (24.1) if $\hat{H}_2^{\text{eff}} = 0$. In general, the results are not symmetrical with respect to $A$ and $B$, respectively. Note, furthermore, that both modes of energy change can eventually be traced back to the same equation of motion, Eq. (24.4). At this stage there is no qualification for the statement that work should be something "better" (more useful) than heat. With respect to the given total scenario both energy parts may enter and leave $A$ in an oscillatory manner.

With reference to (24.16), (24.17) the environment $B$ may be classified (with respect to $A$) as being a (in general finite) heat reservoir, if đ$W_A = 0$, as being a work reservoir, if đ$Q_A = 0$.

### 24.1.2 Applications

We start by noting that the embedding $B$ may be small (e.g., comparable to $A$, say) and thus does not necessarily impart thermal properties on $A$ in the sense of Part II of this book. As a result details of the dynamical behavior of $A$ will sensitively depend on the model used. In general, its state will lack stability with respect to perturbations. We consider three examples; in each the subsystem $A$ is assumed to start in a thermal equilibrium state:

1. Clausius Equality: In the case of a quasi-static evolution, one should expect

$$dS_A = \frac{1}{T_A} đ Q_A. \tag{24.18}$$

Given our generalized expressions for $dS_A$ and đ$Q_A$ it is far from obvious that this equality would indeed hold. Applying our definition for the transferred heat (24.17) we get

$$dS_A = \frac{1}{T_A} \text{Tr}_A \left\{ \hat{H}'_A \mathcal{L}_{\text{inc}}(\hat{\rho}) \right\} dt , \tag{24.19}$$

with $T_A$ being associated with the local temperature of $A$. Using (24.8), (24.9) this temperature is given by

$$T_A = \frac{\text{Tr}_A \left\{ \hat{H}'_A \mathcal{L}_{\text{inc}}(\hat{\rho}) \right\}}{-\text{Tr}_A \left\{ \mathcal{L}_{\text{inc}}(\hat{\rho}) \log \hat{\rho}_A \right\}} . \tag{24.20}$$

For diagonal states $\hat{H}'_A$ commutes with $\hat{\rho}_A$ and $\mathcal{L}_{\text{inc}}(\hat{\rho}_A)$. Thus (24.20) is equivalent to the classical definition

$$T_A = \frac{\partial U_A}{\partial S_A} , \tag{24.21}$$

using (24.15) for the internal energy (with $\dot{\hat{H}}_A^{\text{eff}} \approx 0$). The consistency of assumption (24.18) is thus confirmed. However, $T_A$ is associated with the process induced on $A$ via its embedding and can deviate from the global temperature $T$ of the full system due to the interaction between the subsystems inducing correlations. In the low temperature/strong coupling regime such deviations between $T_A$ and $T$ become most significant.

For general processes (allowing for transient non-equilibrium) the above strict relation between $dS_A$ and $đ Q_A$ has to be relaxed: The famous Clausius inequality, $T_A dS_A \geq đ Q_A$, is obtained [3]. Apparent violations, $T dS_A < đ Q_A$, as derived in [4] for specific processes may be due to the inconsistent use of $T$ instead of $T_A$. There are strong indications that the second law does remain intact also in the quantum limit [5].

2. Coherent Driving: As a second example we consider a two-level atom with local Hamiltonian $\hat{H}_A$ interacting with a coherent laser field (subsystem $B$). Semiclassically, such a scenario can be described by the Hamiltonian

$$\hat{H}_A + \hat{H}_A^{\text{eff}} = \frac{\Delta E}{2}\hat{\sigma}_z + g \sin(\omega t)\hat{\sigma}_x, \tag{24.22}$$

where $g$ is the coupling strength and $\omega$ the laser frequency. Note that $\hat{H}_A^{\text{eff}}$ contains time-dependent Hamilton parameters (control parameters), a typical feature of mechanical control. The Hamiltonian (24.22) can be made time independent in the rotating wave approximation. We investigate the situation where the atom is initially in a thermal state described by

$$\hat{\rho}_A(0) = Z^{-1} \exp(-\beta \hat{H}_A), \tag{24.23}$$

with $Z$ being the partition function and $\beta$ the inverse temperature. The time evolution of the density operator $\hat{\rho}_A(t)$ can be obtained by switching to the rotating frame and diagonalizing the Hamiltonian. Since (24.22) already is an effective description for $A$, work and heat can be computed directly once $\hat{\rho}_A(t)$ is known. As Rabi oscillations occur within the system, we may consider the energy stored in $A$ after a half period ($\Omega t = \frac{\pi}{2}$, where $\Omega = \sqrt{g^2 + \delta^2}$ is the Rabi frequency and $\delta = \omega - \Delta E$ is the detuning from the resonance frequency). Using our definitions, we find

$$đ W_A = \frac{\Delta E g^2}{2\Omega} \tanh\left(\frac{\beta \Delta E}{2}\right) \sin(\Omega t)dt , \tag{24.24}$$

$$đ Q_A = 0 . \tag{24.25}$$

For comparison, the common definition (24.1) for the work would lead to

$$đ W_A = \frac{(\Delta E + \delta)g^2}{2\Omega} \tanh\left(\frac{\beta \Delta E}{2}\right) \sin(\Omega t)dt , \tag{24.26}$$

which means the maximum would neither be at resonance ($\delta = 0$) nor would $đ W_A$ disappear for large $\delta$, which is unphysical. This example shows the importance of explicitly referring to the appropriate measurement basis according to the LEMBAS principle in order to obtain the correct physical result. If $\hat{H}_2^{\text{eff}} = 0$ (as will be the case in Chap. 25) such precautions are not needed and we can base our analysis simply on Eq. (24.1).

3. Heat Transport: In Chap. 21 we have investigated the heat transport through an interacting spin chain. Starting from the respective Hamiltonian (cf. (21.1))

$$\hat{H} = \sum_{\mu=1}^{N} \hat{h}^{(\mu)} + J \sum_{\mu=1}^{N-1} \hat{h}^{(\mu,\mu+1)} \tag{24.27}$$

and assuming weak coupling, the following expression for the pertinent energy current between site $\mu$ and site $\mu + 1$ has been proposed (cf. (21.16)):

$$\hat{J}^{(\mu,\mu+1)} = \mathrm{i} J \operatorname{Tr} \left\{ [\hat{h}^{(\mu,\mu+1)}, \hat{h}^{(\mu)}] \hat{\rho} \right\} . \tag{24.28}$$

Identifying $\mu = A$, $\mu + 1 = B$ the weak coupling approximation amounts to setting $\hat{H}_A^{\text{eff}} = 0$ in the sense of (24.7). As a consequence the transferred work $đW_A$ according to (24.16) is zero under stationary conditions. For the heat current (24.17) we have with $\hat{H}_A' = \hat{H}_A$

$$\frac{dQ_A}{dt} = \mathrm{i} \operatorname{Tr}_{AB} \left\{ [\hat{H}_{AB}, \hat{H}_A, \hat{\mathbb{1}} C_{AB} \right\} . \tag{24.29}$$

This result is identical with (24.28) if

$$\operatorname{Tr}_{AB} \left\{ \hat{H}_{AB} [\hat{H}_A \otimes \hat{\mathbb{1}}_B, \hat{\rho}_A \otimes \hat{\rho}_B] \right\} = 0, \tag{24.30}$$

i.e., if the reduced density operator $\hat{\rho}_A$ commutes with $\hat{H}_A$. This is the case, e.g., for $\hat{\rho}_A$ being canonical, which should be expected for a chain exhibiting a local temperature profile.

### 24.1.3 Remarks

In this chapter we have focussed on two interrelated questions:

(1) How can one define local energy, if the considered subsystem $A$ interacts with its environment?

(2) If specified, how can one split the local energy change into work and heat?

Our answer has been the LEMBAS principle. Contrary to the typical quantum thermodynamic scenarios we have not imposed the condition that the "environment" $B$ should be "large" compared to $A$. As a consequence the dynamics of work flow and heat flow may significantly deviate from that in the thermal domain proper. Here we add two more questions:

(3) Can the LEMBAS scenario further be generalized?

(4) To what extent are those various energy contributions measurable?

The following generalizations are straightforward:

i. By its definition the LEMBAS scenario is not restricted to thermal equilibrium. In fact, it holds for any state $\hat{\rho}$ the dynamics of which is described by (24.3). As a consequence, work and heat could be introduced even far from thermal equilibrium. (This is reminiscent of the von Neumann entropy.)

ii. The bipartite system $\hat{H}$ need not be autonomous: We might include external driving, making $\hat{H}_A$, say, explicitly time dependent. An additional work term $đW_A^{ext}$ results, which is responsible also for the change of the total energy of the bipartite system.

iii. The subsystem $A$ may still be subject to an external bath coupling. Then an additional incoherent contribution had to be included in (24.10) and (24.17).

In motivating the LEMBAS scenario we have assumed measurability of the local energy $E_A'$ of the subsystem $A$ via its effective Hamiltonian $\hat{H}_A'$ (Eq. (24.13)). This means that a projective measurement onto state $|j\rangle$ would yield the eigenvalue $E_j' = E_j + (\hat{H}_A^{eff})_{jj}$, where $E_j$ is the eigenvalue of the isolated subsystem. Insofar as $\hat{C}_{AB}$ only influences the effective dynamics (it appears in the heat term), the reduced density operator $\hat{\rho}_B$ entering $\hat{H}_A^{eff}$ remains constant under the measurement of $E_A'$, i.e., "co-jumps" [6] are not considered here. With respect to the total system these measurements are incomplete.

Heat and work are not observables [7], as, in general, the observed change of energy cannot be split into these two individual contributions. Heat and work are process related. Similar to the intensive thermodynamic variables (cf. Chap. 23) work and heat can be measured at most indirectly, e.g., via temperature changes or in the context of specialized environments: Energy transferred to a heat reservoir is heat, energy transferred to a work reservoir is work (cf. Chap. 25).

## 24.2 Fluctuations of Work[2]

As we have argued, work and heat are associated with processes; these are usually assumed to consist of equilibrium states only. However, if a general process is performed on a system initially in a canonical state, it may be driven out of equilibrium. As soon as the system gets too far away from equilibrium, its behavior can no longer be described by linear response theory or other near-equilibrium approximations and its dynamics may be dominated by fluctuations. Nevertheless, there exist some general theorems about the properties of such non-equilibrium situations [9–14]. Here we want to restrict ourselves on the so-called Jarzynski relation. This relation

---

[2] Based on [8].

connects the work $W$ required for such a non-equilibrium process with the free energy difference $\Delta F$ by

$$\overline{e^{-\beta W}} = e^{-\beta \Delta F} = \frac{Z(t)}{Z(0)}. \tag{24.31}$$

Here $Z(0)$ is the partition sum at the beginning and $Z(t)$ at the end of the process. The initial state is a canonical state with inverse temperature $\beta$. The average is taken over various trajectories of different $W$ provided by a work reservoir.

This relation has been proven for classical systems by Jarzynski [9, 10]. From (24.31) one derives the inequality [1] $\overline{W} \geq \Delta F = \overline{W} - \overline{W}_{\text{diss}}$, which implies for the dissipated work $\overline{W}_{\text{diss}} = T \Delta S - \Delta \overline{Q} \geq 0$, i.e., the Clausius inequality. This inequality is one of several formulations of the second law. As far as applications go, one typically restricts oneself to small mechanical systems characterized by forces and displacements, đ$W = \dot{\lambda} \frac{\partial H}{\partial \lambda} dt$, where $\lambda$ presents a generalized coordinate under external control.

The Jarzynski relation has been shown to be valid also for closed quantum systems by Mukamel [15]. Here the difference $\Delta E_{fi}(t) = \epsilon_f(t) - \epsilon_i(0)$ between the measured system energy eigenvalues at the end and at the beginning, respectively, of a given process is a random variable and interpreted as work. These two measurements are the only source of dissipation here; note that without dissipation there would be no fluctuation at all: "Interactions with the environment are the fundamental source of noise in both classical and quantum systems" [16].

Quantum mechanically the basic fluctuation theorem refers to energy changes as documented by measurement data. The average taken over many such correlated measurement pairs obtained under otherwise identical conditions is interpreted as the change of work đ$W$ introduced by that process:

$$\overline{e^{-\beta \Delta E_{fi}(t)}} = \frac{Z(t)}{Z(0)}. \tag{24.32}$$

In the following we inquire to what extent this theorem might be generalized to more general open quantum systems.

For this purpose we again consider a bipartite system, split into the system of interest $A$ and environment $B$. The process performed by the system is realized by a time-dependent Hamiltonian $\hat{H}_A(t)$. The Hamiltonian of the environment $\hat{H}_B$ and the interaction part $\hat{H}_{AB}$ are assumed to be time independent. Thus, the Hamiltonian of the total system reads

$$\hat{H}(t) = \hat{H}_A(t) \otimes \hat{1}_B + \hat{1}_A \otimes \hat{H}_B + \hat{H}_{AB}. \tag{24.33}$$

The explicit time dependence of $\hat{H}_A$ leads to an extra work term, đ$W_A^{\text{ext}} = \text{Tr}_A\left\{\dot{\hat{H}}_A(t)\hat{\rho}_A\right\} dt$ supplied by the external control. Local two-point measurements on subsystem $A$ yield, according to the LEMBAS definition,

$$\Delta(E'_A(t))_{fi} = E'_f(t) - E'_i(0) ,  \tag{24.34}$$

where $E'_i(0)$ is the initial energy at time 0 and $E'_f(t)$ the final energy at time $t$. The ensemble average over such measurement pairs should give

$$\Delta U_A = \overline{\Delta(E'_A(t))_{fi}} .  \tag{24.35}$$

Considering the total change of local internal energy according to Eqs. (24.15)–(24.17) we get the effective balance

$$\Delta U_A = \Delta W_A^{\text{ext}} + \Delta W_{AB} + \Delta Q_{AB}.  \tag{24.36}$$

$\Delta W_A^{\text{ext}}$ can be interpreted here as the average energy change of the total system:

$$\Delta W_A^{\text{ext}} = \overline{\Delta E_{fi}(t)}.  \tag{24.37}$$

### 24.2.1 Microcanonical Coupling

First we consider microcanonical coupling between system and environment, i.e., neither particle nor energy exchange is allowed. As we have seen in Sect. 10.2, the system is nevertheless influenced by the environment.

For the bipartite system the partition sum is given by

$$Z(t) = \text{Tr} \left\{ e^{-\beta \hat{H}(t)} \right\} = \text{Tr} \left\{ e^{-\beta(\hat{H}_A(t) \otimes \hat{1}_B + \hat{1}_A \otimes \hat{H}_B + \hat{H}_{AB})} \right\} .  \tag{24.38}$$

Since there is no energy transfer between system and environment, the Hamiltonians $\hat{H}_A(t)$ and $\hat{H}_B$ have to commute with $\hat{H}(t)$ at any time $t$:

$$[\hat{H}(t), \hat{H}_A(t)] = 0 , \quad [\hat{H}(t), \hat{H}_B] = 0 .  \tag{24.39}$$

Observing that $[\hat{H}_A(t), \hat{H}_B] = 0$ we get

$$[\hat{H}_A(t), \hat{H}_{AB}] = 0 , \quad [\hat{H}_B, \hat{H}_{AB}] = 0 ,  \tag{24.40}$$

and thus $\hat{H}_2^{\text{eff}} = 0$ (cf. (24.12)). The exponential function in (24.38) can be split, leading to

$$Z(t) = \text{Tr} \left\{ e^{-\beta \hat{H}_A(t)} \otimes e^{-\beta \hat{H}_B} e^{-\beta \hat{H}_{AB}} \right\} .  \tag{24.41}$$

Using the definition of a canonical density operator $\hat{\rho}^{\text{can}} := e^{-\beta \hat{H}}/Z$, we rewrite

$$Z(t) = \text{Tr} \left\{ \hat{\rho}_A^{\text{can}}(t) \otimes \hat{\rho}_B^{\text{can}} e^{-\beta \hat{H}_{AB}} \right\} Z_A(t) Z_B .  \tag{24.42}$$

Up to second order in the coupling strength we get

$$\frac{d}{dt}\left[\text{Tr}\left\{\hat{\rho}_A^{\text{can}}(t)\otimes\hat{\rho}_B^{\text{can}}e^{-\beta\hat{H}_{AB}}\right\}\right]$$

$$\approx\frac{d}{dt}\left[\text{Tr}\left\{\hat{\rho}_A^{\text{can}}(t)\otimes\hat{\rho}_B^{\text{can}}\left(1-\beta\hat{H}_{AB}+\frac{\beta^2}{2}\hat{H}_{AB}^2\right)\right\}\right]$$

$$=\frac{\beta^2}{2}\text{Tr}\left\{\dot{\hat{\rho}}_A^{\text{can}}(t)\otimes\hat{\rho}_B^{\text{can}}\hat{H}_{AB}^2\right\}. \qquad (24.43)$$

Assuming that we can neglect the second-order correction,

$$\frac{d}{dt}\text{Tr}\left\{\hat{\rho}_A^{\text{can}}(t)\otimes\hat{\rho}_B^{\text{can}}e^{-\beta\hat{H}_{AB}}\right\}=0\quad\forall t, \qquad (24.44)$$

we have

$$\text{Tr}\left\{\hat{\rho}_A^{\text{can}}(t)\otimes\hat{\rho}_B^{\text{can}}e^{-\beta\hat{H}_{AB}}\right\}=\text{Tr}\left\{\hat{\rho}_A^{\text{can}}(0)\otimes\hat{\rho}_B^{\text{can}}e^{-\beta\hat{H}_{AB}}\right\}. \qquad (24.45)$$

Thus, using (24.42) and (24.32), we approximately arrive at

$$\overline{e^{-\beta\Delta E_{fi}}}=\frac{Z_A(t)}{Z_A(0)}. \qquad (24.46)$$

We now consider the left-hand side of (24.46), where $\overline{\Delta E_{fi}}$ can be identified with $\overline{\Delta E_{fi}}=\Delta U_A-\Delta W_{AB}-\Delta Q_{AB}$, where $\Delta U_A=\overline{\Delta(E_A')}_{fi}$. According to the LEMBAS definition

$$\text{d}W_{AB}=\text{Tr}_A\left\{\hat{H}_A^{\text{eff}}\dot{\hat{\rho}}_A\right\}dt-i\text{Tr}_A\left\{[\hat{H}_A',\hat{H}_2^{\text{eff}}]\hat{\rho}_A\right\}dt. \qquad (24.47)$$

The first term in $\text{d}W_{AB}$ disappears, if $\hat{\rho}_B$ is taken to be stationary, the second term is zero, as $\hat{H}_2^{\text{eff}}=0$. As a consequence, $\text{d}W_{AB}=0$. Rewriting

$$\text{d}Q_{AB}=-i\text{Tr}\left\{\hat{H}_A'[\hat{H}_{AB},\hat{C}_{AB}]\right\}dt$$
$$=-i\text{Tr}\left\{\hat{C}_{AB}[\hat{H}_A',\hat{H}_{AB}]\right\}dt \qquad (24.48)$$

and observing $[\hat{H}_A,\hat{H}_{AB}]=0$ so that $[\hat{H}^{\text{eff}},\hat{H}_{AB}]=0$, we see that $\text{d}Q_{AB}=0$ for any $\hat{C}_{AB}$. While $\hat{C}_{AB}=0$ certainly implies $\text{d}Q_{AB}=0$, this condition is sufficient but not necessary. With $\text{d}Q_{AB}=0$ and $\text{d}W_{AB}=0$ one may identify $\Delta E_{fi}=\Delta(E_A')_{fi}$, so that (24.46) reduces to a local statement about $A$.

### 24.2.2  Canonical Coupling: Constant Interaction Energy

Now we consider the case of canonical coupling, i.e., energy exchange between the system and environment is now possible. Furthermore, we assume the interaction energy to be a constant of motion. The total system again is described by the Hamiltonian (24.33)

$$[\hat{H}(t), \hat{H}_{AB}] = 0 \Rightarrow [\hat{H}_A(t) + \hat{H}_B, \hat{H}_{AB}] = 0 . \tag{24.49}$$

This condition is slightly less restrictive than (24.40). It can be motivated by the fact that the interaction energy has to be much smaller than the energy of the system and the environment, respectively, as already stated in Sect. 9.3. Thus, the change of interaction energy should also be negligible. The Jarzynski relation can then be proven analogously to the microcanonical case: Under the same weak coupling assumptions using (24.49), the partition sum (24.38) can also be written in the form of (24.42). Together with the Liouville–von Neumann equation and using the cyclic property of the trace, we obtain

$$\frac{d}{dt}\left(\frac{Z(t)}{Z_A(t)Z_B}\right) = -i \operatorname{Tr}\left\{\hat{\rho}_A^{\mathrm{can}}(t) \otimes \hat{\rho}_B^{\mathrm{can}}[\mathrm{e}^{-\beta\hat{H}_{AB}}, \hat{H}(t')]\right\} . \tag{24.50}$$

Because of (24.49) we have $[\mathrm{e}^{-\beta\hat{H}_{AB}}, \hat{H}(t')] = 0 \ \forall t'$. Thus, (24.44) also holds, which again leads us to

$$\overline{\mathrm{e}^{-\beta\Delta E_{fi}(t)}} = \frac{Z_A(t)}{Z_A(0)} = \mathrm{e}^{-\beta\Delta F_A} . \tag{24.51}$$

We now consider the left-hand side of the equation: The total energy change of the system, $\Delta E_{fi}$, is measurable, so this relation can be used experimentally. However, it has not the Jarzynski form: $\Delta E_{fi}$ cannot be identified with $\Delta(E'_A)_{fi}$, as $đW_{AB} \neq 0$, in general; likewise $đQ_{AB} \neq 0$. An alternative scheme for weak coupling, but without constraint (24.49), has been proposed by Talkner et al. [17]. They obtain (24.51) with $\Delta E_{fi} = \Delta(E_A)_{fi} - \Delta(Q_{AB})_{fi}$. Here, the second term is the energy difference of the bath to be measured at the same instants of time as the energy of the system $A$.

### 24.2.3  Canonical Coupling: High-Temperature Limit

Finally, we consider systems at high temperature (i.e., small $\beta$), which allows us to expand the exponent of the partition sum in $\beta$. Neglecting terms of order $O(\beta^3)$, the Baker–Campbell–Hausdorff formula leads to

$$e^{-\beta(A+B)} = e^{-(\beta/2)(A+B)}e^{(\beta^2/4)[A,B]}e^{(\beta^2/4)[B,A]}e^{-(\beta/2)(B+A)}$$
$$= e^{-(\beta/2)A}e^{-(\beta/2)B}e^{-(\beta/2)B}e^{-(\beta/2)A} \ . \tag{24.52}$$

Thus, using the cyclic property of the trace, we have

$$\mathrm{Tr}\left\{e^{-\beta(A+B)}\right\} = \mathrm{Tr}\left\{e^{-(\beta/2)A}e^{-(\beta/2)B}e^{-(\beta/2)B}e^{-(\beta/2)A}\right\}$$
$$= \mathrm{Tr}\left\{e^{-\beta A}e^{-\beta B}\right\} \ . \tag{24.53}$$

Therefore, identifying

$$A := \hat{H}_A(t) \otimes \hat{1}_B + \hat{1}_A \otimes \hat{H}_B \ , \qquad B := \hat{H}^{\mathrm{I}} \ , \tag{24.54}$$

the partition sum reads

$$Z(t) = \mathrm{Tr}\left\{e^{-\beta(\hat{H}_A(t)\otimes\hat{1}_B+\hat{1}_A\otimes\hat{H}_B)}e^{-\beta\hat{H}_{AB}}\right\} \ . \tag{24.55}$$

Now we can proceed as in the above cases and under the same assumptions: The partition sum can then be rewritten as

$$Z(t) = \mathrm{Tr}\left\{\hat{\rho}_A^{\mathrm{can}}(t) \otimes \hat{\rho}_B^{\mathrm{can}}e^{-\beta\hat{H}_{AB}}\right\} Z_A(t)Z_B \ . \tag{24.56}$$

Using the Liouville–von Neumann equation and the cyclic property of the trace again leads to

$$\frac{\mathrm{d}}{\mathrm{d}t}\left(\frac{Z(t)}{Z_A(t)Z_B}\right) = -i\mathrm{Tr}\left\{[\hat{H}_A(t'), \hat{\rho}_A^{\mathrm{can}}(t)] \otimes \hat{\rho}_B^{\mathrm{can}}e^{-\beta\hat{H}_{AB}}\right\} \ . \tag{24.57}$$

For $[\hat{H}_A(t'), \hat{H}_A(t)] = 0 \ \forall t'$ we immediately arrive at

$$\frac{\mathrm{d}}{\mathrm{d}t}\left(\frac{Z(t)}{Z_A(t)Z_B}\right) = 0 \tag{24.58}$$

and therefore

$$\frac{Z(t)}{Z(0)} = \frac{Z_A(t)}{Z_A(0)} = \overline{e^{-\beta\Delta E_{fi}(t)}} \ . \tag{24.59}$$

For $\beta \to 0$ the initial state for the total system approaches the unit operator, which commutes with $\hat{H}(t)$, so that $\hat{C}_{AB} = 0$ for all times. This is approximately true also for finite $\beta$, so that $đQ_{AB} \to 0$. If the driving does not induce coherence, $\hat{H}_2^{\mathrm{eff}} = 0$. As a consequence, $đW_{AB} = 0$ and $đW_A^{\mathrm{ext}} = \overline{\mathrm{d}E_{fi}} = \mathrm{d}U_A$. Thus, the Jarzynski relation should approximately be valid as a local statement on $A$.

From the preceding results one may expect that for the Jarzynski relation to hold, it is sufficient to have $đW_{AB} = 0$ and $đQ_A = 0$: Then $\mathrm{d}E_A = đW_A^{\mathrm{ext}}$. Deviations from the Jarzynski relation should increase as $\beta$ is increased, which can be investigated numerically.

### 24.2.4 Numerical Verification

To check the analytical results of the last sections numerically, we consider a particle in a box (with cutoff at level $n$) coupled to an environment. For microcanonical coupling the environment is chosen to be a single, highly degenerate, energy level (cf. Sect. 16.1). The particle is prepared initially in a canonical state. This gives the probability distribution for the initial energy eigenstates. In order to perform a process, we assume one wall of the box to be movable, i.e., the wall width becomes time dependent $L = L(t)$. Then, we calculate the Schrödinger dynamics of the system, initially in some pure energy eigenstate at $L(0) = L_0$, and determine the energy distribution at the end of the process at $L(t_f) = 2L_0$. From all this the work distribution $P(W)$ follows. This distribution can then be used, to calculate the average $\overline{e^{-\beta W}}$, which finally can be compared with $Z_A(t)/Z_A(0)$ to check the validity of the Jarzynski relation.

Figure 24.1 shows an example for a work distribution $P(W)$ of a microcanonical system compared with a closed system. As one can see, the deviation is very small. Note that possible values for $W$ are discrete and roughly the same for both cases. In the microcanonical case, though, the interaction between system and environment

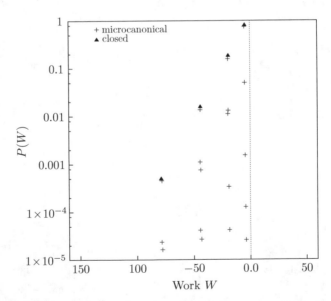

**Fig. 24.1** Distribution of work $P(W)$ for a particle in a box. The process considered is an expansion with constant velocity $v$ and $L(t_{\text{final}}) = 2L(t_0)$. For the microcanonical case a weak coupling is chosen between the system and the environment with degeneracy $N_B = 150$. The cutoff for the spectrum of the system is at $N = 5$. Probabilities smaller than $1 \times 10^{-5}$ are suppressed

causes a tiny splitting of the system's energy levels. Some of these levels can be resolved within numerical accuracy leading to the different values of $P(W)$ for slightly different $W$ depicted in Fig. 24.1. The comparison of the average $\overline{e^{-\beta W}}$ and $Z_A(t)/Z_A(0)$ shows that the Jarzynski relation is valid in both cases within numerical accuracy.

We note in passing that for the quantum Jarzynski relation to hold the set of eigenfunctions of $\hat{H}$ at the end of the process should be expandable into those at the beginning. Obviously this is not the case for the box with changing width. Fortunately, this box model can be approximated by a time-dependent finite potential barrier, for which the above condition is fulfilled [18].

Qualitatively similar conclusions are reached for canonical coupling at high temperatures, whereas for low temperatures the Jarzynski relation is, indeed, found to be violated [8].

# References

1. F. Ritort, Poincaré Seminar **2**, 195 (2003). cond-mat/0401311
2. H. Weimer, M.J. Henrich, F. Rempp, H. Schröder, G. Mahler, Europhys. Lett. **83**, 30008 (2008)
3. S. Abe, S. Thurner, Europhys. Lett. **81**, 10004 (2008)
4. A.E. Allahverdyan, T.M. Nieuwenhuizen, Phys. Rev. Lett. **85**, 1799 (2000)
5. I. Kim, G. Mahler, Eur. Phys. J. B **60**, 401 (2007)
6. C.M. Granzow, A. Liebman, G. Mahler, Eur. Phys. J. B **2**, 131 (1998)
7. P. Talkner, E. Lutz, P. Hänggi, Phys. Rev. E **75**, 050102(R) (2007)
8. J. Teifel, G. Mahler, Phys. Rev. E **76**, 051126 (2007)
9. C. Jarzynski, Phys. Rev. Lett. **78**, 2690 (1997)
10. C. Jarzynski, Phys. Rev. E **56**, 5018 (1997)
11. C. Jarzynski, J. Stat. Phys. **98**, 77 (2000)
12. D.J. Evans, E.G.D. Cohen, G.P. Morriss, Phys. Rev. Lett. **71**, 2401 (1993)
13. G.E. Crooks, Phys. Rev. E **60**, 2721 (1999)
14. U. Seifert, Phys. Rev. Lett. **95**, 040602 (2005)
15. S. Mukamel, Phys. Rev. Lett. **90**, 170604 (2003)
16. M. Nielsen, I. Chuang, *Quantum Computation and Quantum Information* (Cambridge University Press, Cambridge, 2000)
17. P. Talkner, M. Campisi, P. Hänggi, J. Stat. Mech., P02025 (2009)
18. J. Teifel, G. Mahler, submitted (2009)

# Chapter 25
# Quantum Thermodynamic Processes[1]

> *Simple models can be invaluable without being "right" in an*
> *engineering sense. Indeed, by such lights, all the best models*
> *are wrong. But they are fruitfully wrong. They are illuminating*
> *abstractions. I think it was Picasso who said, "Art is a lie that*
> *helps us see the truth".*
>
> — J. M. Epstein [4]

**Abstract** Based on quantum thermodynamic reasoning even small embedded systems $S$ may well be in thermodynamic equilibrium. This equilibrium depends on the system, here formalized by a parameter $\gamma$ controlling the spectrum of $S$ and on the embedding, formalized via a parameter $\alpha$ controlling the resulting attractor state for $S$. Cyclic processes in this $\alpha/\gamma$-control space will be investigated and the effects of non-equilibrium studied.

## 25.1 Model of Control

The study of heat engines has always been an important part of thermodynamics. For acting as a heat engine, a physical system usually has to go through a cyclic process enforced by its environment. Here we want to discuss how this concept of thermodynamic machines can be carried over to small quantum systems. Note that this problem can be dealt with only because – according to quantum thermodynamics – even small systems, single spins, say, can meaningfully be described in thermodynamic terms. We have seen that quantum thermodynamics is able to explain thermodynamical systems by their embedding into some appropriate environment. Here we use a phenomenological short-cut: We postulate the existence of environments that produce desired effects (attractor states and so-called mechanical changes) on the embedded system.

As we have seen in Sect. 14.1, decoherence is an unavoidable consequence of the coupling between a system and its environment and therefore typical for such open quantum systems. The only quantum properties of the embedded system, which could survive such strong decoherence, are the discrete spectrum of the subsystem and the probabilistic character of its state. Both these properties are also influenced

---

[1] Based on [1–3].

Gemmer, J. et al.: *Quantum Thermodynamic Processes*. Lect. Notes Phys. **784**, 291–313 (2009)
DOI 10.1007/978-3-540-70510-9_25 © Springer-Verlag Berlin Heidelberg 2009

by the environment. To describe this effect, we use some kind of control theoretical approach introducing two formal parameters, one related to the spectrum and the other to the state of the quantum system: We assume that the local Hamiltonian $\hat{H}'$ (in the sense of (24.13)) is controlled by the mechanical parameter $\gamma$: $\{E_i^{\mathrm{eff}}(\gamma)\}_{i=1}^N$, where $N$ is the number of energy levels. In the very general case, the $E_i^{\mathrm{eff}}(\gamma)$ would define $N$ independent functions that would make the model hardly tractable. Thus we consider, in the following, the special class of the spectral control:

$$\hat{H}'(\gamma) = \sum_i E_i^{\mathrm{eff}}(\gamma)|i(\gamma)\rangle\langle i(\gamma)|, \tag{25.1}$$

$$E_i^{\mathrm{eff}}(\gamma) = g(\gamma) \cdot \epsilon_i , \tag{25.2}$$

where $g$ is some monotonous function independent of $i$ and $\{\epsilon_i\}_{i=1}^N$ are to be regarded as a set of characteristic constants. One example for this kind of spectrum would be a spin in a magnetic field, where the energy levels are split proportional to the field $B$. Taking the magnetic field as the control parameter ($\gamma = B$) then results in $g(\gamma) = \gamma$. Another example would be the particle in a box, where $E_i \propto 1/L^2$ leads to $g(\gamma) = \gamma^{-2}$ identifying $\gamma = L$. Together with the fast decoherence, this special control allows us to treat the state of the system formally independent of $\gamma$ even if the spectrum explicitly enters the distribution $\{p_i = \rho_{ii}\}_{i=1}^N$ as, e.g., in the canonical case. This distribution $p_i$ is assumed to be controlled by the statistical parameter $\alpha$, i.e., we postulate the existence of an attractor state, $\{\tilde{p}_i(\alpha)\}_{i=1}^N$. Whenever the actual distribution $p_i$ differs from $\tilde{p}_i(\alpha)$ for some given $\alpha$, it will relax toward this attractor. The deviation ("non-equilibrium") will decay on a timescale $\tau_R \gg \tau_{\mathrm{dec}}$, which can be described phenomenologically by the relaxation time approximation

$$\dot{p}_i = -\tau_R^{-1}(p_i - \tilde{p}_i(\alpha)) . \tag{25.3}$$

Note that – from the present point of view – any distribution may be used as attractor $\tilde{p}_i(\alpha)$, as long as all thermodynamic quantities introduced in the next paragraph are well defined. An important example is the canonical attractor

$$\tilde{p}_i(\alpha) = Z_{\mathrm{Can}}^{-1}e^{-\alpha\epsilon_i}, \quad Z_{\mathrm{Can}} = \sum_i e^{-\alpha\epsilon_i} , \tag{25.4}$$

which would result from canonical coupling and an exponential increasing degeneracy in the environment as discussed in Sect. 10.4. As we will see, some typical thermodynamical properties do not depend on the choice of the attractor at all. In a typical process, $\alpha$ and $\gamma$ will be time dependent. This control dynamics must externally be given; it is not subject of the present theory.

## 25.2 Quasi-static Limit

If the characteristic time of enforced $\alpha$-parameter alterations is much larger than the relaxation time $\tau_R$, the system is always in the equilibrium state

$$p_i = \tilde{p}_i(\alpha) . \tag{25.5}$$

This so-called quasi-static limit plays a fundamental role in thermodynamics. In this section we will introduce the thermodynamic quantities in this limit, which then allows us to study quantum thermodynamic processes acting as heat engines or heat pumps.

### 25.2.1 Thermodynamic Quantities

In terms of our control theory the quasi-static limit can be regarded as the limit of perfect control: The spectrum and the distribution are completely determined by the control parameters $\gamma$ and $\alpha$. Thus, the main thermodynamical quantities, namely the internal energy $U$ and the entropy $S$, are also functions of these parameters only:

$$U(\alpha, \gamma) := g(\gamma) \sum \epsilon_i \tilde{p}_i(\alpha) =: g(\gamma)h(\alpha) , \tag{25.6}$$

$$S(\alpha) := - \sum \tilde{p}_i(\alpha) \ln \tilde{p}_i(\alpha) . \tag{25.7}$$

$k_B$ is set to 1 here. The temperature can now be defined in the thermodynamical sense as the conjugate variable to the entropy $S$:

$$T := \left( \frac{\partial U}{\partial S} \right)_\gamma = \left( \frac{\partial U}{\partial \alpha} \right)_\gamma \left( \frac{dS}{d\alpha} \right)^{-1} . \tag{25.8}$$

Using (25.6) and (25.7) yields

$$T(\alpha, \gamma) = g(\gamma)\Theta(\alpha)^{-1} , \tag{25.9}$$

with

$$\Theta(\alpha) := - \frac{\sum \ln \tilde{p}_i (d\tilde{p}_i/d\alpha)}{dh/d\alpha} . \tag{25.10}$$

Note that this temperature, like the spectral temperature introduced in Sect. 13.1, can be defined for any attractor state; a canonical distribution is not mandatory. However, based on (25.4) one finds

$$\Theta_{\mathrm{Can}}(\alpha) = \alpha . \tag{25.11}$$

It is also possible to define the conjugate variable to $\gamma$, which in regard to classical thermodynamics is called $P$ (for pressure):

$$P(\alpha, \gamma) := -\left(\frac{\partial U}{\partial \gamma}\right)_\alpha = -\frac{dg(\gamma)}{d\gamma} h(\alpha) \,. \tag{25.12}$$

One has to keep in mind, though, that the meaning of $P$ strongly depends on the realization of the control. For the particle in a box $P$ indeed corresponds to the classical pressure (i.e., is an intensive variable), while for the spin in the magnetic field $B = \gamma$, $P$ would rather be the magnetization, an extensive variable.

In order to describe thermodynamical heat engines and heat pumps one also has to introduce the notions of heat and work (cf. Chap. 24). Within our model, these definitions are straightforward: We define the work as change of the internal energy at constant entropy, i.e., constant $\alpha$,

$$\text{d} W(\alpha, \gamma) := \left(\frac{\partial U}{\partial \gamma}\right)_\alpha d\gamma = \frac{dg}{d\gamma} h(\alpha)\, d\gamma \,, \tag{25.13}$$

while the heat is given by

$$\text{d} Q(\alpha, \gamma) := T dS = \left(\frac{\partial U}{\partial \alpha}\right)_\gamma d\alpha = g(\gamma)\frac{dh}{d\alpha} d\alpha \,. \tag{25.14}$$

These definitions ensure the validity of the first law

$$dU(\alpha, \gamma) = \left(\frac{\partial U}{\partial \alpha}\right)_\gamma d\alpha + \left(\frac{\partial U}{\partial \gamma}\right)_\alpha d\gamma = \text{d} Q(\alpha, \gamma) + \text{d} W(\alpha, \gamma) \,. \tag{25.15}$$

Because all thermodynamic quantities in the quasi-static limit are functions of the control parameters $\alpha$ and $\gamma$ only, every thermodynamic state is defined as a point in the $\alpha\gamma$-plane. A thermodynamic process can thus be depicted as a line in this control plane.

Combining such processes to a closed loop leads to cyclic processes, which can act as heat engines or heat pumps. One should bear in mind, however, that any machine functionality has to refer to the environment – in terms of input/output – relations. As the environment is not explicitly included here, one has to settle for certain supplementary assumptions. These are also needed if one wants to account for the entropy balance of the system and surrounding. (Note that the entropy change of the system alone is always zero per cycle.) But then this approach allows us, in principle, to transfer any process known from classical thermodynamics to quantum systems, i.e., from the macroscopic to the nanoscopic scale. In the following we want to consider two examples for thermodynamic cycles, namely the Otto and the Carnot cycle.

### 25.2.2 Otto Cycle

The Otto cycle is in some sense the most fundamental cycle in quantum thermo-dynamics: It consists of two isentropic steps, where only the spectrum changes while the distribution stays constant, and two isochoric steps, where the distribution changes at constant spectrum. Most of the theoretically discussed quantum thermo-dynamic cycles are of the Otto type [5–7]. In the control plane the Otto cycle has a very simple form, as depicted in Fig. 25.1. Using the definitions of temperature and pressure, the cycle can also be illustrated as $TS$ or $P\gamma$ diagrams. Figures 25.2 and 25.3 show such diagrams for the particle in a box.

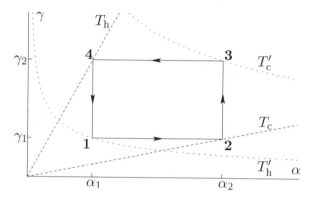

**Fig. 25.1** Otto cycle in the $(\alpha, \gamma)$-plane. The isentropes are given by $\alpha = \texttt{const}$ and the isochors by $\gamma = \texttt{const}$. The *dashed lines* are canonical isotherms with the highest and lowest temperatures of the cycle for $g(\gamma) = \gamma$, while the *dotted ones* hold for $g(\gamma) = \gamma^{-2}$

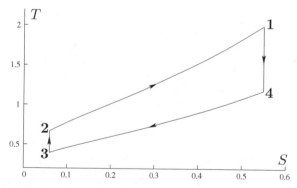

**Fig. 25.2** $TS$-diagram of an Otto cycle for the particle in a box. The chosen parameters are $\alpha_1 = 0.5$, $\alpha_2 = 1.5$, $\gamma_1 = 1$, and $\gamma_2 = 1.3$

To calculate the contributions of heat and work, one has to integrate (25.13) and (25.14) along the respective steps. The sign of these contributions depends on the monotonicity of the functions $g(\gamma)$ and $h(\alpha)$ as well as on the clockwise or anti-clockwise direction of the cycle. Assuming, e.g., the anti-clockwise direction shown in Fig. 25.1 we get

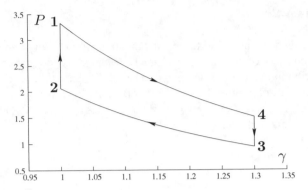

**Fig. 25.3** $P\gamma$-diagram of the Otto cycle in Fig. 25.2

$$Q_{12} = \int_{\alpha_1}^{\alpha_2} g(\gamma_1)\frac{dh}{d\alpha}d\alpha = g(\gamma_1)(h(\alpha_2) - h(\alpha_1)) \ , \tag{25.16}$$

$$W_{23} = \int_{\gamma_1}^{\gamma_2} h(\alpha_2)\frac{dg(\gamma)}{d\gamma}d\gamma = h(\alpha_2)(g(\gamma_2) - g(\gamma_1)) \ , \tag{25.17}$$

$$Q_{34} = \int_{\alpha_2}^{\alpha_1} g(\gamma_2)\frac{dh}{d\alpha}d\alpha = g(\gamma_2)(h(\alpha_1) - h(\alpha_2)) \ , \tag{25.18}$$

$$W_{41} = \int_{\gamma_2}^{\gamma_1} h(\alpha_1)\frac{dg(\gamma)}{d\gamma}d\gamma = h(\alpha_1)(g(\gamma_1) - g(\gamma_2)) \ . \tag{25.19}$$

All other terms are zero. The cycle works as a heat engine, if the total work per cycle

$$W_\circ = (g(\gamma_2) - g(\gamma_1))(h(\alpha_2) - h(\alpha_1)) \tag{25.20}$$

is negative, which is, e.g., the case for $\frac{dh}{d\alpha} < 0$; $\frac{dg}{d\gamma} > 0$. This work is assumed to be taken up by the environment in return for the input $Q_{34}$. The efficiency then is given by

$$\eta^O = \frac{-W_\circ}{Q_{34}} = 1 - \frac{g(\gamma_1)}{g(\gamma_2)} \ . \tag{25.21}$$

If the total work (25.20) is positive one has to run the cycle in the opposite direction to get a heat engine. In general, the efficiency of the Otto cycle reads

$$\eta^O = 1 - \frac{\min(g(\gamma))}{\max(g(\gamma))} \ , \tag{25.22}$$

where $\min(\dots)$ ($\max(\dots)$) stands for the minimal (maximal) value within the cycle. As one can see, this result does not depend on the choice of the attractor state but on the spectral control only. It is reminiscent of the efficiency of the classical Otto cycle, which is determined by the ratio of the minimal and maximal volumes of the gas ("compression").

As a consequence of the second law, the efficiency of a heat engine acting between two heat baths of the temperatures $T_h$ and $T_c$ is bounded from above by the Carnot efficiency $\eta^C = 1 - \frac{T_c}{T_h}$. To check this for the Otto cycle, we have to replace $\min(g(\gamma))$ and $\max(g(\gamma))$ by the highest and lowest temperatures of the cycle. Using (25.9) these are given by

$$T_c = \frac{\min(g(\gamma))}{\max(\Theta(\alpha))} , \quad T_h = \frac{\max(g(\gamma))}{\min(\Theta(\alpha))} . \tag{25.23}$$

Thus the efficiency of the Otto cycle can be rewritten as

$$\eta^O = 1 - \frac{T_c \max(\Theta(\alpha))}{T_h \min(\Theta(\alpha))} \leq 1 - \frac{T_c}{T_h} . \tag{25.24}$$

The Carnot efficiency is only reached in the limit $\max(\Theta(\alpha)) \to \min(\Theta(\alpha)) \Leftrightarrow \alpha_1 \to \alpha_2$, where the total work output vanishes.

### 25.2.3 Carnot Cycle

We now consider a cycle of the Carnot type, which means a cycle consisting of two isothermal and two isentropic steps. This cycle is of fundamental importance in thermodynamics, as it has the highest efficiency of any engine acting between two heat baths of given temperature $T_h > T_c$. Now we want to investigate to what extent this universality carries over to our model.

In contrast to the Otto cycle the form of the Carnot cycle in the control plane depends on the chosen distribution $\tilde{p}_i(\alpha)$ and the control function $g(\gamma)$ (Figs. 25.4 and 25.5). Nevertheless, the following considerations hold for any choice of these functions.

To calculate the efficiency of the Carnot cycle we have again to integrate the heat and work along the different steps. For the isotherm at temperature $T_c$ we get

$$W_{12} = \int_{\gamma_1}^{\gamma_2} h(\alpha) \frac{dg(\gamma)}{d\gamma} d\gamma = T_c \int_{\alpha_1}^{\alpha_2} h(\alpha) \frac{d\Theta}{d\alpha} d\alpha , \tag{25.25}$$

$$Q_{12} = \int_{\gamma_1}^{\gamma_2} g(\gamma) \frac{dh}{d\alpha} d\alpha = T_c \int_{\alpha_1}^{\alpha_2} \Theta(\alpha) \frac{dh}{d\alpha} d\alpha . \tag{25.26}$$

The work on the following adiabatic step is given by

$$W_{23} = \int_{\gamma_2}^{\gamma_3} h(\alpha_2) \frac{dg(\gamma)}{d\gamma} d\gamma = h(\alpha_2)(g(\gamma_3) - g(\gamma_2))$$
$$= h(\alpha_2)\Theta(\alpha_2)(T_h - T_c) . \tag{25.27}$$

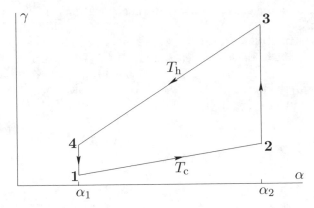

**Fig. 25.4** A Carnot cycle in the control plane for the spin in a magnetic field ($g(\gamma) = \gamma$) and canonical attractor (25.4). According to (25.11) the isotherms are here just *straight lines* with gradient $T$

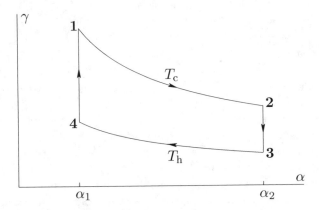

**Fig. 25.5** A Carnot cycle in the control plane for the particle in a box ($g(\gamma) = \gamma^{-2}$) and canonical attractor (25.4). Here, the isotherms are given by $\gamma = 1/\sqrt{T\alpha}$

The contributions on the remaining two parts are analogously given by

$$W_{34} = -T_{\mathrm{h}} \int_{\alpha_1}^{\alpha_2} h(\alpha) \frac{\mathrm{d}\Theta}{\mathrm{d}\alpha} \mathrm{d}\alpha \, , \tag{25.28}$$

$$Q_{34} = -T_{\mathrm{h}} \int_{\alpha_1}^{\alpha_2} \Theta(\alpha) \frac{\mathrm{d}h}{\mathrm{d}\alpha} \mathrm{d}\alpha \, , \tag{25.29}$$

$$W_{41} = h(\alpha_1)\Theta(\alpha_1)(T_{\mathrm{c}} - T_{\mathrm{h}}) \, . \tag{25.30}$$

Integration by parts leads to the total work

$$W_\circ = (T_h - T_c) \left( h(\alpha_2)\Theta(\alpha_2) - h(\alpha_1)\Theta(\alpha_1) - \int_{\alpha_1}^{\alpha_2} h(\alpha)\frac{d\Theta}{d\alpha}d\alpha \right)$$

$$= (T_h - T_c) \int_{\alpha_1}^{\alpha_2} \Theta(\alpha)\frac{dh}{d\alpha}d\alpha . \tag{25.31}$$

So the efficiency of the Carnot heat engine is indeed the well-known Carnot efficiency

$$\eta^C = \frac{-W_\circ}{Q_{34}} = 1 - \frac{T_c}{T_h} , \tag{25.32}$$

independent of the attractor state as well as of the control function $g(\gamma)$. The Carnot efficiency results from the structure of the cycle only. Note that the Carnot efficiency as an upper bound for cyclic machines is a consequence of the second law.

### 25.2.4 Continuum Limit of the Particle in a Box

In this paragraph we want to study the particle in a $D$-dimensional box of adjustable length $L$ in the limit of large $L$. As we will see, the particle in the box then shows, within our model, the typical properties of a classical ideal gas.

In the above limit, the spectrum becomes continuous and the density of states is given by

$$\rho(\epsilon) \propto \epsilon^{\frac{D}{2}-1} . \tag{25.33}$$

For the control parameter $\gamma$ we choose the volume of the box ($\gamma = V = L^D$). Because $E_i^{box} \propto \frac{1}{L^2}$ holds independently of the dimension $D$, the control function is given by

$$g(\gamma) = \gamma^{-\frac{2}{D}} . \tag{25.34}$$

As we have seen before, the function $h(\alpha)$ plays an important role for the thermodynamic quantities. This is why we want to calculate this function in the continuum limit first. For the canonical attractor (25.4) one finds

$$h(\alpha) = \sum_i \epsilon_i \tilde{p}_i = \frac{\sum_i \epsilon_i e^{-\epsilon_i \alpha}}{\sum_i e^{-\epsilon_i \alpha}} \rightarrow \frac{\int_0^\infty \epsilon \rho(\epsilon) e^{-\epsilon \alpha} d\epsilon}{\int_0^\infty \rho(\epsilon) e^{-\epsilon \alpha} d\epsilon} . \tag{25.35}$$

Using (25.33) and some properties of the $\Gamma$-function yields the simple result

$$h(\alpha) = \frac{D}{2\alpha} . \tag{25.36}$$

With this relation we recover the typical properties of the ideal gas: Inserting (25.36) in (25.12) we get

$$P = -\frac{dg(\gamma)}{d\gamma}h(\alpha) = -\frac{Dg'(\gamma)}{2\alpha} \, . \tag{25.37}$$

With the control function (25.34) and $T = \frac{g(\gamma)}{\alpha}$ one ends up with the equation of state

$$P\gamma = T, \tag{25.38}$$

which, because of $k_B = 1$, is the ideal gas law $PV = Nk_BT$ for $N = 1$. For the internal energy one gets

$$U = g(\gamma)h(\alpha) = g(\gamma)\frac{D}{2\alpha} = \frac{D}{2}T \, , \tag{25.39}$$

which is also formally identical with the internal energy of the ideal gas $U = \frac{f}{2}Nk_BT$, with $N = 1$ and the number of degrees of freedom $f$ being equal to the dimension $D$.

Because the equations of state for the particle in a box in the continuum limit are the same as for the classical ideal gas, one expects that all other properties of the ideal gas should also be reproduced for this system. Indeed, this is the case: In particular, all efficiencies, which can be calculated within this model, reduce to the results known from classical thermodynamics. For example, for the Otto cycle discussed before, we get the well-known result

$$\eta^O = 1 - \frac{\min\left(g(\gamma)\right)}{\max\left(g(\gamma)\right)} = 1 - \left(\frac{\gamma_1}{\gamma_2}\right)^{2/f} \, . \tag{25.40}$$

As we will see below (cf. Sect. 25.3.3), we are even able to describe the non-equilibrium behavior of the classical Carnot cycle very well.

## 25.3 Processes Beyond the Quasi-static Limit

The quasi-static limit allows for reversibility but at the cost of making the power of heat engines tend to zero. Real (and useful) machines are irreversible. To deal with such real processes finite-time thermodynamics [8] has been invented in 1975 – entirely from a macroscopic point of view with heat conductances, friction, etc. In this context the model proposed by Curzon and Ahlborn [9] has received much attention.

From a microscopic point of view irreversibility derives from non-equilibrium – an unavoidable aspect of increased process speed. When talking about non-equilibrium, one usually thinks of transport scenarios (based on local equilibrium) as discussed in Part III. Here, the situation is different: The small quantum system considered is driven into global non-equilibrium by the external control. This happens as soon as the change of the parameter $\alpha$ proceeds on a timescale comparable to the relaxation time $\tau_R$ (cf. (25.3)). Such a non-equilibrium state can be written as

$$p_i(\alpha, \Delta p_i) = \tilde{p}_i(\alpha) + \Delta p_i , \qquad (25.41)$$

where the deviations from the equilibrium state $\Delta p_i$ are not directly controllable but determined by the relaxation behavior of the system. Nevertheless, the internal energy and the entropy can be defined like in the quasi-static limit, just by replacing $\tilde{p}_i$ by $p_i$:

$$U^*(\alpha, \gamma) := g(\gamma) \sum \epsilon_i p_i(\alpha) , \qquad (25.42)$$

$$S^*(\alpha) := - \sum p_i(\alpha) \ln p_i(\alpha) . \qquad (25.43)$$

In the following the asterisk "*" always denotes the non-equilibrium quantities. Usually the temperature is an equilibrium property only, but with respect to our control theory it can also be defined for non-equilibrium in a consistent way: The temperature still has to be a control variable indicating the response of the internal energy with respect to changes of $\alpha$, where $\Delta p_i$ and $\gamma$ are kept fixed:

$$
\begin{aligned}
T^*(\alpha, \gamma, \Delta p_i) : &= \left( \frac{\partial U^*}{\partial \alpha} \right)_{\gamma, \Delta p_i} \left( \frac{\partial S^*}{\partial \alpha} \right)^{-1}_{\Delta p_i} \\
&= -g(\gamma) \frac{\sum \epsilon_i (d\tilde{p}_i/d\alpha)}{\sum \ln p_i (d\tilde{p}_i/d\alpha)} .
\end{aligned}
\qquad (25.44)
$$

Because this temperature is no longer determined by the current values of $\alpha$ and $\gamma$ but also depends on the past process via $p_i$, we shall call this quantity a *process temperature*.

The conjugate variable to $\gamma$ in non-equilibrium is just given by

$$P^*(\alpha, \gamma, \Delta p_i) := - \left( \frac{\partial U^*}{\partial \gamma} \right)_{\alpha, \Delta p_i} = - \frac{dg(\gamma)}{d\gamma} \sum \epsilon_i p_i(\alpha) . \qquad (25.45)$$

As we will see later, these quantities are appropriate to qualitatively illustrate the behavior of thermodynamic cycles at finite velocities.

Also the definitions of heat and work in non-equilibrium are straightforward: The complete differential of the internal energy reads

$$dU^* = \left( \frac{\partial U^*}{\partial \gamma} \right)_{\alpha, \Delta p_i} d\gamma + \left( \frac{\partial U^*}{\partial \alpha} \right)_{\gamma, \Delta p_i} d\alpha + \sum \left( \frac{\partial U^*}{\partial \Delta p_i} \right)_{\gamma, \alpha} d(\Delta p_i) . \qquad (25.46)$$

Again work is defined as change of the internal energy at constant entropy:

$$\eth W^* = \left(\frac{\partial U^*}{\partial \gamma}\right)_{\alpha, \Delta p_i} d\gamma$$

$$= \frac{dg}{d\gamma} \sum \epsilon_i p_i \, d\gamma \, , \tag{25.47}$$

while the other terms are considered as heat

$$\eth Q^* = \left(\frac{\partial U^*}{\partial \alpha}\right)_{\gamma, \Delta p_i} d\alpha + \sum \left(\frac{\partial U^*}{\partial \Delta p_i}\right)_{\gamma, \alpha} d(\Delta p_i)$$

$$= g(\gamma) \sum \epsilon_i \left(\frac{d\tilde{p}_i}{d\alpha} d\alpha + d(\Delta p_i)\right)$$

$$= g(\gamma) \sum \epsilon_i d p_i \, . \tag{25.48}$$

These definitions underline our observation of Sect. 24.1.3 that heat and work may indeed be generalized to cover even non-equilibrium scenarios. Their concrete specification, though, requires to know $p_i(\alpha, \Delta p_i)$, i.e., all the $\Delta p_i$. As will be shown below, this is the case along the externally controlled cycles in the $\alpha/\gamma$-control plane.

### 25.3.1 Solution of the Relaxation Equation

Let us consider a process for which $\alpha$ changes at finite, but piecewise constant speed. The behavior of $p_i$ is not completely determined by (25.3), since the time dependence of $\alpha$ has not yet been specified. In the following we assume a linear dependency:

$$\alpha = \alpha_0 + vt \, , \tag{25.49}$$

where $\alpha_0$ is the initial $\alpha$-value and $v$ denotes the velocity of change. The relaxation equation (25.3) then reads

$$(dp_i/d\alpha) = -(v\tau_R)^{-1}(p_i - \tilde{p}_i(\alpha)) \, . \tag{25.50}$$

The solution for this differential equation is given by

$$p_i = p_i^{(0)} e^{-\frac{\alpha - \alpha_0}{v\tau_R}} + (v\tau_R)^{-1} \int_{\alpha_0}^{\alpha} \tilde{p}_i(\alpha') e^{-\frac{\alpha - \alpha'}{v\tau_R}} d\alpha' \, , \tag{25.51}$$

with the distribution $p_i^{(0)}$ at the beginning of the process. Thus, the momentary distribution $p_i$ depends now on the history of the process.

As we are interested in cyclic processes, we consider such a process running between the $\alpha$-values $\alpha_1$ and $\alpha_2$. We allow for two different velocities in either direction, which means

$$\alpha_{1\rightarrow 2}(t) = \alpha_1 + \kappa(t - t_1), \quad t_1 \leq t \leq t_2, \tag{25.52}$$

$$\alpha_{2\rightarrow 1}(t) = \alpha_2 - \lambda\kappa(t - t_3), \quad t_3 \leq t \leq t_4, \tag{25.53}$$

where $\kappa$ and $\lambda$ are both positive parameters, and $t_2 = t_1 + (\alpha_2 - \alpha_1)/\kappa$, $t_4 = t_3 + (\alpha_2 - \alpha_1)/\lambda\kappa$. Using (25.51) and demanding the distribution at the beginning to be the same as at the end of the cycle lead to the distributions at the turning points

$$p_i^{(1)} = \left(e^{\frac{\Delta\alpha}{\lambda\kappa\tau_R}} - e^{-\frac{\Delta\alpha}{\kappa\tau_R}}\right)^{-1} (\kappa\tau_R)^{-1}$$
$$\times \int_{\alpha_1}^{\alpha_2} \tilde{p}_i(\alpha') \left(e^{-\frac{\alpha_2-\alpha'}{\kappa\tau_R}} + \frac{1}{\lambda} e^{\frac{\alpha_2-\alpha'}{\lambda\kappa\tau_R}}\right) d\alpha', \tag{25.54}$$

$$p_i^{(2)} = \left(e^{\frac{\Delta\alpha}{\kappa\tau_R}} - e^{-\frac{\Delta\alpha}{\lambda\kappa\tau_R}}\right)^{-1} (\kappa\tau_R)^{-1}$$
$$\times \int_{\alpha_1}^{\alpha_2} \tilde{p}_i(\alpha') \left(e^{\frac{\alpha'-\alpha_1}{\kappa\tau_R}} + \frac{1}{\lambda} e^{-\frac{\alpha'-\alpha_1}{\lambda\kappa\tau_R}}\right) d\alpha', \tag{25.55}$$

with $\Delta\alpha = \alpha_2 - \alpha_1$. Note that when starting at $\alpha_1$ with a different distribution than (25.54), this would settle down after some cycles ("transient effect", cf. [10]). The non-equilibrium distribution function is attracted toward the time-periodic solution with

$$p_{1\rightarrow 2,i}(\alpha) = p_i^{(1)} e^{-\frac{\alpha-\alpha_1}{\kappa\tau_R}} + (\kappa\tau_R)^{-1} \int_{\alpha_1}^{\alpha} \tilde{p}_i(\alpha') e^{-\frac{\alpha-\alpha'}{\kappa\tau_R}} d\alpha', \tag{25.56}$$

and

$$p_{2\rightarrow 1,i}(\alpha) = p_i^{(2)} e^{-\frac{\alpha_2-\alpha}{\lambda\kappa\tau_R}} - (\lambda\kappa\tau_R)^{-1} \int_{\alpha_2}^{\alpha} \tilde{p}_i(\alpha') e^{\frac{\alpha-\alpha'}{\lambda\kappa\tau_R}} d\alpha'. \tag{25.57}$$

In the following, we want to apply these considerations to the Otto and Carnot cycles at finite driving velocities.

### 25.3.2 Otto Cycle at Finite Velocity

We consider the Otto cycle of Sect. 25.2.2 again, but now driven with finite velocity on the isochors ($\kappa$ and $\lambda\kappa$, respectively). We assume no relaxation on the adiabats, because there is no coupling to a heat bath along these steps. In particular this means that the velocity of changing $\gamma$ during these processes has no influence on the work exchanged with the environment as long as our idealized model assumptions are fulfilled. This velocity only affects the cycle time and therefore only enters the power of the cycle, as will be discussed later. First we want to study the efficiency, by calculating heat and work along the different steps. Because we have a cyclic process with relaxation only on two steps, we can directly apply the result (25.54), (25.55), (25.56) and (25.57). Let us start the process at $\alpha = \alpha_1$ and $g(\gamma) = \min(g(\gamma))$. The distribution at this point is given by (25.54). The isochoric process to $\alpha_2$ is carried out with velocity $\kappa$. Using (25.48) then leads to

$$Q_{12}^* = \min(g(\gamma)) \sum \epsilon_i \left( p_i^{(2)} - p_i^{(1)} \right) < 0 . \tag{25.58}$$

The proof of $Q_{12}^*$ being negative can be found in Ref. [3]. Because there is no relaxation on the adiabats, for the whole next step $p_i = p_i^{(2)}$ holds. Thus, from (25.47) we get

$$W_{23}^* = [\max(g(\gamma)) - \min(g(\gamma))] \sum \epsilon_i p_i^{(2)} . \tag{25.59}$$

Accordingly the contributions by the other two processes read

$$Q_{34}^* = -\max(g(\gamma)) \sum \epsilon_i \left( p_i^{(2)} - p_i^{(1)} \right) > 0 , \tag{25.60}$$

$$W_{41}^* = -[\max(g(\gamma)) - \min(g(\gamma))] \sum \epsilon_i p_i^{(1)} . \tag{25.61}$$

Thus, the total work is given by

$$W_{\circ}^* = [\max(g(\gamma)) - \min(g(\gamma))] \sum \epsilon_i \left( p_i^{(2)} - p_i^{(1)} \right) < 0, \tag{25.62}$$

which leads to the efficiency

$$\eta^{0*} = \frac{-W_{\circ}^*}{Q_{34}^*} = 1 - \frac{\min(g(\gamma))}{\max(g(\gamma))} . \tag{25.63}$$

Hence one gets the same result as in the quasi-static limit. The efficiency of the Otto cycle is independent of the driving velocities! This special property only applies to the Otto cycle. For other cycles like the Carnot cycle the efficiency decreases with increasing velocity, as we will see later.

No change in the efficiency, of course, does not mean that there is no change in heat and work at all: In fact, all contributions tend to vanish with increasing velocity, as depicted in Fig. 25.6. As a consequence, the constancy of the efficiency is valid only in the idealized case. Additional leakage, for example, would gain in importance for increasing velocity, which would cause a decreasing efficiency. The decrease of the total work per cycle can also be visualized by means of the $T^*S^*$- or $P^*\gamma$-diagrams using the renormalized non-equilibrium definitions (25.43), (25.44), and (25.45). Figures 25.7, 25.8, and 25.9, 25.10 show an example for such diagrams and different velocities. As one can see, the enclosed area and thus the total work decreases with increasing velocity.

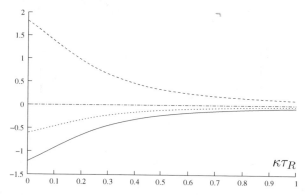

**Fig. 25.6** Behavior of $W_\circ^*$ (*solid line*), $Q_{34}^*$ (*dashed line*), and $Q_{12}^*$ (*dotted line*) with increasing velocity $\kappa$ for the Otto cycle. The chosen parameters are $\{\epsilon_i\}_{i=1}^N = \{-3/2; -1/2; 1/2; 3/2\}$, $\lambda = 0.5$, $\alpha_1 = 0.2$, $\alpha_2 = 0.8$, $\gamma_1 = 1$, $\gamma_2 = 3$. As one can see, the ratio $\frac{W_\circ^*}{Q_{34}^*}$, which defines the efficiency, always stays the same

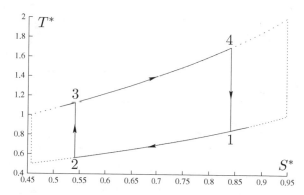

**Fig. 25.7** $T^* S^*$-diagram of an Otto cycle with $g(\gamma) = \gamma$ for $\kappa \tau_R = 0.2$. The other parameters are $\{\epsilon_i\}_{i=1}^N = \{-3/2; -1/2; 1/2; 3/2\}$, $\lambda = 1$, $\alpha_1 = 1$, $\alpha_2 = 2$, $\gamma_1 = 1$, $\gamma_2 = 2$. The points **1** bis **4** correspond to those in Fig. 25.1. The *dotted lines* show the quasi-static limit for comparison

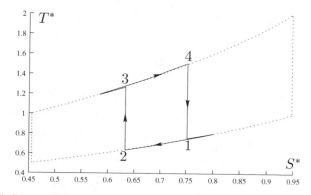

**Fig. 25.8** $T^* S^*$-diagram of the same Otto cycle as in Fig. 25.7, now with $\kappa \tau_R = 0.5$

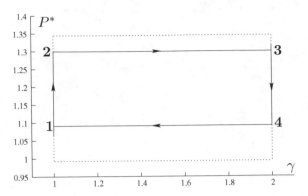

**Fig. 25.9** $P^*\gamma$-diagram of the Otto cycle in Fig. 25.7 for $\kappa\tau_R = 0.2$. The *dotted lines* show the quasi-static limit again

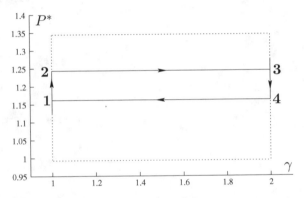

**Fig. 25.10** $P^*\gamma$-diagram of the Otto cycle with $\kappa\tau_R = 0.5$

The power $\mathcal{P} = \frac{-W^*_\circ}{t_\circ}$ of the heat engine is of more practical importance than the efficiency. The duration of the isochoric process steps is given by $\Delta t_{12} = \frac{\alpha_2 - \alpha_1}{\kappa}$ and $\Delta t_{34} = \frac{\alpha_2 - \alpha_1}{\lambda\kappa}$, respectively. We assume that the relative part of time per cycle spent on the isochors is given by $\delta$. Thus, the total duration for one cycle is

$$t_\circ^O = \frac{\alpha_2 - \alpha_1}{\delta\kappa}\frac{\lambda + 1}{\lambda} . \tag{25.64}$$

Together with the work per cycle (25.62) the power of the Otto cycle is then given by

$$\mathcal{P}^O = \frac{-W^*_\circ}{t_\circ^O} = \frac{\delta\lambda\kappa\left[\max\left(g(\gamma)\right) - \min\left(g(\gamma)\right)\right]\sum\epsilon_i\left(p_i^{(1)} - p_i^{(2)}\right)}{(\alpha_2 - \alpha_1)(1 + \lambda)} . \tag{25.65}$$

Figure 25.11 shows an example of the typical dependency of the power on the velocities: A well-defined maximum exists here for a certain choice of $\kappa$ and $\lambda$. In

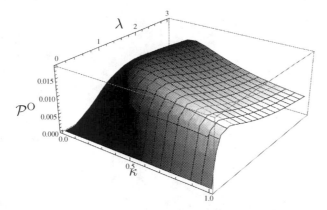

**Fig. 25.11** Power of the Otto cycle depending on $\kappa$ and $\lambda$. The chosen parameters are $\{\epsilon_i\}_{i=1}^{N} = \{-3/2; -1/2; 1/2; 3/2\}$, $\tau_R = 1$, $\delta = 0.8$, $\alpha_1 = 1$, $\alpha_2 = 2$, $\gamma_1 = 1$, $\gamma_2 = 2$

order to calculate this maximum analytically, we consider the limit of slow driving ($\kappa\tau_R/\Delta\alpha \ll 1$). In this limit the relaxation equation (25.50) can be solved iteratively, which leads to the distributions at the turning points:

$$p_i^{(1)} = \tilde{p}_i(\alpha_1) + \lambda\kappa\tau_R \left(\frac{d\tilde{p}_i}{d\alpha}\right)_{\alpha_1} + o(\kappa\tau_R) , \qquad (25.66)$$

$$p_i^{(2)} = \tilde{p}_i(\alpha_2) - \kappa\tau_R \left(\frac{d\tilde{p}_i}{d\alpha}\right)_{\alpha_2} + o(\kappa\tau_R) . \qquad (25.67)$$

The power then reads

$$\mathcal{P}^O = \frac{\delta\lambda\kappa \left[a - \kappa\tau_R (b_2 + \lambda b_1)\right]}{(\alpha_2 - \alpha_1)(1 + \lambda)} , \qquad (25.68)$$

where we have introduced the abbreviations

$$a := \left[\max\left(g(\gamma)\right) - \min\left(g(\gamma)\right)\right] (h(\alpha_1) - h(\alpha_2)) > 0 , \qquad (25.69)$$
$$b_1 := -\left[\max\left(g(\gamma)\right) - \min\left(g(\gamma)\right)\right] h'(\alpha_1) > 0 , \qquad (25.70)$$
$$b_2 := -\left[\max\left(g(\gamma)\right) - \min\left(g(\gamma)\right)\right] h'(\alpha_2) > 0 . \qquad (25.71)$$

The necessary conditions for maximizing the power $\frac{\partial\mathcal{P}^O}{\partial\kappa} = \frac{\partial\mathcal{P}^O}{\partial\lambda} = 0$ finally lead to the optimum values for the velocities

$$\lambda_{\mathrm{maxP}} = \sqrt{\frac{b_2}{b_1}} = \sqrt{\frac{h'(\alpha_2)}{h'(\alpha_1)}} , \qquad (25.72)$$

$$\kappa_{\mathrm{maxP}}\tau_R = \frac{a}{2(b_2 + \sqrt{b_1 b_2})} = \frac{h(\alpha_1) - h(\alpha_2)}{2\left(-h'(\alpha_2) + \sqrt{h'(\alpha_1)h'(\alpha_2)}\right)} , \qquad (25.73)$$

and to the maximum power

$$\mathcal{P}^{\text{O}}_{\text{max}} = \frac{\delta a^2}{4(\sqrt{b_1} + \sqrt{b_2})^2(\alpha_2 - \alpha_1)}$$

$$= \frac{\delta \left[ \max\left( g(\gamma) \right) - \min\left( g(\gamma) \right) \right] (h(\alpha_1) - h(\alpha_2))^2}{4 \left( \sqrt{-h'(\alpha_1)} + \sqrt{-h'(\alpha_2)} \right)^2 (\alpha_2 - \alpha_1)} . \tag{25.74}$$

Because the efficiency of the Otto cycle is independent of the velocities, the efficiency at maximum power is, of course, also given by (25.63).

### 25.3.3  Carnot Cycle at Finite Velocity

The calculations for the Carnot cycle are similar to those for the Otto cycle done in the last section. We allow for two different velocities: $\kappa$ on the low-temperature isotherm and $\lambda\kappa$ along the high-temperature isotherm and again assume no relaxation on the adiabats. This allows us to apply (25.54), (25.55), (25.56) and (25.57) for the distributions. Integration along these processes then leads to

$$W^*_{12} = W_{12} - \kappa \tau_R T_{\text{c}} \sum_i \epsilon_i \left( p_i^{(2)} - p_i^{(1)} \right) , \tag{25.75}$$

$$Q^*_{12} = T_{\text{c}} \sum_i \epsilon_i \left( \alpha_2 p_i^{(2)} - \alpha_1 p_i^{(1)} \right) - W^*_{12} , \tag{25.76}$$

$$W^*_{23} = \alpha_2 \left( T_{\text{h}} - T_{\text{c}} \right) \sum_i \epsilon_i p_i^{(2)} , \tag{25.77}$$

$$W^*_{34} = W_{34} - \lambda \kappa \tau_R T_{\text{h}} \sum_i \epsilon_i \left( p_i^{(2)} - p_i^{(1)} \right) , \tag{25.78}$$

$$Q^*_{34} = T_{\text{h}} \sum_i \epsilon_i \left( \alpha_1 p_i^{(1)} - \alpha_2 p_i^{(2)} \right) - W^*_{34} , \tag{25.79}$$

$$W^*_{41} = \alpha_1 \left( T_{\text{c}} - T_{\text{h}} \right) \sum_i \epsilon_i p_i^{(1)} , \tag{25.80}$$

where $W_{12}$ and $W_{34}$ denote the quantities in the quasi-static limit. We are interested the efficiency, the power and, in particular, the efficiency at maximum power. To enable an analytical discussion we consider the limit of slow velocities again ($\kappa \tau_R / \Delta\alpha \ll 1$). This results in the efficiency

$$\eta^{\text{C}*} = \frac{-W_{\circ} - \kappa \tau_R \left( (T_{\text{c}} + \lambda T_{\text{h}})\tilde{a} + \lambda \Delta T \tilde{b}_1 + \Delta T \tilde{b}_2 \right)}{Q_{34} - \kappa \tau_R T_{\text{h}} \left( \lambda(\tilde{a} + \tilde{b}_1) + \tilde{b}_2 \right)} , \tag{25.81}$$

with the abbreviations

$$\tilde{a} := h(\alpha_1) - h(\alpha_2) > 0 , \tag{25.82}$$

$$\tilde{b}_1 := -\alpha_1 h'(\alpha_1) > 0 , \tag{25.83}$$

$$\tilde{b}_2 := -\alpha_2 h'(\alpha_2) > 0 . \tag{25.84}$$

One easily convinces oneself that this efficiency is bounded from above by the quasi-static Carnot efficiency $\eta^C = \frac{-W_\circ}{Q_{34}} = 1 - \frac{T_c}{T_h}$ being reached only for $\kappa = 0$. Thus, unlike the Otto cycle, the efficiency of the Carnot cycle depends on the driving velocities.

As for the Otto cycle we assume the relative duration of the isotherms to be $\delta$, which again leads to the duration per cycle $t_\circ^C = \frac{\alpha_2 - \alpha_1}{\delta\kappa} \frac{\lambda+1}{\lambda}$ and therefore to the power

$$\mathcal{P}^C = \frac{\delta\kappa\lambda}{\Delta\alpha(\lambda+1)} \cdot \left[-W_\circ - \kappa\tau_R\big((T_c + \lambda T_h)\tilde{a} + \lambda\Delta T\tilde{b}_1 + \Delta T\tilde{b}_2\big)\right] . \quad (25.85)$$

From the necessary conditions for maximizing the power $\frac{\partial\mathcal{P}^C}{\partial\kappa} = \frac{\partial\mathcal{P}^C}{\partial\lambda} = 0$, we get the optimum velocities

$$\lambda_{\mathrm{maxP}} = \sqrt{\frac{T_c}{T_h}} \cdot \sqrt{\frac{1 + \Delta T\tilde{b}_2/T_c\tilde{a}}{1 + \Delta T\tilde{b}_1/T_h\tilde{a}}} \quad (25.86)$$

and

$$\kappa_{\mathrm{maxP}}\tau_R = \frac{-W_\circ}{2\sqrt{T_c\tilde{a} + \Delta T\tilde{b}_2}\left(\sqrt{T_c\tilde{a} + \Delta T\tilde{b}_2} + \sqrt{T_h\tilde{a} + \Delta T\tilde{b}_1}\right)} . \quad (25.87)$$

Inserting these velocities into (25.81) yields the efficiency at maximum power

$$\eta_{\mathrm{maxP}}^{C*} = \eta^C \left(1 + \frac{1}{\sqrt{(1 + \Delta T\tilde{b}_1/T_h\tilde{a})(1 + \Delta T\tilde{b}_2/T_c\tilde{a})}\sqrt{\frac{T_c}{T_h}}}\right)^{-1} . \quad (25.88)$$

This efficiency is bounded from above by the Carnot efficiency $\eta^C$. Interestingly, there also exists a lower bound, which is reached for $\Delta T\tilde{b}_1/T_h\tilde{a} \ll 1$ and $\Delta T\tilde{b}_2/T_c\tilde{a} \ll 1$ and given by the famous Curzon–Ahlborn efficiency

$$\eta_{\mathrm{maxP}}^{C*} \approx 1 - \sqrt{\frac{T_c}{T_h}} = \eta^{CA} . \quad (25.89)$$

According to Curzon and Ahlborn $\eta^{CA}$ should be the efficiency at maximum power for the classical Carnot cycle [9]. We want to check this, using the analogy of the particle in a box in the continuum limit with the classical ideal gas as discussed in Sect. 25.2.4. With (25.36) we get

$$\tilde{a} = \frac{D}{2}\left(\frac{1}{\alpha_1} - \frac{1}{\alpha_2}\right), \tag{25.90}$$

$$\tilde{b}_1 = \frac{D}{2\alpha_1}, \tag{25.91}$$

$$\tilde{b}_2 = \frac{D}{2\alpha_2}, \tag{25.92}$$

which leads us to the efficiency at maximum power for the continuum limit

$$\eta_{\mathrm{maxP}}^{\mathrm{C*}} = \eta^{\mathrm{C}}\left[1 + \sqrt{\frac{T_c}{T_h}}\left(\left(1 + \frac{T_h/T_c - 1}{\alpha_2/\alpha_1 - 1}\right)\left(1 + \frac{1 - T_c/T_h}{1 - \alpha_1/\alpha_2}\right)\right)^{-1/2}\right]^{-1}. \tag{25.93}$$

As one can see, also in this limit $\eta_{\mathrm{maxP}}^{\mathrm{C*}} > \eta^{\mathrm{CA}}$ always holds. $\eta^{\mathrm{CA}}$ is reached only for $T_c/T_h \to 1$. This obviously differs from the result of Curzon and Ahlborn. So, where does this difference come from?

In their derivation Curzon and Ahlborn assumed the temperatures to be constant along the isothermal steps also in the non-equilibrium case. They assumed a constant shift of these temperatures compared to the bath temperatures and then maximized the power with respect to these temperature differences, which finally led to $\eta^{\mathrm{CA}}$ for the efficiency at maximum power. But is it possible to realize a Carnot cycle at finite velocity for which the temperatures stay constant on the processes which were isothermal in the quasi-static limit? Within our model the answer is simply "no." We want to illustrate this for the example of a two-level system. Figure 25.12 shows the $T^*S^*$-diagram for such a system at finite velocities. As one can see, the temperature no longer stays constant on the originally isothermal steps: For the high-temperature isotherm the process starts at $T > T_h$ and then decreases to a value smaller than $T_h$. For the cold isotherm it is exactly the other way round. To understand this behavior

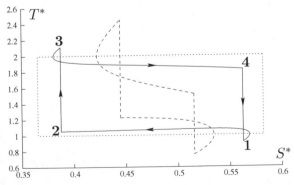

**Fig. 25.12** $T^*S^*$-diagram of a Carnot cycle with $g(\gamma) = \gamma$ for $\kappa\tau_R = 0.1$ (*solid line*) and $\kappa\tau_R = 0.4$ (*dashed line*) compared with the quasi-static limit (*dotted line*). The chosen parameters are $\{\epsilon_i\}_{i=1}^N = \{-1/2; 1/2\}, T_c = 1, T_h = 2, \alpha_1 = 1, \alpha_2 = 2, \lambda = 1$

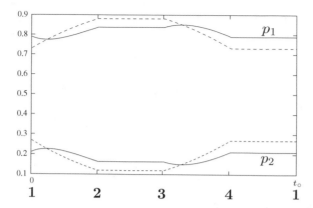

**Fig. 25.13** Evolution of the probabilities $p_1$ and $p_2$ of a two-level system during a cycle with two adiabats ($2 \to 3$ and $4 \to 1$). On the processes ($1 \to 2$ and $3 \to 4$) which are subject to relaxation, the probabilities are given by (25.56) and (25.57). The *dashed lines* show the attractor state. The chosen parameters are $\{\epsilon_i\}_{i=1}^{N} = \{-1/2; 1/2\}$, $\alpha_1 = 1$, $\alpha_2 = 2$, $\lambda = 1$, and $\kappa\tau_R = 0.4$

we need to consider the evolution of the occupation probabilities during the cycle: On the one hand, the process has to be cyclic, which means the probabilities have to return to their starting values at the end of the cycle. On the other hand, the probabilities have to show relaxation during the non-adiabatic steps. These requirements can only be fulfilled by an evolution as shown in Fig. 25.13: At point **1** the $p_1$ has to be larger than in the quasi-static limit and therefore initially decreases due to the relaxation. Since the attractor $\tilde{p}_1$ increases at the same time, there exists a point where $p_1$ and $\tilde{p}_1$ intersect. After this point the relaxation causes an increase of $p_1$. Because large $p_1$ means low temperature and vice versa, this describes exactly the behavior of the temperature during the isothermal step **1** $\to$ **2**. An analogous explanation can be made for the step **3** $\to$ **4**. Thus, a temperature pattern as shown in Fig. 25.12 is a necessary consequence of a cyclic process under relaxation.

Of course, the question arises, whether this behavior can also be observed in classical systems, as expected due to the analogy with the particle in a box. Indeed, this seems to be the case. Recently molecular dynamics simulations have been done on finite-time Carnot cycles [11]. The temperature observed within these numerical experiments qualitatively confirms the behavior discussed before. Furthermore the efficiency at maximum power was found there to be always larger than the Curzon–Ahlborn efficiency, which is, despite the completely different starting point, in good agreement with our result.

As one can see in Fig. 25.12, the behavior of temperature during the isothermal steps causes loops in the diagram, which increase with increasing velocities. These loops also occur in the $P^*\gamma$-diagrams. At high velocities, they even dominate the diagrams, as shown in Figs. 25.14 and 25.15. Because the loops are performed with opposite orientation, this indicates a change of sign of the total work per cycle. Indeed, when we look at the velocity dependency of work and heat given by (25.75),

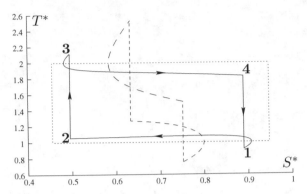

**Fig. 25.14** $T^*S^*$-diagram of a Carnot cycle with $g(\gamma) = \gamma$ at different velocities: $\kappa\tau_R = 0.1$ (*solid line*) and $\kappa\tau_R = 0.5$ (*dashed line*). The $TS$-diagram in the quasi-static limit is also shown for comparison (*dotted line*). The other parameters are $\{\epsilon_i\}_{i=1}^N = \{-3/2; -1/2; 1/2; 3/2\}$, $T_c = 1$, $T_h = 2$, $\alpha_1 = 1$, $\alpha_2 = 2$, $\lambda = 1$

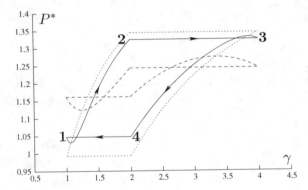

**Fig. 25.15** $P^*\gamma$-diagram of the Carnot cycle in Fig. 25.14 for $\kappa\tau_R = 0.1$ (*solid line*) and $\kappa\tau_R = 0.5$ (*dashed line*). The *dotted lines* correspond to the quasi-static case

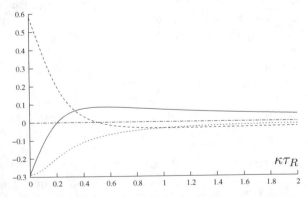

**Fig. 25.16** Behavior of $W_o^*$ (*solid lines*), $Q_{34}^*$ (*dashed lines*), and $Q_{12}^*$ (*dotted lines*) with increasing velocity. The chosen parameters are $\{\epsilon_i\}_{i=1}^N = \{-3/2; -1/2; 1/2; 3/2\}$ and $\alpha_1 = 0.2$, $\alpha_2 = 0.8$, $T_c = 1$, $T_h = 2$. As one can see, both $W_o^*$ and $Q_{34}^*$ change sign at certain finite velocities, while $Q_{12}^*$ always stays negative

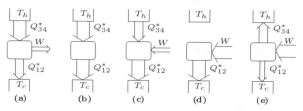

**Fig. 25.17** The development of the Carnot cycle with increasing cycle velocity: For small $\kappa \tau_R$, the cycle works as a heat engine (**a**). At a certain velocity, the total work vanishes, and there is only heat flowing from the hot to the cold bath (**b**). After that point, work has changed its sign (**c**). Then there exists a velocity, where $Q^*_{34}$ vanishes. The work therefore is completely transformed into heat, flowing into the cold bath (**d**). Finally, heat is flowing into the cold as well as into the hot bath (**e**). Note that the size of the arrows is rescaled at each picture: All components of heat and work tend to decrease for increasing velocity as shown in Fig. 25.16

(25.76), (25.77), (25.78), (25.79), and (25.80), we find a change of sign for the total work as well as for $Q^*_{34}$ at certain velocities (Fig. 25.16). Thus, the Carnot cycle looses its function of a heat engine already at finite velocity, even in the case of our idealized model. The development of the Carnot cycle with increasing velocity can be illustrated schematically as done in Fig. 25.17. Three different regimes can be distinguished, (a), (c), (e); they are separated by the special velocity scenarios (b) and (d), respectively.

# References

1. T. Jahnke, J. Birjukov, G. Mahler, Eur. Phys. J. ST **151**, 167 (2007)
2. T. Jahnke, J. Birjukov, G. Mahler, Ann. Phys. **17**, 88 (2008)
3. J. Birjukov, T. Jahnke, G. Mahler, Eur. Phys. J. B **64**, 105 (2008)
4. J.M. Epstein, SFI Working Papers (2008). DOI: SFI-WP 08-09-040
5. T. Feldmann, R. Kosloff, Phys. Rev. E **68**(1), 016101 (2003)
6. H.E.D. Scovil, E.O. Schulz-Du Bois, Phys. Rev. Lett. **2**, 262 (1959)
7. F. Rempp, M. Michel, G. Mahler, Phys. Rev. A **76**, 032325 (2007)
8. P. Salamon, B. Andresen, R.S. Berry, Phys. Rev. A **15**, 2094 (1977)
9. F. Curzon, B. Ahlborn, Am. J. Phys. **43**, 22 (1975)
10. W.K.A. Salem, J. Fröhlich, J. Stat. Phys. **126**, 431 (2007)
11. Y. Izumida, K. Okuda, Eur. Phys. Lett. **83**, 60003 (2008)

# Appendix A
# Hyperspheres

**Abstract** The mathematical tool to evaluate the Hilbert space averages and variances that play a crucial role throughout this book is the averaging of polynomial functions over hyperspheres. In this appendix we explain the corresponding mathematics in some detail and thus provide this tool.

## A.1 Averaging of Functions over Hyperspheres

As explained in Sects. 7.3 and 8.1 the accessible regions (ARs) we are considering essentially confine the real and imaginary parts of the amplitudes $\psi_i$ (which define the state with respect to a given basis) to hyperspheres of given radii. Thus the Hilbert space averages we have to do are eventually averages over such hyperspheres. As stated below (see Appendix B.1) the respective functions are not too complicated. They simply consist of products of powers of the coordinates (real and imaginary parts). The largest products we will have to consider contain four factors. Hence in this section we abstractly compute the averages of such functions over hyperspheres of given dimension $d$ and given radius $R$.

We start by considering the integral of the above function over a hypersphere which we denote by $Z(R, d, l, k, u, v)$ as follows:

$$Z(R, d, l, k, u, v) \equiv \int_{-\infty}^{\infty} \prod_{i=1}^{d} dx_i \quad \delta\left(\sqrt{\textstyle\sum_j^d x_j^2} - R\right) x_a^l x_b^k x_g^u x_h^v. \qquad (A.1)$$

The arguments of $Z$ denote radius, dimension, and the respective powers of the four coordinate factors. Note that $Z(R, d, 0, 0, 0, 0)$ simply is the "surface area" of the hypersphere. Thus the evaluation of (A.1) will eventually suffice to compute not only the interval but the wanted average over the hypersphere, cf. (A.10). Since it is not obvious how to evaluate $Z$ in a straightforward manner, we instead analyze a function $G$ which is related to $Z$ in the following way:

Gemmer, J. et al.: *Appendices*. Lect. Notes Phys. **784**, 315–346 (2009)
DOI 10.1007/978-3-540-70510-9_BM2      © Springer-Verlag Berlin Heidelberg 2009

$$\int_0^\infty dR \; e^{-R^2} Z(R, d, l, k, u, v) \tag{A.2}$$

$$= \int_0^\infty dR \int_{-\infty}^\infty \prod_{i=1}^d dx_i \delta\left(\sqrt{\sum_j^d x_j^2} - R\right) e^{-\sum_m^d x_m^2} x_a^l x_b^k x_g^u x_h^v$$

$$= \int_{-\infty}^\infty \prod_{i=1}^d dx_i \; e^{-\sum_m^d x_m^2} x_a^l x_b^k x_g^u x_h^v$$

$$\equiv G(d, l, k, u, v).$$

Here we first replaced $R^2$ in the exponent of the integrand by $\sum_j^d x_j^2$ which is in order since the $\delta$-function requires them to be equal anyway. Then we performed the integration over $R$ which simply makes the $\delta$-function vanish. The remaining integral factorizes; it may be evaluated using any standard textbook on basic integrals:

$$G(d, l, k, u, v) = \begin{cases} \pi^{\frac{d-4}{2}} \Gamma\left(\frac{l+1}{2}\right) \Gamma\left(\frac{k+1}{2}\right) \Gamma\left(\frac{u+1}{2}\right) \Gamma\left(\frac{v+1}{2}\right) & \text{for } * \\ 0 & \text{for } ** \end{cases}$$

$* \; (l, k, u, v)$ all even,        $** \; (l, k, u, v)$ any of them odd.

$$\tag{A.3}$$

To further pursue the evaluation of $Z$ we now perform a coordinate transformation $x_i \rightarrow y_i$ according to

$$x_i = R \, y_i. \tag{A.4}$$

Applying this transformation we may "extract" the $R$-dependence from $Z$:

$$Z(R, d, l, k, u, v) \tag{A.5}$$

$$= R^{d+l+k+u+v} \int_{-\infty}^\infty \prod_{i=1}^d dy_i \; \delta\left(R\left(\sqrt{\sum_j^d y_j^2} - 1\right)\right) y_a^l y_b^k y_g^u y_h^v$$

$$= R^{d+l+k+u+v-1} \underbrace{\int_{-\infty}^\infty \prod_{i=1}^d dy_i \; \delta\left(\sqrt{\sum_j^d y_j^2} - 1\right) y_a^l y_b^k y_g^u y_h^v}_{z(d, l, k, u, v)}.$$

Here we simply made use of the fact that prefactors in the arguments of $\delta$-functions may be (inversely) pulled out of the arguments. The function defined by the underbrace, $z$, is essentially the wanted function $Z$, only with the $R$-dependence factored out. Using this formulation we may write the above function $G$ from (A.2) in an yet other form:

$$\underbrace{\int_0^\infty dR\, e^{-R^2}\, R^{d+l+k+u+v-1}}_{F(R,d,l,k,u,v)}\, z(d,l,k,u,v) = G(d,l,k,u,v). \qquad (A.6)$$

Again, the function defined by the underbrace, $F$, may be evaluated using any standard textbook on basic integrals:

$$F(R,d,l,k,u,v) = \frac{\Gamma\left(\frac{d+l+k+u+v}{2}\right)}{2}. \qquad (A.7)$$

Now, eventually plugging (A.5) and (A.6) together yields

$$Z(R,d,l,k,u,v) = \frac{G(d,l,k,u,v)}{F(R,d,l,k,u,v)}\, R^{d+l+k+u+v-1}. \qquad (A.8)$$

The different functions needed to write out (A.8) explicitly may be read off from (A.3) and (A.7), thus we find

$$Z(R,d,l,k,u,v) = \begin{cases} 2\pi^{\frac{d-4}{2}} \dfrac{\Gamma(\frac{l+1}{2})\Gamma(\frac{k+1}{2})\Gamma(\frac{u+1}{2})\Gamma(\frac{v+1}{2})}{\Gamma(\frac{d+l+k+u+v}{2})}\, R^{d+l+k+u+v-1} & \text{for } * \\ 0 & \text{for } ** \end{cases}$$

$*\ (l,k,u,v)$ all even,      $**\ (l,k,u,v)$ any of them odd.

$$(A.9)$$

As mentioned above we are eventually not primarily interested in integrals but in averages over hyperspheres. However, since those are linked in a rather obvious way we may define the average of a function of the above type over the above hypersphere by

$$E(R,d,l,k,u,v) \equiv \frac{Z(R,d,l,k,u,v)}{Z(R,d,0,0,0,0)}. \qquad (A.10)$$

Note that the averages of all functions containing odd powers of any coordinate vanish. As turns out below, cf. Appendices B.1, B.2, apart from this general statement, we only need the concrete values of three of those averages. They all may be computed from (A.10), (A.9), thus we simply list them here

$$E(R,d,2,0,0,0) = \frac{R^2}{d}, \qquad (A.11)$$

$$E(R,d,2,2,0,0) = \frac{R^4}{d^2+2d}, \qquad (A.12)$$

$$E(R,d,4,0,0,0) = \frac{3R^4}{d^2+2d}. \qquad (A.13)$$

# Appendix B
# General Hilbert Space Averages and Variances

**Abstract** In this appendix we concretely calculate the Hilbert space averages and variances of some quantities which are imperative for the general reasoning given in Chap. 8. To calculate these averages we exploit the averages of polynomial functions over hyperspheres in high-dimensional spaces as given in Appendix A.

## B.1 Hilbert Space Averages of Polynomial Terms

To account for the Hilbert space averages and variances of the physically relevant quantities (observables, etc.) discussed in Sects. 8.2 and 8.3 we eventually need to evaluate Hilbert space averages of products of different $\psi_{\alpha i}$ or their complex conjugates. Thus, schematically these products look like $\psi_{\alpha i} \psi_{\beta j}^* \psi_{\gamma k} \ldots$. With $\psi_{\alpha i} = \eta_{\alpha i} + i\xi_{\alpha i}$ the averages of those $\psi$-products may be expanded into linear combinations of averages over hyperspheres of polynomial functions of their coordinates $\eta_{\alpha i}, \xi_{\alpha i}$. Thus, identifying $\eta_{\alpha i}, \xi_{\alpha i}$ with $x_i, x_j, \ldots$ we may use the results from Sect. A.1. (The largest product we need to consider contains four $\psi$-factors; thus the general results on the integration of polynomial functions over hyperspheres discussed in Sect. A.1 suffice.) Two features facilitate the evaluation of those averages significantly: (i) according to (7.24) the average of such a $\psi$-product decomposes into a product of averages of $\psi$-products, the factors of which correspond to the same subspace $\alpha, \beta, \ldots$. (Those products may possibly consist of a single $\psi$-factor.) (ii) The averages of $\psi$-products corresponding to some individual subspace will be zero at any rate unless the $\psi$-factors are pairwise equal. Otherwise the corresponding averages over hyperspheres will be averages of polynomial functions in which at least one coordinate appears in an odd power. According to (A.10) those averages are all zero. As a consequence of these two features the averages over all original $\psi$-products are zero at any rate unless the $\psi$-factors are pairwise equal. Equal here means equal with respect to the subspace $\alpha, \beta, \ldots$ as well as with respect to the individual level within the subspace, i.e., $i, j, \ldots$.

In the following we evaluate the specific $\psi$-product averages that we eventually need. Based on these general considerations the average of a single $\psi$ may directly be given as

$$[\![\psi_{\alpha i}]\!] = 0. \tag{B.1}$$

Furthermore we may immediately write

$$[\![\psi_{\alpha i}^* \psi_{\beta i}]\!] = \delta_{\alpha\beta}\, \delta_{ij} [\![|\psi_{\alpha i}|^2]\!]. \tag{B.2}$$

In order to evaluate $[\![|\psi_{\alpha i}|^2]\!]$ we now explicitly exploit (A.11):

$$[\![|\psi_{\alpha i}|^2]\!] = [\![\eta_{\alpha i}^2 + \xi_{\alpha i}^2]\!] = 2E(W_\alpha, 2N_\alpha, 2, 0, 0, 0) = \frac{W_\alpha}{N_\alpha}. \tag{B.3}$$

In the same fashion we now evaluate the following collection of Hilbert space averages which suffices to compute the Hilbert space averages and variances of expectation values of observables which are needed in Chaps. 8 and 10. The respective Hilbert space averages and variances are then concretely evaluated on the basis of the below list in Sect. B.2:

$$[\![\psi_{\alpha i}^2]\!] = [\![\eta_{\alpha i}^2 - \xi_{\alpha i}^2 + 2i\eta_{\alpha i}\xi_{\alpha i}]\!] = 2iE(W_\alpha, 2N_\alpha, 1, 1, 0, 0) = 0, \tag{B.4}$$

$$\begin{aligned}
[\![|\psi_{\alpha i}|^2|\psi_{\alpha j}|^2]\!] &= [\![(\eta_{\alpha i}^2 + \xi_{\alpha i}^2)(\eta_{\alpha j}^2 + \xi_{\alpha j}^2)]\!] \\
&= 4E(W_\alpha, 2N_\alpha, 2, 2, 0, 0) \\
&= \frac{1}{N_\alpha(N_\alpha + 1)},
\end{aligned} \tag{B.5}$$

$$\begin{aligned}
[\![|\psi_{\alpha i}|^4]\!] &= [\![\eta_{\alpha i}^4 + \xi_{\alpha i}^4 + 2\eta_{\alpha i}^2\xi_{\alpha i}^2]\!] \\
&= 2(E(W_\alpha, 2N_\alpha, 4, 0, 0, 0) + E(W_\alpha, 2N_\alpha, 2, 2, 0, 0)) \\
&= \frac{2}{N_\alpha(N_\alpha + 1)},
\end{aligned} \tag{B.6}$$

$$\begin{aligned}
[\![(\psi_{\alpha i}^*)^2 \psi_{\alpha j}^2]\!] &= [\![\eta_{\alpha i}^2\eta_{\alpha j}^2 - \eta_{\alpha i}^2\xi_{\alpha j}^2 - \xi_{\alpha i}^2\eta_{\alpha j}^2 + \xi_{\alpha i}^2\xi_{\alpha j}^2 + \\
&\quad + i(2\eta_{\alpha i}\xi_{\alpha i}(\xi_{\alpha j}^2 - \eta_{\alpha j}^2) + 2\eta_{\alpha j}\xi_{\alpha j}(\eta_{\alpha i}^2 - \xi_{\alpha i}^2))]\!] \\
&= 0.
\end{aligned} \tag{B.7}$$

## B.2 Hilbert Space Variance of an Observable

In the following we derive the result quoted in (8.18). Expanded according to sub-spaces the corresponding Hilbert space variance reads

$$\llbracket \langle \hat{A} \rangle^2 \rrbracket = \sum_{\alpha\beta\alpha'\beta'} \llbracket \langle \psi_\alpha | \hat{A} | \psi_\beta \rangle \langle \psi_{\alpha'} | \hat{A} | \psi_{\beta'} \rangle \rrbracket . \tag{B.8}$$

As explained in Sect. B.1 such a Hilbert space average can only be non-zero if the full indices of the corresponding $\psi$s are either pairwise equal or all equal. This can only occur if, regardless of the "intra-subspace index," the corresponding subspaces are either pairwise equal or all equal. Thus, the sum decomposes into four different terms:

$$\llbracket \langle \hat{A} \rangle^2 \rrbracket = P_1 + P_2 + P_3 + P_4, \tag{B.9}$$

with

$$
\begin{aligned}
P_1 &: \alpha = \beta = \alpha' = \beta', \\
P_2 &: \quad \alpha = \alpha', \qquad \beta = \beta', \\
P_3 &: \quad \alpha = \beta, \qquad \alpha' = \beta', \\
P_4 &: \quad \alpha - \beta', \qquad \beta = \alpha'.
\end{aligned}
\tag{B.10}
$$

$$\tag{B.11}$$

In the following we discuss those terms separately. The first term reads

$$P_1 = \sum_\alpha \llbracket (\langle \psi_\alpha | \hat{A} | \psi_\alpha \rangle)^2 \rrbracket = \sum_{\alpha ijkl} A_{ij} A_{kl} \llbracket \psi_{\alpha i}^* \psi_{\alpha j} \psi_{\alpha k}^* \psi_{\alpha l} \rrbracket . \tag{B.12}$$

Again as explained in Sect. B.1 not only the subspaces but also the "intra-subspace indices" have to be either pairwise equal or all equal for the Hilbert space average to be non-zero at all. Thus, similar to the scheme described in (B.10) one may single out the addends that contribute to $P_1$. These are found to be

$$
\begin{aligned}
P_1 =& \sum_{\substack{\alpha ij \\ i \neq j}} \llbracket |\psi_{\alpha i}|^2 |\psi_{\alpha j}|^2 \rrbracket (A_{\alpha i,\alpha i} A_{\alpha j,\alpha j} + A_{\alpha i,\alpha j} A_{\alpha j,\alpha i}) \\
&+ \sum_{\substack{\alpha ij \\ i \neq j}} \llbracket (\psi_{\alpha i}^*)^2 (\psi_{\alpha j})^2 \rrbracket A_{\alpha i,\alpha j}^2 + \sum_{\alpha i} A_{\alpha i,\alpha i}^2 \llbracket |\psi_{\alpha i}|^4 \rrbracket ,
\end{aligned}
\tag{B.13}
$$

where we again used the notation introduced in (8.11). Exploiting (B.5), (B.6), and (B.7) we find

$$P_1 = \frac{W_\alpha^2}{N_\alpha(N_\alpha + 1)} \left( \sum_{\substack{\alpha ij \\ i \neq j}} (A_{\alpha i, \alpha i} A_{\alpha j, \alpha j} + A_{\alpha i, \alpha j} A_{\alpha j, \alpha i}) + 2 \sum_{\alpha i} A_{\alpha i, \alpha i}^2 \right). \quad \text{(B.14)}$$

The terms that are left out in the first sum due to $i \neq j$ are exactly the terms in the second sum. Thus one may simply simultaneously drop $i \neq j$ and the second sum. Without $i \neq j$ the sum may be performed in a straightforward manner. The result may be written as

$$P_1 = \sum_\alpha \frac{W_\alpha^2}{N_\alpha(N_\alpha + 1)} \left( \text{Tr}\left\{ \hat{A}_{\alpha\alpha}^2 \right\} + (\text{Tr}\left\{ \hat{A}_{\alpha\alpha} \right\})^2 \right). \quad \text{(B.15)}$$

The second term yields

$$P_2 = \sum_{\substack{\alpha\beta \\ \alpha \neq \beta}} \llbracket ((\psi_\alpha | \hat{A} | \psi_\beta))^2 \rrbracket$$

$$= \sum_{\substack{\alpha\beta \\ \alpha \neq \beta}} \sum_{ijkl} \llbracket A_{\alpha i, \beta j} \psi_{\alpha i}^* \psi_{\alpha k}^* A_{\alpha k, \beta l} \psi_{\beta j} \psi_{\beta l} \rrbracket$$

$$= \sum_{\substack{\alpha\beta \\ \alpha \neq \beta}} \sum_{ijkl} A_{\alpha i, \beta j} A_{\alpha k, \beta l} \llbracket \psi_{\alpha i}^* \psi_{\alpha k}^* \psi_{\beta j} \psi_{\beta l} \rrbracket. \quad \text{(B.16)}$$

Since $\alpha \neq \beta$ this term can only be non-zero for $i = k$ and $j = l$ finding

$$P_2 = \sum_{\substack{\alpha\beta \\ \alpha \neq \beta}} \sum_{ij} A_{\alpha i, \beta j}^2 \llbracket (\psi_{\alpha i}^*)^2 (\psi_{\beta j})^2 \rrbracket. \quad \text{(B.17)}$$

According to (7.24) this factorizes as

$$P_2 = \sum_{\substack{\alpha\beta \\ \alpha \neq \beta}} \sum_{ij} A_{\alpha i, \beta j}^2 \llbracket (\psi_{\alpha i}^*)^2 \rrbracket \llbracket (\psi_{\beta j})^2 \rrbracket = 0, \quad \text{(B.18)}$$

which is zero according to (B.4). The third term may also be factorized and evaluated on the basis of (8.9) as follows:

$$P_3 = \sum_{\substack{\alpha\beta \\ \alpha \neq \beta}} [\![ \langle \psi_\alpha | \hat{A} | \psi_\alpha \rangle \langle \psi_\beta | \hat{A} | \psi_\beta \rangle ]\!]$$

$$= \sum_{\substack{\alpha\beta \\ \alpha \neq \beta}} [\![ \langle \psi_\alpha | \hat{A} | \psi_\alpha \rangle ]\!] [\![ \langle \psi_\beta | \hat{A} | \psi_\beta \rangle ]\!]$$

$$= \sum_{\substack{\alpha\beta \\ \alpha \neq \beta}} \frac{W_\alpha W_\beta}{N_\alpha N_\beta} \text{Tr} \left\{ \hat{A}_{\alpha\alpha} \right\} \text{Tr} \left\{ \hat{A}_{\beta\beta} \right\} . \tag{B.19}$$

For the last term we find

$$P_4 = \sum_{\substack{\alpha\beta \\ \alpha \neq \beta}} [\![ \langle \psi_\alpha | \hat{A} | \psi_\beta \rangle \langle \psi_\beta | \hat{A} | \psi_\alpha \rangle ]\!]$$

$$= \sum_{\substack{\alpha\beta \\ \alpha \neq \beta}} \sum_{ijkl} [\![ A_{\alpha i,\beta j} A_{\beta k,\alpha l} \psi_{\alpha i}^* \psi_{\alpha l} \psi_{\beta j}^* \psi_{\beta k} ]\!]. \tag{B.20}$$

Again, since $\alpha \neq \beta$ this term can only be non-zero for $i = l$ and $j = k$. The remaining addends read and factorize as follows:

$$P_4 = \sum_{\substack{\alpha\beta \\ \alpha \neq \beta}} \sum_{ij} A_{\alpha l,\beta j} A_{\beta j,\alpha i} [\![ |\psi_{\alpha i}|^2 |\psi_{\beta j}|^2 ]\!]$$

$$= \sum_{\substack{\alpha\beta \\ \alpha \neq \beta}} \sum_{ij} A_{\alpha i,\beta j} A_{\beta j,\alpha i} [\![ |\psi_{\alpha i}|^2 ]\!] [\![ |\psi_{\beta j}|^2 ]\!]$$

$$= \sum_{\substack{\alpha\beta \\ \alpha \neq \beta}} \frac{W_\alpha W_\beta}{N_\alpha N_\beta} \text{Tr} \left\{ \hat{A}_{\alpha\beta} \hat{A}_{\alpha\beta}^\dagger \right\} , \tag{B.21}$$

where we have used (B.3). Plugging all those contributions together we find for the first term of the variance

$$[\![ \langle \hat{A} \rangle^2 ]\!] = \sum_{\alpha\beta} \frac{W_\alpha W_\beta}{N_\alpha (N_\beta + \delta_{\alpha\beta})} \left( \text{Tr} \left\{ \hat{A}_{\alpha\beta} \hat{A}_{\alpha\beta}^\dagger \right\} + \text{Tr} \left\{ \hat{A}_{\alpha\alpha} \right\} \text{Tr} \left\{ \hat{A}_{\beta\beta} \right\} \right). \tag{B.22}$$

This result is quoted in (8.18).

# Appendix C
# Special Hilbert Space Averages and Variances

**Abstract** In this part of the appendix various Hilbert space averages and variances are evaluated which are needed in the context of typicality of reduced local states in composite quantum systems, especially in Chaps. 10 and 11.

## C.1 General Distances Between States

First we derive the relation between the mean-squared distance of states from some average state and the Hilbert space variances of certain observables as expressed in (10.13). Starting from (10.12) we have to compute first the trace of the average of the squared density operator

$$
\mathrm{Tr}\left\{[\![\hat{\rho}^2]\!]\right\} = \mathrm{Tr}\Bigg\{ \Bigg[\!\Bigg[ \sum_{AC}\sum_{ac}(X^{Aa,Cc} - \mathrm{i}Y^{Aa,Cc})|A,a\rangle\langle C,c|
$$
$$
\sum_{A'C'}\sum_{a'c'}(X^{A'a',C'c'} - \mathrm{i}Y^{A'a',C'c'})|A',a'\rangle\langle C',c'|\Bigg]\!\Bigg] \Bigg\}, \tag{C.1}
$$

which is after exchanging trace and average and performing the trace operation

$$
\mathrm{Tr}\left\{[\![\hat{\rho}^2]\!]\right\} = \Bigg[\!\Bigg[ \sum_{AC}\sum_{ac}(X^{Aa,Cc} - \mathrm{i}Y^{Aa,Cc})(X^{Cc,Aa} - \mathrm{i}Y^{Cc,Aa})\Bigg]\!\Bigg]. \tag{C.2}
$$

The term in the second bracket is just the complex conjugate of the first one, and thus we find

$$
\mathrm{Tr}\left\{[\![\hat{\rho}^2]\!]\right\} = \sum_{AC}\sum_{ac}\left([\![(X^{Aa,Cc})^2]\!] + [\![(Y^{Aa,Cc})^2]\!]\right). \tag{C.3}
$$

The trace over the squared Hilbert space average (HA) of the density operator, using (10.10) and (10.11) yields

$$\mathrm{Tr}\left\{\hat{\omega}^2\right\} = \mathrm{Tr}\left\{\sum_{AC}\sum_{ac}([X^{Aa,Cc}] - i[Y^{Aa,Cc}])|A,a\rangle\langle C,c|\right.$$

$$\left.\sum_{A'C'}\sum_{a'c'}([X^{A'a',C'c'}] - i[Y^{A'a',C'c'}])|A',a'\rangle\langle C',c'|\right\}, \qquad (C.4)$$

which reduces to

$$\mathrm{Tr}\left\{\hat{\omega}^2\right\} = \sum_{AC}\sum_{ac}([X^{Aa,Cc}] - i[Y^{Aa,Cc}])([X^{Cc,Aa}] - i[Y^{Cc,Aa}])$$

$$= \sum_{AC}\sum_{ac}([X^{Aa,Cc}]^2 + [Y^{Aa,Cc}]^2), \qquad (C.5)$$

where we used again that the second bracket is the complex conjugate of the first one. Plugging this into (10.12) and exploiting furthermore (10.10) and (10.11) yields

$$D^2(\hat{\rho},\hat{\omega}) = \sum_{AC}\sum_{ac}\left(\Delta_{\mathrm{H}}^2(X^{Aa,Cc}) + \Delta_{\mathrm{H}}^2(Y^{Aa,Cc})\right). \qquad (C.6)$$

This result is the one that is referred to below (10.13).


## C.2 Microcanonical Hilbert Space Averages

Here we compute the HA which is required for the result quoted in (10.17): The average of expectation value of the operator $\hat{Y}^{Aa,Cc}$ can be computed in the very same way as the average of expectation value of the operator $\hat{X}^{Aa,Cc}$. The latter computation is given in Sect. 10.2.1. Starting with the operator

$$\hat{Y}^{Aa,Cc}_{A'B,A'B}$$

$$= \frac{i}{2}\hat{\Pi}_{A'}\hat{\Pi}_B\left((|A,a\rangle\langle C,c| - |C,c\rangle\langle A,a|)\otimes\hat{1}\right)\hat{\Pi}_{A'}\hat{\Pi}_B$$

$$= \frac{i}{2}\delta_{A'A}\,\delta_{A'C}(|A,a\rangle\langle C,c| - |C,c\rangle\langle A,a|)\otimes\hat{\Pi}_B, \qquad (C.7)$$

we find the trace

$$\mathrm{Tr}\left\{\hat{Y}^{Aa,Cc}_{A'B,A'B}\right\} = 0 \qquad (C.8)$$

and thus its Hilbert space average of $\hat{Y}^{Aa,Cc}$ vanishes, $[\langle\hat{Y}^{Aa,Cc}\rangle] = 0$. Thus exploiting (10.10) and (10.16) we find for the Hilbert space average of a density matrix element

$$\llbracket \rho^{Aa,Cc} \rrbracket = \sum_B \frac{W_{AB}}{N_A} \delta_{AC} \delta_{ac} . \tag{C.9}$$

This result is quoted in (10.17).

## C.3 Microcanonical Hilbert Space Variances

Here we are going to compute the Hilbert space variances that are quoted in (10.20) and (10.21). The pertinent accessible region (AR) is given by (9.15). The general formula for the Hilbert space variance of the expectation value of a Hermitian operator is given in (8.19). Using this, we first need the operator

$$\hat{X}^{Aa,Cc}_{A'B,C'D}$$

$$= 1/2 \hat{\Pi}_{A'} \hat{\Pi}_B \Big( (|A,a\rangle\langle C,c| + |C,c\rangle\langle A,a|) \otimes \hat{1} \Big) \hat{\Pi}_{C'} \hat{\Pi}_D$$

$$= 1/2 \delta_{BD} \big( \delta_{A'A} \delta_{C'C} |A,a\rangle\langle C,c|$$

$$+ \delta_{A'C} \delta_{C'A} |C,c\rangle\langle A,a| \big) \otimes \hat{\Pi}_B . \tag{C.10}$$

The adjoint operator is given by

$$(\hat{X}^{Aa,Cc}_{A'B,C'D})^\dagger = 1/2 \delta_{BD} \big( \delta_{A'A} \delta_{C'C} |C,c\rangle\langle A,a|$$

$$+ \delta_{A'C} \delta_{C'A} |A,a\rangle\langle C,c| \big) \otimes \hat{\Pi}_B . \tag{C.11}$$

The product of (C.10) and (C.11) is

$$\hat{X}^{Aa,Cc}_{A'B,C'D}(\hat{X}^{Aa,Cc}_{A'B,C'D})^\dagger = 1/4 \delta_{BD} \big( \delta_{A'A} \delta_{C'C} |A,a\rangle\langle A,a|$$

$$+ \delta_{A'C} \delta_{C'A} |C,c\rangle\langle C,c| \big) \otimes \hat{\Pi}_B , \tag{C.12}$$

and its trace

$$\mathrm{Tr} \left\{ \hat{X}^{Aa,Cc}_{A'B,C'D}(\hat{X}^{Aa,Cc}_{A'B,C'D})^\dagger \right\}$$

$$= 1/4 \delta_{BD} (\delta_{A'A} \delta_{C'C} + \delta_{A'C} \delta_{C'A}) N_B . \tag{C.13}$$

Plugging this trace into (8.19), realizing that $N_{AB} = N_A N_B$ and using (10.15) for the second term, we find

$$\Delta^2_{\mathrm{H}}(\langle \hat{X}^{Aa,Cc}\rangle)$$

$$= \frac{1}{4}\sum_{A'BC'D}\frac{W_{A'B}W_{C'D}}{N_{A'B}(N_{C'D}+\delta_{A'C'}\,\delta_{BD})}\,\delta_{BD}\big(\delta_{A'A}\,\delta_{C'C}+\delta_{A'C}\,\delta_{C'A}\big)N_B$$

$$-\,\delta_{AC}\delta_{ac}\sum_B \frac{W^2_{AB}N^2_B}{N^2_{AB}(N_{AB}+1)}$$

$$= \frac{1}{4}\sum_B \frac{W_{AB}W_{CB}}{N_A N_B N_C + N_A\delta_{AC}} + \frac{1}{4}\sum_B \frac{W_{AB}W_{CB}}{N_A N_B N_C + N_C\delta_{AC}}$$

$$-\,\delta_{AC}\delta_{ac}\sum_B \frac{W^2_{AB}}{N^2_A(N_A N_B +1)}\,. \tag{C.14}$$

Because of the delta function $\delta_{AC}$ we may change the dimension $N_C$ in the denominator of the second term in front of the delta function to a dimension $N_A$. Thus, the first two terms are equivalent and the variance simplifies to

$$\Delta^2_{\mathrm{H}}(\langle \hat{X}^{Aa,Cc}\rangle) = \frac{1}{2}\sum_B \frac{W_{AB}W_{CB}}{N_A(N_C N_B + \delta_{AC})} - \delta_{AC}\delta_{ac}\sum_B \frac{W^2_{AB}}{N^2_A(N_A N_B + 1)}\,. \tag{C.15}$$

This result is quoted in (10.20). Now we do the same analysis for the corresponding $\hat{Y}$-operator

$$\hat{Y}^{Aa,Cc}_{A'B,C'D}$$

$$= \frac{\mathrm{i}}{2}\hat{\Pi}_{A'}\hat{\Pi}_B\Big((|A,a\rangle\langle C,c| - |C,c\rangle\langle A,a|)\otimes \hat{1}\Big)\hat{\Pi}_{C'}\hat{\Pi}_D$$

$$= \frac{\mathrm{i}}{2}\delta_{BD}\big(\delta_{A'A}\,\delta_{C'C}|A,a\rangle\langle C,c|$$

$$- \delta_{A'C}\,\delta_{C'A}|C,c\rangle\langle A,a|\big)\otimes \hat{\Pi}_B\,, \tag{C.16}$$

and its adjoint reads

$$(\hat{Y}^{Aa,Cc}_{A'B,C'D})^\dagger = -\frac{\mathrm{i}}{2}\delta_{BD}\big(\delta_{A'A}\,\delta_{C'C}|C,c\rangle\langle A,a|$$

$$- \delta_{A'C}\,\delta_{C'A}|A,a\rangle\langle C,c|\big)\otimes \hat{\Pi}_B\,. \tag{C.17}$$

The product of (11.13) and (11.14) yields

$$\hat{Y}^{Aa,Cc}_{A'B,C'D}(\hat{Y}^{Aa,Cc}_{A'B,C'D})^\dagger = 1/4\,\delta_{BD}\big(\delta_{A'A}\,\delta_{C'C}|A,a\rangle\langle A,a|$$

$$+ \delta_{A'C}\,\delta_{C'A}|C,c\rangle\langle C,c|\big)\otimes \hat{\Pi}_B\,, \tag{C.18}$$

and its trace is

$$\mathrm{Tr}\left\{\hat{Y}^{Aa,Cc}_{A'B,C'D}(\hat{Y}^{Aa,Cc}_{A'B,C'D})^{\dagger}\right\}$$
$$= 1/4\,\delta_{BD}\left(\delta_{A'A}\,\delta_{C'C} + \delta_{A'C}\,\delta_{C'A}\right)N_B\,. \qquad (C.19)$$

Since this result is equivalent to (C.13) the first term of the variance is identical to the first two terms of (C.14). However, because the trace of $\hat{Y}^{Aa,Cc}$ vanishes according to (C.8) the second term does not exist here. The Hilbert space variance thus reads

$$\Delta^2_{\mathrm{H}}(\langle\hat{Y}^{Aa,Cc}\rangle) = \frac{1}{2}\sum_B \frac{W_{AB}W_{CB}}{N_A(N_C N_B + \delta_{AC})}\,. \qquad (C.20)$$

This result is quoted in (10.21).


## C.4 Energy Exchange Hilbert Space Averages

Here we derive the result quoted in (10.27). The pertinent AR is given by (9.19). Again, we use the general result (8.12) to compute the Hilbert space average. Therefore we need the operator

$$\hat{X}^{Aa,Cc}_{EE}$$
$$= \frac{1}{2}\hat{\Pi}_E\left(|A,a\rangle\langle C,c| + |C,c\rangle\langle A,a|\right)\hat{\Pi}_E$$
$$= \frac{1}{2}\sum_{A'B}\hat{\Pi}_{A'}\hat{\Pi}_B M_{E,A'B}\left(|A,a\rangle\langle C,c| + |C,c\rangle\langle A,a|\right)\sum_{C'D}\hat{\Pi}_{C'}\hat{\Pi}_D M_{E,C'D}$$
$$= \frac{1}{2}\sum_{A'BC'} M_{E,A'B}\,M_{E,C'D}\hat{\Pi}_{A'}\left(|A,a\rangle\langle C,c| + |C,c\rangle\langle A,a|\right)\hat{\Pi}_{C'}\hat{\Pi}_B\,. \qquad (C.21)$$

Realizing that $M_{E,A'B}\,M_{E,C'D} = M_{E,A'B}\,\delta_{A'C'}$ we get

$$\hat{X}^{Aa,Cc}_{EE} = \frac{1}{2}\sum_{A'B} M_{E,A'B}\hat{\Pi}_{A'}\left(|A,a\rangle\langle C,c| + |C,c\rangle\langle A,a|\right)\hat{\Pi}_{A'}\hat{\Pi}_B$$
$$= \frac{1}{2}\sum_B M_{E,AB}\,\delta_{AC}\left(|A,a\rangle\langle C,c| + |C,c\rangle\langle A,a|\right)\hat{\Pi}_B\,. \qquad (C.22)$$

The trace over this operator yields

$$\mathrm{Tr}\left\{\hat{X}^{Aa,Cc}_{EE}\right\} = \sum_B M_{E,AB}\,\delta_{AC}\,\delta_{ac}\,N_B\,, \qquad (C.23)$$

and thus the average reads

$$[\![\langle \hat{X}^{Aa,Cc}\rangle]\!] = \sum_E \frac{W_E}{N_E} \mathrm{Tr}\left\{ \hat{X}^{Aa,Cc}_{EE} \right\}$$

$$= \sum_E \frac{W_E}{N_E} M_{E,AB}\, \delta_{AC}\, \delta_{ac}\, N_B \,. \qquad\qquad (\text{C.24})$$

Again the average of $\hat{Y}^{Aa,Cc}$ vanishes, and finally, the Hilbert space average of the density matrix element is

$$[\![\rho^{Aa,Cc}]\!] = \sum_E \frac{W_E}{N_E} M_{E,AB}\, \delta_{AC}\, \delta_{ac}\, N_B \,. \qquad\qquad (\text{C.25})$$

This result is quoted in (10.27).

## C.5 Energy Exchange Hilbert Space Variance

Here we derive the result quoted in (10.30) on the basis of the AR as given in (9.19). Again we use the general formula for the Hilbert space variance of the expectation value of a Hermitian operator as given in (8.19). Using this formula we first need the operator

$$\hat{X}^{Aa,Cc}_{EE'}$$

$$= \frac{1}{2}\hat{\Pi}_E\Big(|A,a\rangle\langle C,c| + |C,c\rangle\langle A,a|\Big)\hat{\Pi}'_E$$

$$= \frac{1}{2}\sum_{A'B}\hat{\Pi}_{A'}\hat{\Pi}_B M_{E,A'B}\Big(|A,a\rangle\langle C,c| + |C,c\rangle\langle A,a|\Big)\sum_{C'D}\hat{\Pi}_{C'}\hat{\Pi}_D M_{E',C'D}$$

$$= \frac{1}{2}\sum_{A'BC'}\hat{\Pi}_{A'}\Big(|A,a\rangle\langle C,c| + |C,c\rangle\langle A,a|\Big)\hat{\Pi}_{C'} M_{E,A'B} M_{E',C'B}\hat{\Pi}_B$$

$$= \frac{1}{2}\sum_{A'BC'}\Big(\delta_{A'A}\delta_{C'C}|A,a\rangle\langle C,c| + \delta_{A'C}\delta_{AC'}|C,c\rangle\langle A,a|\Big) M_{E,A'B} M_{E',C'B}\hat{\Pi}_B$$

$$= \frac{1}{2}\sum_B \Big( M_{E,AB} M_{E',CB}|A,a\rangle\langle C,c| + M_{E,CB} M_{E',AB}|C,c\rangle\langle A,a|\Big)\hat{\Pi}_B \,. \quad (\text{C.26})$$

The adjoint operator reads

$$(\hat{X}^{Aa,Cc}_{EE'})^\dagger$$

$$= \frac{1}{2}\sum_B \Big( M_{E,AB} M_{E',CB}|C,c\rangle\langle A,a| + M_{E,CB} M_{E',AB}|A,a\rangle\langle C,c|\Big)\hat{\Pi}_B \,, \quad (\text{C.27})$$

and the product yields

$$\hat{X}_{EE'}^{Aa,Cc}(\hat{X}_{EE'}^{Aa,Cc})^\dagger$$

$$= \frac{1}{4}\sum_B \left( M_{E,AB}M_{E',CB}|A,a\rangle\langle A,a| + M_{E,CB}M_{E',AB}|C,c\rangle\langle C,c| \right)\hat{\Pi}_B, \quad (C.28)$$

where we have used the identity $M_{E,AB}M_{E,AD} = M_{E,AB}\delta_{BD}$. The trace over this operator yields

$$\mathrm{Tr}\left\{\hat{X}_{EE'}^{Aa,Cc}(\hat{X}_{EE'}^{Aa,Cc})^\dagger\right\} = \frac{1}{4}\sum_B \left( M_{E,AB}M_{E',CB} + M_{E,CB}M_{E',AB}\right)N_B, \quad (C.29)$$

and thus, we finally find for the variance

$$\Delta_{\mathrm{H}}^2(\langle\hat{X}^{Aa,Cc}\rangle)$$

$$= \sum_{EE'} \frac{W_E W_{E'} N_B}{N_E(N_{E'} + \delta_{EE'})} \sum_B \left[ \frac{N_B}{4}(M_{E,AB}M_{E',CB} + M_{E,CB}M_{E',AB}) \right.$$

$$\left. - \frac{\delta_{EE'}}{N_E}M_{E,AB}\,\delta_{AC}\,\delta_{ac}N_B^2 \right]$$

$$= \frac{1}{2}\sum_{EE'}\sum_B \frac{W_E W_{E'} N_B}{N_E(N_{E'} + \delta_{EE'})}M_{E,AB}M_{E',CB}$$

$$- \delta_{AC}\,\delta_{ac}\sum_E\sum_B \frac{W_E^2 N_B^2}{N_E^2(N_E + 1)}M_{E,AB}. \quad (C.30)$$

As in the microcanonical situation the Hilbert space variance of the expectation value $\langle\hat{Y}^{Aa,Cc}\rangle$ is similar to $\Delta_{\mathrm{H}}^2(\langle\hat{X}^{Aa,Cc}\rangle)$. However, since the trace of $\hat{Y}^{Aa,Cc}$ vanishes the second term is zero again. Using (10.13) mean-squared distance yields

$$D^2(\hat{\rho},\hat{\omega}) = \sum_{AC}\sum_{ac}\left(\sum_{EE'}\sum_B \frac{W_E W_{E'} N_B}{N_E(N_{E'} + \delta_{EE'})}M_{E,AB}M_{E',CB}\right.$$

$$\left. - \delta_{AC}\,\delta_{ac}\sum_E\sum_B \frac{W_E^2 N_B^2}{N_E^2(N_E + 1)}M_{E,AB}\right)$$

$$= \sum_{EE'}\sum_{ABC} \frac{W_E W_{E'} N_B N_A N_C}{N_E(N_{E'} + \delta_{EE'})}M_{E,AB}M_{E',CB}$$

$$- \sum_E\sum_{AB} \frac{W_E^2 N_B^2 N_A}{N_E^2(N_E + 1)}M_{E,AB}. \quad (C.31)$$

This result is quoted in (10.30).

# Appendix D
# Power of a Function

**Abstract** In this appendix we show that the $k$th power of any function with a global maximum will essentially be a Gaussian.

For this purpose we consider a function $f(x)$ with a global maximum at $x = 0$. Because of the positivity of $f(x) > 0$ we can rewrite the function

$$f(x) = e^{g(x)}, \quad \text{with} \quad g(x) = \ln f(x) . \tag{D.1}$$

Since the logarithm is a monotonous function, we consider instead of $f(x)$ the expansion of the function $g(x)$ around the global maximum $x = 0$,

$$g(x) = \sum_i C_i x^i = C_0 - C_2 x^2 + C_3 x^3 + \cdots , \tag{D.2}$$

with $C_2 > 0$ and thus

$$f(x) = e^{C_0} e^{-C_2 x^2} e^{C_3 x^3} \cdots , \tag{D.3}$$

with some constants $C_i$. Since multiplying the function with itself will amplify the maximum in the center, we can truncate the decomposition in this way. Multiplying the function $k$ times with itself we get

$$\left( f(x) \right)^k = e^{k C_0} e^{-k C_2 x^2} e^{k C_3 x^3} \cdots . \tag{D.4}$$

The value of $x$, for which the quadratic part will have reduced the function to half maximum, i.e., for which

$$\exp\left(-k C_2 x_h^2\right) = 1/2, \tag{D.5}$$

is

$$x_h = \pm \sqrt{\frac{\ln 2}{k C_2}} \ . \tag{D.6}$$

Evaluating the third-order part of $(f(x))^k$ at $x_h$ then yields

$$\exp\left(k C_3 x_h^3\right) = \exp\left(\frac{C_3 \left(\frac{\ln 2}{C_2}\right)^{3/2}}{\sqrt{k}}\right) , \tag{D.7}$$

which tends to 1 if $k$ approaches infinity. For the relevant region, in which the function is peaked, we can thus approximate

$$\left(f(x)\right)^k \approx e^{k C_0} e^{-k C_2 x^2}, \tag{D.8}$$

which is essentially a Gaussian.

# Appendix E
# Local Temperature Conditions for a Spin Chain[1]

**Abstract** In this appendix we present the technical details of the application of the local temperature conditions to a spin chain.

The entire chain with periodic boundary conditions may be diagonalized via successive Jordan–Wigner, Fourier, and Bogoliubov transformations [4]. The relevant energy scale is introduced via the thermal expectation value (without the ground state energy)

$$\overline{E} = \frac{N N^G}{2\pi} \int_{-\pi}^{\pi} dk \, \frac{\omega_k}{\exp(\beta \, \omega_k) + 1} \,, \tag{E.1}$$

where

$$\omega_k = 2\Delta E \sqrt{(1 - K \cos k)^2} \,, \tag{E.2}$$

with $K = \lambda / \Delta E$. The ground state energy $E_0$ is given by

$$E_0 = -\frac{N N^G}{2\pi} \int_{-\pi}^{\pi} dk \, \frac{\omega_k}{2} \,. \tag{E.3}$$

Since $N^G \gg 1$, the sums over all modes have been replaced by integrals.

If one partitions the chain into $N^G$ groups of $N$ subsystems each, the groups may also be diagonalized via a Jordan–Wigner and a Fourier transformation [4] and the energy $E_a$ reads

$$E_a = 2\Delta E \sum_{\nu=1}^{N^G} \sum_k (1 - K \cos(k)) \left( n_k^a(\nu) - \frac{1}{2} \right) \,, \tag{E.4}$$

---

[1] Based on [1–4] Hartmann et al.

where $k = \pi l/(N+1)$ $(l = 1, \ldots, N)$ and $n_k^a(\nu)$ is the fermionic occupation number of mode $k$ of group number $\nu$ in the state $|a\rangle$. It can take on the values 0 and 1.

For the model at hand one has $\varepsilon_a = 0$ for all states $|a\rangle$, while the squared variance $\Delta_a^2$ reads

$$\Delta_a^2 = \sum_{\nu=1}^{N^G} \Delta_\nu^2 \,, \tag{E.5}$$

with

$$\Delta_\nu^2 = \frac{\Delta E^2\, K^2}{2} - $$
$$- \frac{8\, \Delta E^2\, K^2}{(N+1)^2} \sum_k \sin^2(k) \left( n_k^a(\nu) - \frac{1}{2} \right) \sum_p \sin^2(p) \left( n_p^a(\nu+1) - \frac{1}{2} \right) . \tag{E.6}$$

We now turn to analyze conditions (23.16) and (23.18). According to (E.6), $\Delta_\nu^2$ cannot be expressed in terms of $E_{\nu-1}$ and $E_\nu$. We therefore approximate (23.16) and (23.18) by simpler expressions.

Let us first analyze condition (23.16). Since it cannot be checked for every state $|a\rangle$, we make the following approximations.

For the present model with $|K| < 1$ all occupation numbers $n_k^a(\nu)$ are zero in the ground state and thus $\Delta_a^2 = 0$ as well as $E_a - E_0 = 0$. Therefore (23.16) cannot hold for this state. However, if one occupation number is changed from 0 to 1, $\Delta_a^2$ changes at most by $4\Delta E^2 K^2/(n+1)$ and $E_a$ changes at least by $2\Delta E(1 - |K|)$. Therefore (23.16) will hold for all states except the ground state if

$$N > 2\Delta E \beta \frac{K^2}{1 - |K|} . \tag{E.7}$$

If $|K| > 1$, occupation numbers of modes with $\cos(k) < 1/|K|$ are 0 in the ground state and occupation numbers of modes with $\cos(k) > 1/|K|$ are equal to 1. $\Delta_a^2$ for the ground state then is $\left[\Delta_a^2\right]_{gs} \approx \left[\Delta_a^2\right]_{max}/2$ (in this entire chapter, $[x]_{min}$ and $[x]_{max}$ denote the minimal and maximal values $x$ can take on). We therefore approximate (23.16) with the stronger condition

$$\frac{E_{min} - E_0}{N^G} > \beta \frac{\left[\Delta_a^2\right]_{max}}{N^G} \,, \tag{E.8}$$

which implies that (23.16) holds for all states $|a\rangle$ in the energy range $[E_{min}, E_{max}]$ (see (23.20) and (23.21)). Equation (E.8) can be rewritten as a condition on the group size $N$

$$N > \beta \frac{\left[\delta_a^2\right]_{max}}{e_{min} - e_0} \, , \tag{E.9}$$

where $e_{min} = E_{min}/(NN^G)$, $e_0 = E_0/(NN^G)$, and $\delta_a^2 = \Delta_a^2/N^G$.

We now turn to analyze condition (23.18). Equation (E.6) shows that the $\Delta_a^2$ do not contain terms that are proportional to $E_a$. One thus has to determine when the $\Delta_a^2$ are approximately constant, which is the case if

$$\beta \frac{\left[\Delta_a^2\right]_{max} - \left[\Delta_a^2\right]_{min}}{2} \ll [E_a]_{max} - [E_a]_{min} \, . \tag{E.10}$$

As a direct consequence, we get $|c_1| \ll 1$ which means that temperature is intensive.

Defining the quantity $e_a = E_a/(NN^G)$, we can rewrite (E.10) as a condition on $N$,

$$N \geq \frac{\beta}{2\varepsilon} \frac{\left[\delta_a^2\right]_{max} - \left[\delta_a^2\right]_{min}}{[e_a]_{max} - [e_a]_{min}} \, , \tag{E.11}$$

where the accuracy parameter $\varepsilon \ll 1$ is equal to the ratio of the left-hand side and the right-hand side of (E.10).

Since (E.10) does not take into account the energy range (23.20), its application needs some further discussion.

If the occupation number of one mode of a group is changed, say from $n_k^a(\nu) = 0$ to $n_k^a(\nu) = 1$, the corresponding $\Delta_a^2$ differ at most by $4\Delta E^2 K^2/(n+1)$. On the other hand, $\left[\Delta_a^2\right]_{max} - \left[\Delta_a^2\right]_{min} = N^G \Delta E^2 K^2$. The state with the maximal $\Delta_a^2$ and the state with the minimal $\Delta_a^2$ thus differ in nearly all occupation numbers and, therefore, their difference in energy is close to $[E_a]_{max} - [E_a]_{min}$. On the other hand, states with similar energies $E_a$ also have a similar $\Delta_a^2$. Hence the $\Delta_a^2$ only change quasi-continuously with energy and (E.10) ensures that the $\Delta_a^2$ are approximately constant even locally, i.e., on any part of the possible energy range.

To compute the required group size $N_{min}$, we need to know the maximal and minimal values $E_a$ and $\Delta_a^2$ can take on. For $E_a$ they are given by

$$\left\{ \begin{matrix} [E_a]_{max} \\ [E_a]_{min} \end{matrix} \right\} = \left\{ \begin{matrix} + \\ - \end{matrix} \right\} N^G N \Delta E \, , \tag{E.12}$$

for $|K| \leq 1$, and by

$$\left\{ \begin{matrix} [E_a]_{max} \\ [E_a]_{min} \end{matrix} \right\} = \left\{ \begin{matrix} + \\ - \end{matrix} \right\} N^G N \Delta E \frac{2}{\pi} \left[ \sqrt{K^2 - 1} + \arcsin\left(\frac{1}{|K|}\right) \right] \, , \tag{E.13}$$

for $|K| > 1$, where the sum over all modes $k$ has been approximated by an integral. The maximal and minimal values of $\Delta_a^2$ are given by

$$\left\{ \begin{bmatrix} \Delta_a^2 \end{bmatrix}_{\text{max}} \\ \begin{bmatrix} \Delta_a^2 \end{bmatrix}_{\text{min}} \right\} = \left\{ \begin{matrix} N^G \, K^2 \, \Delta E^2 \\ 0 \end{matrix} \right\} . \tag{E.14}$$

Plugging these results into (E.11) as well as (E.1) and (E.3) (taking into account (23.20) and (23.21)) into (E.9) for $|K| > 1$ and using (E.7) for $|K| < 1$, the minimal number of systems per group can be calculated.

# References

1. M. Hartmann, J. Gemmer, G. Mahler, O. Hess, Euro. Phys. Lett. **65**, 613 (2004)
2. M. Hartmann, G. Mahler, O. Hess, Lett. Math. Phys. **68**, 103 (2004)
3. M. Hartmann, G. Mahler, O. Hess, Phys. Rev. Lett. **93**, 080402 (2004)
4. M. Hartmann, G. Mahler, O. Hess, Phys. Rev. E **70**, 066148 (2004)

# Index